COMPLEXITY
AND
CRITICALITY

Imperial College Press Advanced Physics Texts

Forthcoming:

Elements of Quantum Information
  by Martin Plenio & Vlatko Vedral

Physics of the Environment
  by Andrew Brinkman

Modern Astronomical Instrumentation
  by Adrian Webster

Symmetry, Groups and Representations in Physics
  by Dimitri Vvedensky & Tim Evans

Imperial College Press Advanced Physics Texts – Vol. 1

COMPLEXITY
AND
CRITICALITY

Kim Christensen
Nicholas R. Moloney
Imperial College London, UK

Imperial College Press

Published by

Imperial College Press
57 Shelton Street
Covent Garden
London WC2H 9HE

Distributed by

World Scientific Publishing Co. Pte. Ltd.
5 Toh Tuck Link, Singapore 596224
USA office: 27 Warren Street, Suite 401-402, Hackensack, NJ 07601
UK office: 57 Shelton Street, Covent Garden, London WC2H 9HE

British Library Cataloguing-in-Publication Data
A catalogue record for this book is available from the British Library.

COMPLEXITY AND CRITICALITY

Copyright © 2005 by Imperial College Press

All rights reserved. This book, or parts thereof, may not be reproduced in any form or by any means, electronic or mechanical, including photocopying, recording or any information storage and retrieval system now known or to be invented, without written permission from the Publisher.

For photocopying of material in this volume, please pay a copying fee through the Copyright Clearance Center, Inc., 222 Rosewood Drive, Danvers, MA 01923, USA. In this case permission to photocopy is not required from the publisher.

ISBN 1-86094-504-X
ISBN 1-86094-517-1 (pbk)

To Krista and Emil Christensen

Preface

The aim of this book is to introduce the concepts of critical phenomena and explore the common ground between complexity and criticality.

The word 'complexity' takes on a variety of meanings depending on the context, and its official definition is continuously being revised. This is because the study of complexity is in its infancy and is a rapidly developing field at the forefront of many areas of science including mathematics, physics, geophysics, economics and biology, to name just a few. Institutes and departments have been formed, conferences and workshops organised, books and countless articles written, all in the name of complexity. And yet, nobody agrees on a clear and concise theoretical formalism with which to study complexity. The danger is therefore that complexity research may become unstructured or even misleading. For our purposes, complexity refers to the repeated application of simple rules in systems with many degrees of freedom that gives rise to emergent behaviour not encoded in the rules themselves.

The word 'criticality', on the other hand, is well defined among statistical physicists. Criticality refers to the behaviour of extended systems at a phase transition where observables are scale free, that is, no characteristic scales exist for these observables. At a phase transition, the many constituent microscopic 'parts' give rise to macroscopic phenomena that cannot be understood by considering the laws obeyed by a single part alone. Criticality is therefore a cooperative feature emerging from the repeated application of the microscopic laws of a system of interacting 'parts'. The phenomenology of phase transitions is well developed and there exists a sound theoretical formalism for its description.

The book is divided into three chapters. In the first two chapters, we carefully introduce the reader to the concepts of critical phenomena using percolation and the Ising model as paradigmatic examples of isolated equilibrium systems. These systems undergo a phase transition only if an external agent finely tunes certain external parameters to particular values.

Percolation is one of the simplest models displaying a phase transition. The phase transition in percolation is purely geometrical and enables the reader to become intuitively familiar with important concepts such as fractals, scaling, and renormalisation.

The celebrated Ising model further develops the reader's intuition of emergent cooperative phenomena at a phase transition.

In the final chapter we consider the emergence of complexity in Nature. We argue that systems in Nature are neither isolated nor in equilibrium. We investigate a class of non-equilibrium systems where the constraint of having to tune external parameters to obtain critical behaviour is relaxed. We invite the reader to speculate on whether self-organisation in non-equilibrium systems may be a unifying concept for disparate fields such as statistical mechanics, geophysics and atmospheric physics.

Although mathematical methods have been developed to describe complexity and criticality, it is our experience that these methods are unfamiliar to scientists outside the field. Our hope is that this book will help students and researchers to treat complexity and criticality more quantitatively. Therefore, throughout the book we emphasise the mathematical quantitative techniques available.

The book is self-contained and therefore accessible to readers not familiar with the concepts of complexity and criticality. The text can form the basis for advanced undergraduate or graduate courses, and serve as an introductory reference for researchers in various fields. Each chapter is accompanied by exercises, full solutions to which can be obtained by contacting the authors via the book's associated website, http://www.worldscibooks.com/physics/p365.html. On this site, readers will also find animation codes to visualise the behaviour of the models considered. The bibliography is an attempt to cite the primary sources and most relevant publications but we readily acknowledge that it is by no means complete and might even be biased. We apologise if the reader is upset by any omissions or prejudice.

The book is based on a lecture course in statistical mechanics given at Imperial College London since 2000. We wish to pay tribute to all the students who attended the course. Their constructive feedback has been vital for the adapted presentation. In particular, we would like to thank Arno Proeme and Georg Schusteritsch for spotting many mistakes and typos. We urge readers with an equally keen eye for details to contact the authors on encountering any errors.

We are grateful to Dimitri Vvedensky for suggesting the project in the first place. Furthermore, we thank Imperial College Press for being efficient and highly professional during the writing process and editor Katie Lydon for her seemingly infinite patience waiting for the final manuscript.

We are grateful to a vast number of people who have helped in one way or another throughout the writing process. In particular, present and former Ph.D. students at Imperial College London have been much involved: Nadia Farid, Vera Pancaldi, Ole Peters, Gunnar Pruessner, and Matthew Stapleton. Without their constant support and enthusiasm, we would never have finished the book. Indeed, the fantastic atmosphere in the condensed matter theory group at Imperial College London has been a fountain of inspiration. We also greatly appreciate the hospitality of PGP, University of Oslo, Norway, where parts of the book were written. Various insightful suggestions have been provided by Álvaro Corral and Jens G. Feder.

K.C. would like to pay a special tribute to Hugo S. Jensen, Hans C. Fogedby, Henrik J. Jensen, Per Bak, Zeev Olami, and Amnon Aharony. Hugo S. Jensen is a sublime high school physics teacher and he is really to blame for my becoming a physicist. Hans Fogedby's and Henrik Jensen's enthusiasm and insight into physics continues to amaze me. Besides being great physicists, they are great friends, as was Per Bak. Per Bak's passion and insight into science was second to none and Zeev Olami and I had some unforgettable years together with Per at Brookhaven National Laboratory. Thank you, Per and Zeev, for your inspiration, your limitless generosity and friendship. You are sorely missed and will always be remembered. K.C. was fortunate to attend a post-graduate course taught by Amnon Aharony on cooperative phenomena. The chapter on percolation is an attempt to live up to his high scientific standard and his pedagogical and formidable lectures. This book would never have materialised were it not for the great influence of these first rate physicists in my scientific life.

Finally, K.C. would like to mention his wonderful family, who have been tremendously supportive during this marathon; they never missed a chance to remind me that the manuscript was long overdue.

Kim Christensen and Nicholas R. Moloney

Contents

Preface	vii
1. Percolation	1
1.1 Introduction	1
1.1.1 Definition of site percolation	3
1.1.2 Quantities of interest	3
1.2 Percolation in $d=1$	5
1.2.1 Cluster number density	6
1.2.2 Average cluster size	8
1.2.3 Transition to percolation	13
1.2.4 Correlation function	13
1.2.5 Critical occupation probability	14
1.3 Percolation on the Bethe Lattice	15
1.3.1 Definition of the Bethe lattice	15
1.3.2 Critical occupation probability	16
1.3.3 Average cluster size	16
1.3.4 Transition to percolation	19
1.3.5 Cluster number density	22
1.3.6 Correlation function	29
1.4 Percolation in $d=2$	30
1.4.1 Transition to percolation	31
1.4.2 Average cluster size	33
1.4.3 Cluster number density – exact	35
1.4.4 Cluster number density – numerical	35
1.5 Cluster Number Density – Scaling Ansatz	39
1.5.1 Scaling function and data collapse	41

      1.5.2 Scaling function and data collapse in $d=1$ 42
      1.5.3 Scaling function and data collapse on the Bethe lattice 44
      1.5.4 Scaling function and data collapse in $d=2$ 49
  1.6 Scaling Relations 51
  1.7 Geometric Properties of Clusters 55
      1.7.1 Self-similarity and fractal dimension 55
      1.7.2 Mass of a large but finite cluster at $p=p_c$ 58
      1.7.3 Correlation length 62
      1.7.4 Mass of the percolating cluster for $p > p_c$ 63
  1.8 Finite-Size Scaling 69
      1.8.1 Order parameter 70
      1.8.2 Average cluster size and higher moments 73
      1.8.3 Cluster number density 75
  1.9 Non-Universal Critical Occupation Probabilities 78
  1.10 Universal Critical Exponents 81
  1.11 Real-Space Renormalisation 82
      1.11.1 Self-similarity and the correlation length 82
      1.11.2 Self-similarity and fixed points 83
      1.11.3 Coarse graining and rescaling 85
      1.11.4 Real-space renormalisation group procedure 87
      1.11.5 Renormalisation in $d=1$ 91
      1.11.6 Renormalisation in $d=2$ on a triangular lattice 95
      1.11.7 Renormalisation in $d=2$ on a square lattice 98
      1.11.8 Approximation via the truncation of parameter space 100
  1.12 Summary 102
  Exercises 104

2. Ising Model   115

  2.1 Introduction 115
      2.1.1 Definition of the Ising model 116
      2.1.2 Review of equilibrium statistical mechanics 119
      2.1.3 Thermodynamic limit 123
  2.2 System of Non-Interacting Spins 124
      2.2.1 Partition function and free energy 125
      2.2.2 Magnetisation and susceptibility 127
      2.2.3 Energy and specific heat 129
  2.3 Quantities of Interest 131
      2.3.1 Magnetisation 131
      2.3.2 Response functions 133

	2.3.3	Correlation length and spin-spin correlation function	134
	2.3.4	Critical temperature and external field	135
	2.3.5	Symmetry breaking	138
2.4	Ising Model in $d=1$		140
	2.4.1	Partition function	141
	2.4.2	Free energy	143
	2.4.3	Magnetisation and susceptibility	145
	2.4.4	Energy and specific heat	149
	2.4.5	Correlation function	151
	2.4.6	Critical temperature	154
2.5	Mean-Field Theory of the Ising Model		156
	2.5.1	Partition function and free energy	157
	2.5.2	Magnetisation and susceptibility	158
	2.5.3	Energy and specific heat	166
2.6	Landau Theory of the Ising Model		169
	2.6.1	Free energy	170
	2.6.2	Magnetisation and susceptibility	172
	2.6.3	Specific heat	175
2.7	Landau Theory of Continuous Phase Transitions		175
2.8	Ising Model in $d=2$		179
	2.8.1	Partition function	179
	2.8.2	Magnetisation and susceptibility	180
	2.8.3	Energy and specific heat	184
	2.8.4	Critical temperature	186
2.9	Widom Scaling Ansatz		188
	2.9.1	Scaling ansatz for the free energy	190
	2.9.2	Scaling ansatz for the specific heat	191
	2.9.3	Scaling ansatz for the magnetisation	191
	2.9.4	Scaling ansatz for the susceptibility	192
	2.9.5	Scaling ansatz for the spin-spin correlation function .	192
2.10 Scaling Relations			193
2.11 Widom Scaling Form and Critical Exponents in $d=1$ . . .			195
2.12 Non-Universal Critical Temperatures			198
2.13 Universal Critical Exponents			199
2.14 Ginzburg Criterion			200
2.15 Real-Space Renormalisation			202
	2.15.1 Kadanoff's block spin transformation		202
	2.15.2 Kadanoff's block spin and the free energy		206
	2.15.3 Kadanoff's block spin and the correlation function .		209

    2.15.4 Renormalisation in $d = 1$ 211
    2.15.5 Renormalisation in $d = 2$ on a square lattice 215
 2.16 Wilson's Renormalisation Group Theory 222
    2.16.1 Coupling space and renormalisation group flow ... 222
    2.16.2 Self-similarity and fixed points 227
    2.16.3 Basin of attraction of fixed points 229
    2.16.4 RG flow in coupling and configurational space 230
    2.16.5 Universality and RG flow near fixed point 231
    2.16.6 Widom scaling form 235
 2.17 Summary 237
 Exercises 241

3. Self-Organised Criticality   249

 3.1 Introduction 249
    3.1.1 Sandpile metaphor 250
 3.2 BTW Model in $d = 1$ 255
    3.2.1 Algorithm of the BTW model in $d = 1$ 256
    3.2.2 Transient and recurrent configurations 257
    3.2.3 Avalanche time series 259
    3.2.4 Avalanche-size probability 260
 3.3 Mean-Field Theory of the BTW Model 264
    3.3.1 Random neighbour BTW model 264
    3.3.2 Algorithm of the random neighbour BTW model .. 264
    3.3.3 Steady state and the average avalanche size 265
 3.4 Branching Process 267
    3.4.1 Branching ratio 267
    3.4.2 Avalanche-size probability – exact 268
    3.4.3 Avalanche-size probability – scaling form 270
 3.5 Avalanche-Size Probability – Scaling Ansatz 273
 3.6 Scaling Relations 275
 3.7 Moment Analysis of Avalanche-Size Probability 276
 3.8 BTW Model in $d = 2$ 278
    3.8.1 Algorithm of the BTW model in $d = 2$ 278
    3.8.2 Steady state and the average avalanche size 279
    3.8.3 Avalanche time series 280
    3.8.4 Avalanche-size probability 281
 3.9 Ricepile Experiment and the Oslo Model 285
    3.9.1 Ricepile experiment 285
    3.9.2 Ricepile avalanche time series 287

3.9.3   Ricepile avalanche-size probability density 289
3.9.4   Ricepile modelling 290
3.9.5   Algorithm of the Oslo model 291
3.9.6   Transient and recurrent configurations 292
3.9.7   Avalanche time series 293
3.9.8   Avalanche-size probability 294
3.10 Earthquakes and the OFC Model 302
   3.10.1 Earthquake mechanism 302
   3.10.2 Earthquake time series 303
   3.10.3 Earthquake-size frequency 304
   3.10.4 Earthquake modelling 305
   3.10.5 Algorithm of the OFC model 309
   3.10.6 Steady state and the average avalanche size 310
   3.10.7 Avalanche time series 314
   3.10.8 Avalanche-size probability 318
3.11 Rainfall 325
   3.11.1 Rainfall mechanism 325
   3.11.2 Rainfall time series 326
   3.11.3 Rainfall-size number density 329
3.12 Summary 331
Exercises 334

Appendix A   Taylor Expansion   341

Appendix B   Hyperbolic Functions   343

Appendix C   Homogeneous and Scaling Functions   345

Appendix D   Fractals   351

Appendix E   Data Binning   355

Appendix F   Boltzmann Distribution   359

Appendix G   Free Energy   361

Appendix H   Metropolis Algorithm   363

Bibliography 365

List of Symbols 371

Index 381

Chapter 1

Percolation

1.1 Introduction

Take some squared paper and black out a portion of the squares randomly. Consider the clusters of adjacent black squares. If only a small portion of squares are blacked out, it is unlikely that a cluster extends across opposite sides of the paper. However, if a large portion of squares are blacked out, it is likely that a cluster 'percolates' across the paper. Indeed, for a paper of infinite size, there exists a unique portion of randomly blacked out squares that mark a phase transition from paper with no percolating infinite cluster to paper with a percolating infinite cluster. Percolation is the study of the clusters as a function of the portion of black squares. It is a purely geometrical problem and is arguably the simplest model that undergoes a phase transition. The challenge in percolation lies in describing its emergent structures rather than understanding its defining rules. Apart from investigating the geometrical properties of the percolating infinite cluster, we will be specifically interested in the statistical properties of finite clusters. One approach to studying these clusters is simply to calculate their size and number explicitly from the rules of percolation, but we will find that in general this is a hopeless task even for clusters of moderate size. In the vicinity of the phase transition, another approach to describing cluster statistics suggests itself, and this is related to the notion of scale invariance. We will focus on the phase transition and dwell upon scale invariance, since these are unifying themes throughout the book. We will develop a mathematical framework for expressing scale invariance, applicable not only to percolation, but to other models undergoing a phase transition.

Percolation is, in fact, highly relevant in a variety of physical settings: oil recovery from porous media [King et al., 1999], epidemic modelling [Cardy and Grassberger, 1985], networks [Cohen *et al.*, 2002], fragmentation [Herrmann and Roux, 1990], metal-insulator transition [Ball *et al.*, 1994], ionic transport in glasses and composites [Roman *et al.*, 1986], fracture patterns and earthquakes in rocks and ground water flow [Sahimi, 1994] among others. Therefore, the concepts in percolation are not only of academic interest but also of considerable practical value. For an introduction to percolation and some of its applications at the level of our presentation, we can recommend the popular text [Stauffer and Aharony, 1994].

After defining percolation and its main observables, our programme is as follows. First, we will investigate one-dimensional percolation, which is simple and exactly solvable. Although trivial, this exercise is not entirely fruitless because it will expose some of the features of a phase transition. Then we will consider percolation on a tree-like lattice where we will also be able to make considerable analytic progress. This so-called mean-field percolation gives a qualitatively more accurate description of a phase transition, but leaves out important features, both qualitative and quantitative, that are present in percolation in general. In two dimensions, we will initially appear to reach an impasse, since the possibility that clusters 'interact' with themselves will force us to abandon explicit calculations even for clusters of moderate size. Fortunately, however, one-dimensional and mean-field percolation provide clues suggesting that something generic happens to the clusters near a phase transition. We will identify and exploit this generic phenomenon by invoking the powerful concept of scale invariance. This will allow us to propose a particular functional form for the statistics of clusters. While this so-called scaling ansatz will not be able to give us the exact statistics of clusters, we will argue that the concept of scale invariance nonetheless gives us something far more preferable: a statistical framework for unifying the description of clusters in percolation in general.

When studying the geometrical properties of percolation in greater detail, we will find that the root of the simplification of scale invariance lies in the fractal nature of percolation at a phase transition. Loosely speaking, this means that at a phase transition percolating systems look alike on different scales. This in turn will refine our intuition of scale invariance, and we will be in a position to introduce the so-called renormalisation group approach. Such an approach is designed to exploit scale invariance in percolation at the phase transition, by preserving only large-scale features of the clusters. We will illustrate this approach with specific examples of renormalisation group transformations.

1.1.1 Definition of site percolation

We now define site percolation, one of the simplest examples of a disordered system. Consider for now a two-dimensional square lattice composed of $L \times L$ sites. We refer to L, measured in units of the lattice spacing, as the lattice size. Note that L is not a length scale as such, but a dimensionless number, L'/a, where L' is the lattice length and a the lattice spacing. Throughout, all lengths are measured in units of the lattice spacing.

To introduce disorder, we occupy sites randomly and independently with occupation probability p, treating all sites equally, see Figure 1.1. The foremost entity of interest is a cluster, which we define as a group of nearest-neighbouring occupied sites. The cluster size, s, is the number of occupied sites in the group.

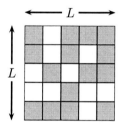

s	$N(s, p; L)$
1	2
2	1
4	1
7	1

Fig. 1.1 A realisation of two-dimensional site percolation on a square lattice of size $L = 5$. Sites are occupied with probability p and are thus empty with probability $(1-p)$. Occupied sites are dark grey while unoccupied sites are white. A cluster is defined as a group of nearest-neighbouring occupied sites. In a square lattice, a bulk site has four nearest neighbours: north, south, east, and west and four next-nearest-neighbours: north-east, north-west, south-east, and south-west. The table to the right displays the cluster size frequency, $N(s, p; L)$.

1.1.2 Quantities of interest

We shall introduce the quantities of interest on a two-dimensional square lattice. For a fixed lattice size L measured in units of the lattice spacing, there is only one parameter in the problem, namely the occupation probability p.

For $p = 0$ the lattice is empty of clusters, while for $p = 1$ there is only one cluster of size L^2. For intermediate values $0 < p < 1$, each realisation will, in general, be different. But qualitatively, we expect the size of the largest cluster to increase with p. Figure 1.2 supports this trend, showing six realisations of increasing p on a lattice of size $L = 150$.

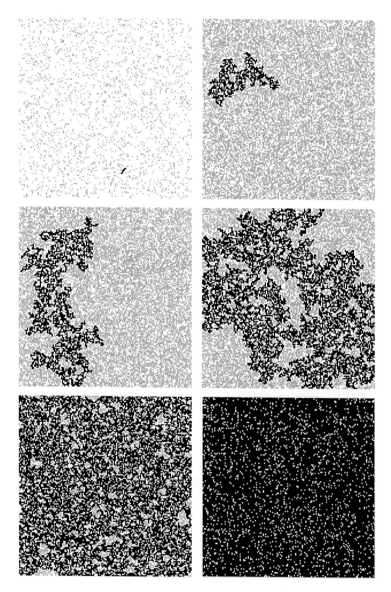

Fig. 1.2 Six realisations of two-dimensional site percolation on a square lattice of size $L = 150$ for occupation probabilities $p = 0.10, 0.55, 0.58, 0.59274621, 0.65, 0.90$, from left to right and top to bottom, respectively. Occupied sites are dark grey while unoccupied sites are white. For each realisation, the largest cluster has been shaded black.

For each realisation, the largest cluster has been shaded black. In fact, for a given p, there is a whole distribution of cluster sizes. It is one of our main aims to determine this distribution, since we would be able to calculate the probability of clusters of any size, including the average cluster size, for example.

By definition, a cluster is percolating if and only if it is infinite. Clearly, clusters that span a finite lattice from left to right or top to bottom are candidates for percolating clusters in infinite systems. In Figure 1.2, for example, the largest clusters in the lattices corresponding to $p = 0.59274621, 0.65$, and 0.90 are spanning clusters whereas the largest clusters in the lattices corresponding to $p = 0.10, 0.55$ and 0.58 are not. The largest cluster in the lattice corresponding to $p = 0.58$, however, is almost spanning.

In an infinite system, there exists a critical occupation probability, p_c, such that for $p < p_c$ there is no percolating infinite cluster, while for $p > p_c$ there is a percolating infinite cluster. At $p = p_c$, it has been conjectured and shown rigorously for $d = 2$, that there is no percolating infinite cluster [Grimmett, 1999]. The large but finite clusters that emerge at $p = p_c$ are called 'incipient infinite clusters'. Throughout, we will often refer to the incipient infinite clusters at $p = p_c$ as percolating clusters, even though they are, strictly speaking, finite.

In an infinite system, how much space does the percolating cluster occupy of the lattice or, in other words, what is the probability that a site belongs to the percolating cluster? What type of geometry does the percolating cluster exhibit? For example, is it fractal? Excluding the percolating infinite cluster, what is the typical 'radius' of the largest finite cluster? As a closely related question, what is the probability that two sites belong to the same finite cluster? Similarly, excluding the percolating infinite cluster, what is the typical 'size' of the largest finite cluster? How is the typical size of the largest finite cluster related with its typical radius? These are some of the questions we will now try to answer.

1.2 Percolation in $d = 1$

We start with one-dimensional percolation, because this is one of the few cases where we are able to calculate quantities exactly. Although the conclusions we reach in one dimension are not general to higher dimensions, some of the main features of percolation are nevertheless present.

1.2.1 Cluster number density

Figure 1.3 is a realisation of site percolation in a one-dimensional lattice of size $L = 20$ where sites are occupied with probability p.

Fig. 1.3 A realisation of one-dimensional site percolation on a lattice of size $L = 20$. Occupied sites are dark grey while unoccupied sites are white. There are three clusters of size $s = 1$, two clusters of size $s = 2$, and one cluster of size $s = 5$.

We can readily construct the cluster size frequency, $N(s, p; L)$. There are three clusters of size $s = 1$, two clusters of size $s = 2$, and one cluster of size $s = 5$. This cluster size frequency pertains to a particular realisation on a given lattice of size L, but we would like to be far more general. We therefore proceed by accounting for the lattice size and calculating cluster size frequencies probabilistically. In doing so, we encounter our first complication with L being finite. This is because the two boundary sites have only one nearest neighbour each, while bulk sites have two nearest neighbours each. For example, consider the cluster of size $s = 5$ in Figure 1.3. If the cluster is in the bulk of the lattice, five consecutive sites would be occupied, while the two sites on either side of the cluster would be unoccupied. If the cluster touches the boundary of the lattice, there would still be five consecutive sites occupied, but only one site unoccupied. However, the ratio of boundary to bulk sites tends to zero for hypercubic lattices as L tends to infinity. Therefore, such a complication would play a diminishing role with increasing L, and completely disappears for an infinite system. From now on, we will therefore ignore the effect of boundary sites.

We are now ready to calculate the probability that a given site belongs to a cluster of size s. The site must be part of a consecutive group of s occupied sites bounded by two unoccupied sites. The probability for this is $s(1-p)p^s(1-p)$, where the factor s comes about because the given site can be any of the s occupied sites. This expression is a simple product because each site is occupied independently with probability p and therefore unoccupied with probability $(1-p)$.

How is this quantity related to the cluster size frequency $N(s, p; L)$? Since $s(1-p)^2 p^s$ is the probability that a given site belongs to a cluster of size s, we might try to identify $Ls(1-p)^2 p^s$ with $N(s, p; L)$. But this reasoning would be wrong, because an s-cluster takes up s sites, so there is only space for approximately L/s such clusters. Therefore, to calculate the expected frequency of s-clusters in a lattice of size L, we must correct our estimate by dividing by s, that is,

$$N(s, p; L) = L(1-p)^2 p^s. \qquad (1.1)$$

The cluster number frequency, however, is not a desirable quantity, because it depends on the lattice size L. Therefore, we normalise by the number of sites in the lattice, and work instead with the cluster number density,

$$n(s, p) = \frac{N(s, p; L)}{L} = (1-p)^2 p^s. \qquad (1.2)$$

In this notation, $sn(s, p)$ is the probability that a given site belongs to an s-cluster. Figure 1.4 displays $n(s, p)$ versus cluster size s for five different probabilities p.

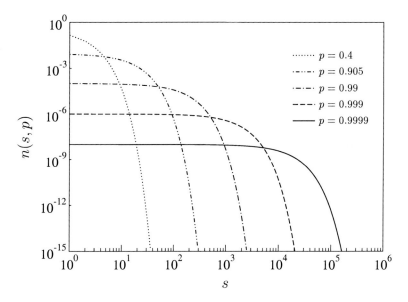

Fig. 1.4 Exact solution of the cluster number density, $n(s, p)$, versus the cluster size, s, for one-dimensional percolation. The five curves correspond to occupation probabilities $p = 0.4, 0.905, 0.99, 0.999, 0.9999$ with characteristic cluster sizes $s_\xi(p) \approx 1, 10, 100, 1\,000, 10\,000$, respectively. For $p \to 1^-$, the characteristic cluster size diverges.

For a given p, the cluster number density $n(s, p)$ decreases with increasing cluster size s, thus large clusters are rare. For large p, the curve remains relatively flat for a large range of s, before decaying rapidly. This motivates the introduction of a 'characteristic' cluster size, $s_\xi(p)$, signalling the onset of the rapid decay. A commonly accepted convention is to mark where the curve has decreased by a factor $1/e$, see Figure 1.4. We can recast the cluster number density in a form revealing this behaviour:

$$\begin{aligned} n(s,p) &= (1-p)^2 p^s \\ &= (1-p)^2 \exp(\ln p^s) \\ &= (1-p)^2 \exp(s \ln p) \\ &= (1-p)^2 \exp(-s/s_\xi), \end{aligned} \quad (1.3)$$

with the definition of the characteristic cluster size

$$s_\xi(p) = -\frac{1}{\ln p}, \quad (1.4)$$

which is a function of p, plotted in Figure 1.5(a). The characteristic cluster size is an increasing function of p and diverges when p tends to one from below. To investigate how $s_\xi(p)$ diverges as p approaches one, we use the Taylor expansion $\ln(1-x) \to -x$ for $x \to 0$, see Appendix A, for the denominator in the neighbourhood of $p = 1$ to find

$$s_\xi(p) = -\frac{1}{\ln(1-[1-p])} \to (1-p)^{-1} \quad \text{for } p \to 1^-. \quad (1.5)$$

Thus, the characteristic cluster size diverges as a power law with exponent -1 in terms of $(1-p)$, the distance of p below one, see Figure 1.5(b).

1.2.2 Average cluster size

What is the average cluster size? For example, in Figure 1.3 there are three clusters of size one, two clusters of size two, and one cluster of size five. Therefore, the average size of the $N_{\text{clu}} = 6$ clusters is

$$\frac{1}{N_{\text{clu}}} \sum_{k=1}^{N_{\text{clu}}} s_k = \frac{1+1+1+2+2+5}{6} = 2. \quad (1.6)$$

Here, each cluster has equal weight. However, there is an alternative form of averaging where each cluster is weighted according to its size, giving more weight to large clusters. To understand this type of averaging, imagine choosing an occupied site at random. How large, on average, is the cluster to which this occupied site belongs? In our example, there are twelve occupied sites. The probability of choosing a particular occupied site is $1/N_{\text{occ}}$, where $N_{\text{occ}} = 12$ is the total number of occupied sites. The probability that the

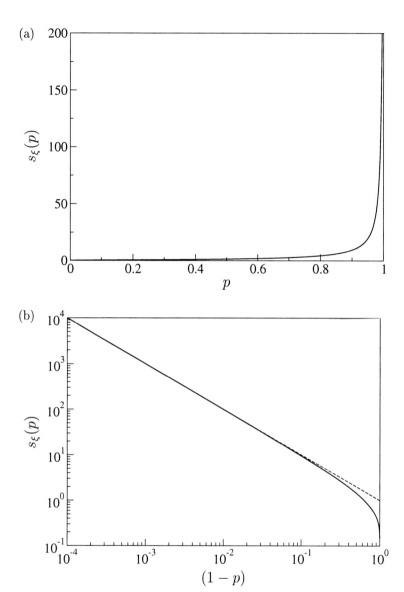

Fig. 1.5 Exact solution of the characteristic cluster size for one-dimensional percolation. (a) The characteristic cluster size, $s_\xi(p)$, versus the occupation probability, p. For p approaching one, the characteristic cluster size diverges. For p tending to zero, $s_\xi(p)$ tends to zero. (b) The characteristic cluster size, $s_\xi(p)$, versus $(1-p)$, the distance of p below one (solid line). For $p \to 1^-$, the characteristic cluster size diverges as a power law with exponent -1 in terms of $(1-p)$. The dashed straight line has slope -1.

chosen site belongs to a particular s-cluster is s/N_{occ}. Weighting each of the N_{clu} clusters by its size, the average size of a cluster to which a randomly chosen occupied site belongs is

$$\chi(p) = \frac{1}{N_{\text{occ}}} \sum_{k=1}^{N_{\text{clu}}} s_k^2 = \frac{1^2 + 1^2 + 1^2 + 2^2 + 2^2 + 5^2}{12} = 3. \quad (1.7)$$

From now on, we will take the average cluster size to be $\chi(p)$. We would like to relate $\chi(p)$ to the cluster number density $n(s, p)$. First we convert the sum running over all clusters indexed by k, to a sum running over all cluster sizes s. This is achieved by replacing a sample of clusters with their frequency,

$$\chi(p) = \frac{1}{N_{\text{occ}}} \sum_{k=1}^{N_{\text{clu}}} s_k^2 = \frac{1}{N_{\text{occ}}} \sum_{s=1}^{\infty} s^2 N(s, p; L). \quad (1.8)$$

In a lattice of size L, the expected number of occupied sites, $N_{\text{occ}} = pL$, and therefore

$$\begin{aligned} \chi(p) &= \frac{\sum_{s=1}^{\infty} s^2 N(s, p; L)}{pL} \\ &= \frac{\sum_{s=1}^{\infty} s^2 n(s, p)}{p} \\ &= \frac{\sum_{s=1}^{\infty} s^2 n(s, p)}{\sum_{s=1}^{\infty} sn(s, p)} \quad \text{for } p < 1. \end{aligned} \quad (1.9)$$

To justify the last step in Equation (1.9), we point out that the probability that an arbitrary site in the lattice belongs to any finite cluster is simply the probability p of it being occupied. This is true for all $p < 1$. The probability that an arbitrary site belongs to an s-cluster is given by $sn(s, p)$. Therefore, the sum of $sn(s, p)$ over all possible s-clusters equals p, the probability that a site is occupied, and we have the sum rule

$$\sum_{s=1}^{\infty} sn(s, p) = p \quad \text{for } p < 1. \quad (1.10)$$

Notice that at $p = 1$, the above identity breaks down: according to Equation (1.2), $n(s, 1) = 0$, making the left-hand side of Equation (1.10) equal to zero, while the right hand-side equals one. The inconsistency occurs because, at $p = 1$, there is only one cluster present and it is infinite in size. We can demonstrate the identity in Equation (1.10) for $p < 1$ rigorously by summing up a geometric series, that is,

$$\sum_{s=1}^{\infty} sn(s,p) = \sum_{s=1}^{\infty} s(1-p)^2 p^s$$

$$= (1-p)^2 \sum_{s=1}^{\infty} p\frac{d}{dp} p^s$$

$$= (1-p)^2 \left(p\frac{d}{dp}\right) \left(\sum_{s=1}^{\infty} p^s\right)$$

$$= (1-p)^2 \left(p\frac{d}{dp}\right) \left(\frac{p}{1-p}\right)$$

$$= p. \qquad (1.11)$$

We will take the latter form in Equation (1.9) as the definition of the average cluster size $\chi(p)$, also in dimensions $d > 1$ [Stauffer and Aharony, 1994]. The average cluster size is therefore

$$\chi(p) = \frac{\sum_{s=1}^{\infty} s^2 n(s,p)}{\sum_{s=1}^{\infty} sn(s,p)}$$

$$= \frac{1}{p}(1-p)^2 \sum_{s=1}^{\infty} s^2 p^s$$

$$= \frac{1}{p}(1-p)^2 \left(p\frac{d}{dp}\right) \left(p\frac{d}{dp}\right) \left(\sum_{s=1}^{\infty} p^s\right). \qquad (1.12)$$

After cautioning the reader that

$$\left(p\frac{d}{dp}\right) \left(p\frac{d}{dp}\right) \neq p^2 \frac{d^2}{dp^2}, \qquad (1.13)$$

we finally arrive at

$$\chi(p) = \frac{1+p}{1-p} \quad \text{for } 0 < p < 1. \qquad (1.14)$$

As expected, the average cluster size is an increasing function of p, see Figure 1.6(a). Like the characteristic cluster size, when p approaches one, the average cluster size also diverges as a power law with exponent -1 in terms of $(1-p)$, the distance of p below one,

$$\chi(p) = \frac{1+p}{1-p} \to 2(1-p)^{-1} \quad \text{for } p \to 1^-, \qquad (1.15)$$

see Figure 1.6(b).

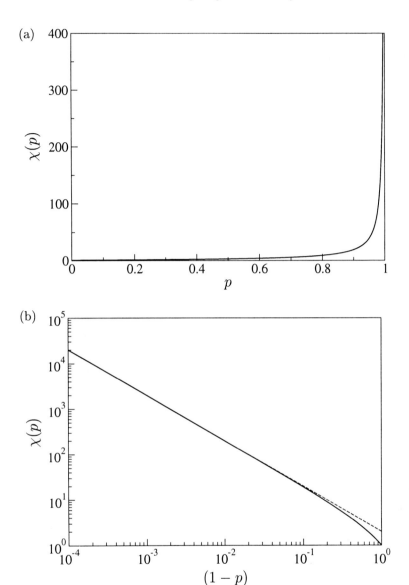

Fig. 1.6 Exact solution of the average cluster size for one-dimensional percolation. (a) The average cluster size, $\chi(p)$, versus the occupation probability, p. For p approaching one, the average cluster size diverges. For p tending to zero, $\chi(p)$ tends to one. (b) The average cluster size, $\chi(p)$, versus $(1-p)$, the distance of p below one (solid line). For $p \to 1^-$, the average cluster size diverges as a power law with exponent -1 in terms of $(1-p)$. The dashed straight line has slope -1.

1.2.3 Transition to percolation

The probability that a site belongs to any finite cluster is p. In one dimension, the percolating infinite cluster must include every site in the lattice for it to span from left to right. What is the probability, $P_\infty(p)$, that a site belongs to the percolating cluster? For $p < 1$, there is no percolating cluster, so the probability that a site belongs to the percolating cluster is zero. However, at $p = 1$, there is a percolating cluster and all sites belong to it. Thus the probability that a site belongs to the percolating cluster is one, that is,

$$P_\infty(p) = \begin{cases} 0 & \text{for } p < 1 \\ 1 & \text{for } p = 1. \end{cases} \qquad (1.16)$$

This quantity allows us to generalise Equation (1.10) to

$$P_\infty(p) + \sum_{s=1}^{\infty} sn(s,p) = p \qquad (1.17)$$

valid for all p. Put differently, Equation (1.17) states that the probability that a site is occupied, p, is equal to the probability that the site belongs to the percolating infinite cluster, $P_\infty(p)$, added to the probability that the site belongs to a finite cluster, $\sum_{s=1}^{\infty} sn(s,p)$. In fact, Equation (1.17) is valid in higher dimensions as well.

1.2.4 Correlation function

The final quantity to be discussed in one dimension is the probability that two sites belong to the same finite cluster. More precisely, let $\mathbf{r}_i$ denote the position vector of site i in the one-dimensional lattice. Given that the site at $\mathbf{r}_i$ is occupied, the probability that a site at position $\mathbf{r}_j$ belongs to the same finite cluster is known as the site-site correlation function. Since the site at position $\mathbf{r}_j$ must be occupied as well as all the $r-1$ intermediate sites, the site-site correlation function

$$g(\mathbf{r}_i, \mathbf{r}_j) = p^r \quad \text{for } 0 < p < 1, \qquad (1.18)$$

where $r = |\mathbf{r}_i - \mathbf{r}_j|$ is the distance between the sites measured in units of the lattice spacing. By definition, $g(\mathbf{r}_i, \mathbf{r}_i) = 1$ when $r = 0$. The correlation function decays with increasing distance r. In the same spirit as for the characteristic cluster size $s_\xi(p)$, we can characterise the decay of the site-site correlation function,

$$g(\mathbf{r}_i, \mathbf{r}_j) = \exp(\ln p^r) = \exp(r \ln p) = \exp(-r/\xi), \quad (1.19)$$

where we have defined a characteristic length scale, $\xi(p)$, the correlation length (measured in units of the lattice spacing) by

$$\xi(p) = -\frac{1}{\ln p} = -\frac{1}{\ln(1 - [1 - p])} \to (1 - p)^{-1} \quad \text{for } p \to 1^-. \quad (1.20)$$

Again we have used the Taylor expansion $\ln(1 - x) \to -x$ for $x \to 0$, see Appendix A. The correlation length diverges as a power law with exponent -1 in terms of $(1-p)$, the distance of p below one. Note that the correlation length is the typical radius of the largest cluster. For this reason we have used the notation $s_\xi(p)$ to denote the characteristic cluster size.

In fact, the correlation function $g(\mathbf{r}_i, \mathbf{r}_j)$ is related to the average cluster size $\chi(p)$. Consider an occupied site at position $\mathbf{r}_i$. The probability that a site at position $\mathbf{r}_j$ is occupied and belongs to the same cluster is $g(\mathbf{r}_i, \mathbf{r}_j)$, hence

$$\begin{aligned}\sum_{\mathbf{r}_j} g(\mathbf{r}_i, \mathbf{r}_j) &= \sum_{\mathbf{r}_j} p^{|\mathbf{r}_i - \mathbf{r}_j|} \\ &= \cdots + p^2 + p^1 + p^0 + p^1 + p^2 + \cdots \\ &= \frac{1+p}{1-p} \\ &= \chi(p), \end{aligned} \quad (1.21)$$

which the reader can show by summing the geometric series.

1.2.5 Critical occupation probability

We conclude our analysis of one-dimensional percolation by remarking on the role played by the occupation probability $p = 1$. Approaching this value, the characteristic cluster size $s_\xi(p)$, the average cluster size $\chi(p)$, and the correlation length $\xi(p)$, all diverge. This is intimately related to the onset of percolation at $p = 1$. To mark the transition to a percolating phase, the probability $p = 1$ is referred to as the critical occupation probability, p_c. The divergence of the above characteristic quantities may be expressed as a power law in terms of $(p_c - p)$, the distance of p below p_c.

1.3 Percolation on the Bethe Lattice

In one-dimensional percolation, it is not hard to calculate the cluster number density $n(s,p)$, thereby obtaining various quantities of interest such as the characteristic cluster size $s_\xi(p)$ and the average cluster size $\chi(p)$, as well as determining the critical occupation probability p_c. In the Bethe lattice, a mathematical construct also known as the Cayley tree, it is also possible to obtain analytical results for the above quantities [Fisher and Essam, 1961]. As we shall see, one-dimensional and Bethe lattices share the simplifying property of having no loops: any two sites in such lattices have a unique path between them.

1.3.1 *Definition of the Bethe lattice*

A Bethe lattice is a tree where each site has z neighbours, see Figure 1.7.

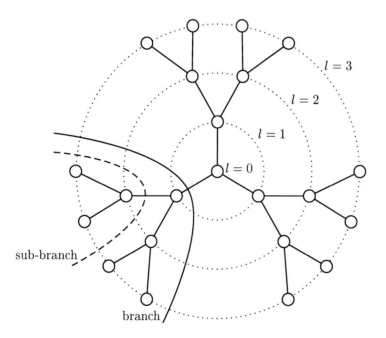

Fig. 1.7 The Bethe lattice with coordination number $z = 3$ and four generations $l = 0, 1, 2, 3$, where the parent in the centre is the 0th generation. One of the $z = 3$ branches originating from the parent in the centre is enclosed with a solid line. One of the $z-1 = 2$ sub-branches within this branch is enclosed with a dashed line. Note that for $z = 2$, the Bethe lattice is equivalent to a one-dimensional lattice.

Topologically, a Bethe lattice with $z > 2$ is different from a hyperlattice. For example, the usual Euclidean distance cannot be naturally carried over to the Bethe lattice. However, the so-called chemical distance, that is, the length of the unique path between two sites, does provide a notion of distance that is applicable in the site-site correlation function, see Section 1.3.6. In the following, it will be convenient to refer to branches and sub-branches of a given site. In Figure 1.7, the centre site has z branches, and each branch contains $z - 1$ sub-branches. However, in infinite lattices, all sites are equivalent. For example, there is nothing special about the centre site, or, for that matter, any of the other sites displayed in Figure 1.7. Thus a sub-branch is itself a branch, and we only make the distinction for the sake of clarifying our discussion.

1.3.2 Critical occupation probability

The percolating cluster is equivalent to a cluster which extends indefinitely. Imagine performing a walk on the percolating infinite cluster where the retracing of steps is forbidden. At each step, there are $(z - 1)$ branches of which only $p(z - 1)$ are accessible on average. In order to continue walking on the cluster, there must be at least one accessible branch to walk along, that is,

$$p(z - 1) \geq 1. \tag{1.22}$$

Therefore, we identify the transition to percolation at

$$p_c = \frac{1}{z - 1}. \tag{1.23}$$

The critical occupation probability depends on the coordination number, z. We say that p_c is non-universal because it depends on the lattice details. Note that for $z = 2$, we recover the one-dimensional critical occupation probability $p_c = 1$. However, for $z > 2$, the critical occupation probability $p_c < 1$. In contrast to percolation in one dimension, p_c can be approached both from below and above.

1.3.3 Average cluster size

What is the average cluster size $\chi(p)$, to which an occupied site belongs on the Bethe lattice? Let us consider this occupied site to be the centre site of the lattice in Figure 1.7. We make use of the fact that a Bethe lattice has no loops, and that, in an infinite lattice, all sites are equivalent, to write closed-form equations for the average cluster size. Let B denote the contribution to the average cluster size from a given branch. Then, the average cluster size, to which the occupied centre site belongs, is

$$\chi(p) = 1 + zB, \qquad (1.24)$$

where the first term is the contribution from the centre site itself and the second term is the contribution from the z branches. If the parent of a branch is unoccupied, there is no contribution. If, however, the parent of a branch is occupied, that parent contributes its own weight together with a contribution B from each of its $z - 1$ sub-branches. The contribution from a sub-branch is identical to the contribution from a branch because all sites are equivalent. Thus

$$B = (1 - p)\, 0 + p\left[1 + (z - 1)B\right] \quad \text{for } 0 < p < p_c, \qquad (1.25)$$

therefore

$$B = \frac{p}{1 - (z - 1)p} \quad \text{for } 0 < p < p_c. \qquad (1.26)$$

Note that we restrict ourselves to $0 < p < p_c$ to ensure that the cluster, to which the centre belongs, is finite. Substituting B into Equation (1.24), we find that

$$\chi(p) = \frac{1 + p}{1 - (z - 1)p} \quad \text{for } 0 < p < p_c, \qquad (1.27)$$

which is consistent with the average cluster size for one-dimensional percolation when $z = 2$, see Equation (1.14). As p tends to p_c from below, the average cluster size diverges. As written in Equation (1.27), however, the way in which the average cluster size diverges is concealed. But identifying $p_c = 1/(z - 1)$ and multiplying the numerator and denominator on the right-hand side of Equation (1.27) by p_c, we arrive at

$$\chi(p) = \frac{p_c(1 + p)}{p_c - p} \quad \text{for } 0 < p < p_c. \qquad (1.28)$$

Figure 1.8 displays the average cluster size as a function of occupation probability p for $p < 1/2$ in a Bethe lattice with $z = 3$. The average cluster size on the Bethe lattice is an increasing function of p and diverges as p

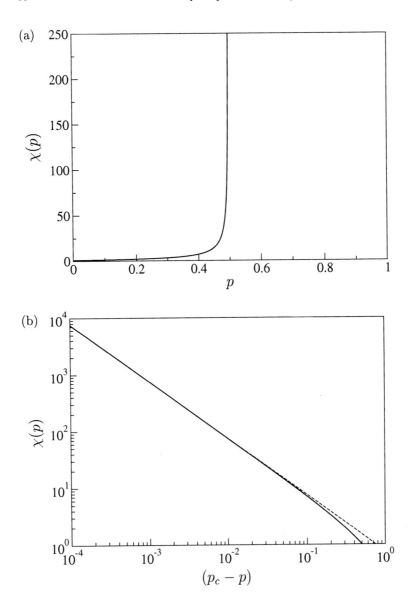

Fig. 1.8 Exact solution of the average cluster size for percolation on the Bethe lattice with $z = 3$ and $p < p_c$, where $p_c = 1/2$. (a) The average cluster size, $\chi(p)$, versus the occupation probability, p. For p approaching p_c from below, the average cluster size diverges. For p tending to zero, $\chi(p)$ tends to one. (b) The average cluster size, $\chi(p)$, versus $(p_c - p)$, the distance of p below p_c (solid line). For $p \to p_c^-$, the average cluster size diverges as a power law with exponent -1 in terms of $(p_c - p)$. The dashed straight line has slope -1.

approaches p_c from below as a power law with exponent -1 in terms of the distance of p below p_c:

$$\chi(p) = \frac{p_c(1+p)}{p_c - p} \to p_c(1+p_c)(p_c - p)^{-1} \quad \text{for } p \to p_c^-. \quad (1.29)$$

The exponent, characterising the divergence of the average cluster size as p approaches p_c is independent of the coordination number of the Bethe lattice. Thus, in contrast to p_c, the exponent is insensitive to the underlying lattice details. We say that the exponent is universal.

1.3.4 Transition to percolation

What is the probability $P_\infty(p)$ that a site belongs to the percolating infinite cluster? When $p \leq p_c$, there is no percolating infinite cluster, so $P_\infty(p) = 0$ for $p \leq p_c$, by definition. In order to calculate $P_\infty(p)$ for $p > p_c$, let $Q_\infty(p)$ denote the probability that a given branch does not connect to the percolating cluster. Thus, the probability that an arbitrarily selected site, for example the centre site in Figure 1.7, belongs to the percolating infinite cluster is the probability that the site is occupied, p, multiplied by the probability $1 - Q_\infty^z(p)$ that at least one of the z branches originating from this site connects to the percolating infinite cluster, that is,

$$P_\infty(p) = p\left[1 - Q_\infty^z(p)\right]. \quad (1.30)$$

A given branch does not connect to the percolating infinite cluster if the parent of the branch is unoccupied or, if the parent is occupied, none of the $z-1$ sub-branches connect to the percolating infinite cluster. Again, using the fact that all sites are equivalent, $Q_\infty(p)$ is also the probability that a sub-branch does not connect to the percolating infinite cluster, that is,

$$Q_\infty(p) = (1-p) + p\, Q_\infty^{z-1}(p). \quad (1.31)$$

To avoid the tedious algebra associated with solving Equation (1.31) for general z (see Exercise 1.7) we restrict ourselves to $z = 3$, where the solution to the resulting quadratic equation is

$$Q_\infty(p) = \frac{1 \pm \sqrt{(2p-1)^2}}{2p} = \begin{cases} 1 & \text{for } p \leq p_c \\ \frac{1-p}{p} & \text{for } p > p_c. \end{cases} \quad (1.32)$$

When $p \leq p_c$, there is no percolating infinite cluster, corresponding to the solution $Q_\infty(p) = 1$. When $p > p_c$, there is a percolating infinite cluster, corresponding to the solution $Q_\infty(p) = (1-p)/p < 1$. With these identifications in mind, we substitute $Q_\infty(p)$ into Equation (1.30) to obtain

$$P_\infty(p) = \begin{cases} 0 & \text{for } p \leq p_c \\ p\left[1 - \left(\frac{1-p}{p}\right)^3\right] & \text{for } p > p_c. \end{cases} \quad (1.33)$$

The probability $P_\infty(p)$ that a site belongs to the percolating infinite cluster is a continuous function of p with $P_\infty(p) = 0$ for $p \leq p_c$ and $P_\infty(p) > 0$ for $p > p_c$, see Figure 1.9(a).

The continuous function $P_\infty(p)$, however, is not differentiable at $p = p_c$. The value p_c, at which $P_\infty(p)$ picks up, marks a transition from a non-percolating to a percolating phase, and physicists describe this phenomenon as a continuous phase transition. The parameter $P_\infty(p)$ describing the phase transition is equivalent to the fraction of lattice sites belonging to the percolating infinite cluster. It becomes non-zero for $p > p_c$ when a finite fraction of the lattice sites belong to the percolating cluster. Thus the transition gives rise to a cluster of macroscopic size.

Although $P_\infty(p)$ is not diverging (after all, probabilities cannot exceed one), we nevertheless investigate how it picks up at $p = p_c$. To extract this information, we Taylor expand $P_\infty(p)$ around $p = p_c = 1/2$, see Appendix A. The first two non-zero terms in the Taylor expansion are

$$P_\infty(p) = 6(p - p_c) - 24(p - p_c)^2 + \cdots \quad \text{for } p > p_c. \quad (1.34)$$

For p approaching p_c from above, the higher order terms become negligible and, to leading order, $P_\infty(p)$ is proportional to the distance of p above p_c, that is,

$$P_\infty(p) \propto (p - p_c) \quad \text{for } p \to p_c^+. \quad (1.35)$$

Figure 1.9(b) shows $P_\infty(p)$ versus $p - p_c$ with logarithmic axes. The graph has been divided into two portions, the left portion being well approximated by Equation (1.35), while the right portion would require higher-order terms to be reproduced faithfully. Since $P_\infty(p)$ is proportional to $(p - p_c)$ raised to the power of $+1$ for $p \to p_c^+$, the left portion is well approximated by a straight line with slope $+1$. The above analysis on the behaviour of $P_\infty(p)$ as p approaches p_c from above is not only valid for $z = 3$. Rather like the behaviour of the average cluster size for p approaching p_c, the exponent of $+1$ characterising the pick-up of $P_\infty(p)$ is universal because it is independent of the coordination number z.

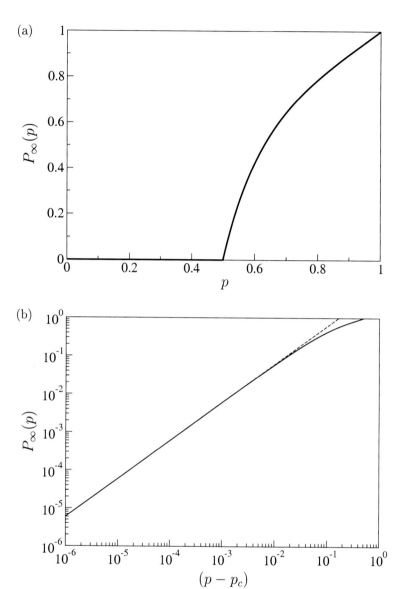

Fig. 1.9 Exact solution of the order parameter for percolation on the Bethe lattice with $z = 3$ where $p_c = 1/2$. (a) The probability, $P_\infty(p)$, that a site belongs to the percolating infinite cluster versus the occupation probability, p. For $p \leq p_c$, $P_\infty(p) = 0$ but then picks up abruptly at $p = p_c$. For $p > p_c$, a finite fraction of the sites belongs to the percolating cluster. (b) The probability, $P_\infty(p)$, versus $(p - p_c)$, the distance of p above p_c (solid line). For $p \to p_c^+$, the probability $P_\infty(p)$ picks up as a power law with exponent $+1$ in terms of $(p - p_c)$. The dashed straight line has slope $+1$.

1.3.5 Cluster number density

We now discuss the cluster number density for the Bethe lattice. We recall from one-dimensional percolation that $n(s,p) = (1-p)^2 p^s$, see Equation (1.2), corresponding to s occupied sites bounded by two unoccupied sites. This result can be formally generalised for clusters of any geometry.

Consider a cluster of size s and let t denote the size of the perimeter, that is, the number of unoccupied nearest neighbours of the cluster. In general, the number density of an s-cluster is not just $(1-p)^t p^s$, since s-clusters may neither have a unique geometry (possibly leading to a different number of perimeter sites t) nor a unique orientation. Introducing a degeneracy factor, $g(s,t)$, to count the number of different clusters of size s and perimeter t, the cluster number density can be expressed as

$$n(s,p) = \sum_{t=1}^{\infty} g(s,t)\,(1-p)^t p^s, \qquad (1.36)$$

where the sum is performed over all possible perimeters t.

In one dimension, an s-cluster has a unique geometry and orientation with $t = 2$ for all s and

$$g(s,t) = \begin{cases} 0 & \text{for } t \neq 2 \\ 1 & \text{for } t = 2, \end{cases} \qquad (1.37)$$

thus recovering Equation (1.2). In the Bethe lattice, $g(s,t)$ is not as trivial.

Figure 1.10 shows two clusters of size $s = 4$ with different geometries. We note that both clusters have $t = 6$ perimeter sites. The Bethe lattice enjoys the simplification that all clusters of size s have the same number of perimeter sites $t = 2 + s(z-2)$. The one-to-one correspondence between the cluster size s and perimeter sites t on the Bethe lattice reduces the sum in Equation (1.36) to a single term, so that

$$n(s,p) = g[s, 2 + s(z-2)]\,(1-p)^{2+s(z-2)} p^s. \qquad (1.38)$$

For coordination number $z = 3$, Equation (1.38) becomes

$$n(s,p) = g(s, 2+s)\,(1-p)^{2+s} p^s \quad \text{for } z = 3. \qquad (1.39)$$

It is, in fact, possible to enumerate $g(s, 2+s)$ by the use of generating functions, see e.g. [Fisher and Essam, 1961], but for our purposes we proceed more simply. To avoid enumerating $g(s, 2+s)$, we study the relative changes of the cluster number density around p_c. In this way, the degeneracy factor

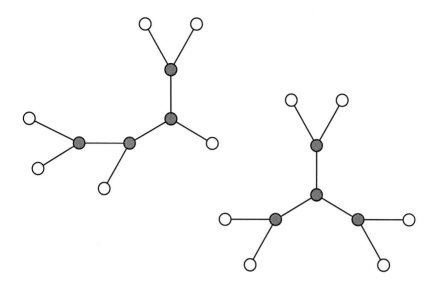

Fig. 1.10 Two different clusters of size $s = 4$ for site percolation on a Bethe lattice with $z = 3$. Occupied sites are dark grey while unoccupied sites are white. The perimeter, t, is the number of unoccupied nearest neighbours of a cluster. For both clusters, the perimeter $t = 2 + s = 6$.

$g(s, 2+s)$ cancels out, and we find

$$\begin{aligned} \frac{n(s,p)}{n(s,p_c)} &= \left[\frac{1-p}{1-p_c}\right]^2 \left[\frac{(1-p)\,p}{(1-p_c)\,p_c}\right]^s \\ &= \left[\frac{1-p}{1-p_c}\right]^2 \exp\left(s \ln\left[\frac{(1-p)\,p}{(1-p_c)\,p_c}\right]\right) \\ &= \left[\frac{1-p}{1-p_c}\right]^2 \exp\left(-s/s_\xi\right), \end{aligned} \quad (1.40)$$

where we have defined the characteristic cluster size

$$s_\xi(p) = -\frac{1}{\ln\left[\frac{(1-p)\,p}{(1-p_c)\,p_c}\right]}. \quad (1.41)$$

Figure 1.11(a) displays the characteristic cluster size $s_\xi(p)$ versus occupation probability p on the Bethe lattice with $z = 3$ where $p_c = 1/2$. When p approaches p_c, the characteristic cluster size diverges. The characteristic cluster size decreases with p when $p > p_c$ since it is a measure of the typical size of the largest finite cluster.

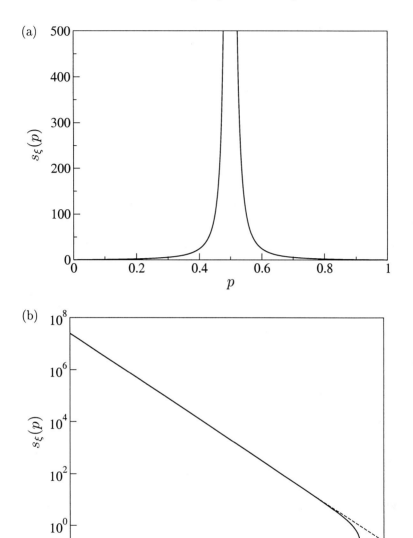

Fig. 1.11 Exact solution of the characteristic cluster size for percolation on the Bethe lattice with $z = 3$ where $p_c = 1/2$. (a) The characteristic cluster size, $s_\xi(p)$, versus the occupation probability, p. For p approaching p_c, the characteristic cluster size diverges. For p tending to zero or one, $s_\xi(p)$ tends to zero. (b) The characteristic cluster size, $s_\xi(p)$, versus $|p - p_c|$, the distance of p from p_c (solid line). For $p \to p_c$, the characteristic cluster size diverges as a power law with exponent -2 in terms of $(p - p_c)$. The dashed straight line has slope -2.

To extract how the characteristic cluster size $s_\xi(p)$ diverges for p approaching p_c (both from below and above), we rework the argument of the logarithm, using $p_c = 1/2$:

$$s_\xi(p) = -\frac{1}{\ln\left[\frac{(1-p)\,p}{(1-p_c)\,p_c}\right]}$$

$$= -\frac{1}{\ln[4p - 4p^2]}$$

$$= -\frac{1}{\ln[1 - 4(p-p_c)^2]}$$

$$\to \frac{1}{4}(p-p_c)^{-2} \qquad \text{for } p \to p_c. \qquad (1.42)$$

Thus the characteristic cluster size $s_\xi(p)$ diverges as a power law with exponent -2 in terms of $(p - p_c)$, the distance of p from $p_c = 1/2$, see Figure 1.11(b).

Rearranging Equation (1.40), we find that the cluster number density at occupation probability p is

$$n(s,p) = \left[\frac{1-p}{1-p_c}\right]^2 n(s,p_c)\exp(-s/s_\xi), \qquad (1.43\text{a})$$

$$s_\xi(p) = -\frac{1}{\ln\left[\frac{(1-p)\,p}{(1-p_c)\,p_c}\right]}, \qquad (1.43\text{b})$$

where the characteristic cluster size is given as in Equation (1.41). Note that for p approaching p_c, the characteristic cluster size $s_\xi(p)$ diverges so that the exponential becomes equal to one, guaranteeing the consistency of Equation (1.43a) at $p = p_c$.

Now we have to consider the behaviour of $n(s,p_c)$, the cluster number density at p_c. For the final word on $n(s,p_c)$, we refer the reader to [Fisher and Essam, 1961]; however, there are strong constraints that can be imposed on $n(s,p_c)$. First, according to Equation (1.17), the probability that a site belongs to a finite cluster, $\sum_{s=1}^{\infty} sn(s,p)$, is finite, see Figure 1.12. Second, the average cluster size

$$\chi(p) = \frac{\sum_{s=1}^{\infty} s^2 n(s,p)}{\sum_{s=1}^{\infty} sn(s,p)} \qquad \text{for } 0 < p < 1 \qquad (1.44)$$

diverges as p approaches p_c from below, see Equation (1.29). But since the denominator is finite, the numerator itself must diverge.

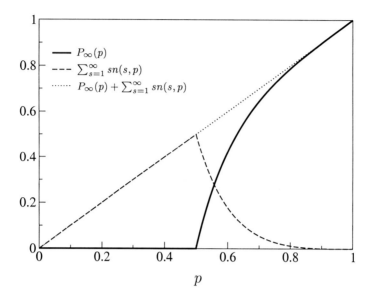

Fig. 1.12 Exact results for percolation on the Bethe lattice with $z = 3$ where $p_c = 1/2$. The probability, $P_\infty(p)$, that a site belongs to the percolating infinite cluster (solid line), added to the probability, $\sum_{s=1}^{\infty} sn(s,p)$, that the site belongs to a finite cluster (dashed line) equals the probability, p, that the site is occupied (dotted line). Note that when $p \leq p_c$, the sum over finite clusters $\sum_{s=1}^{\infty} sn(s,p_c) = p$, since $P_\infty(p) = 0$.

In summary, $n(s, p_c)$ must satisfy the constraints

$$\sum_{s=1}^{\infty} sn(s, p_c) = p_c - P_\infty(p_c), \tag{1.45a}$$

$$\sum_{s=1}^{\infty} s^2 n(s, p) \to \infty \quad \text{for } p \to p_c. \tag{1.45b}$$

Although these two constraints do not uniquely determine the form of $n(s, p_c)$, a power-law decay, $n(s, p_c) \propto s^{-\tau}$ for $s \gg 1$, is not only consistent with the two constraints in Equation (1.45) if the exponent $2 < \tau \leq 3$, but is also consistent with the exact cluster number density displayed in Figure 1.13(a). For large cluster sizes, the power-law decay of the cluster number density with exponent $\tau = 5/2$ can be proved rigorously, as we shall see in Section 3.4.

Note that in one dimension, the cluster number density equals zero at p_c and does not decay like a power law. Therefore, the lower constraint $2 < \tau$, originating from Equation (1.45a), does not apply. Collecting all these

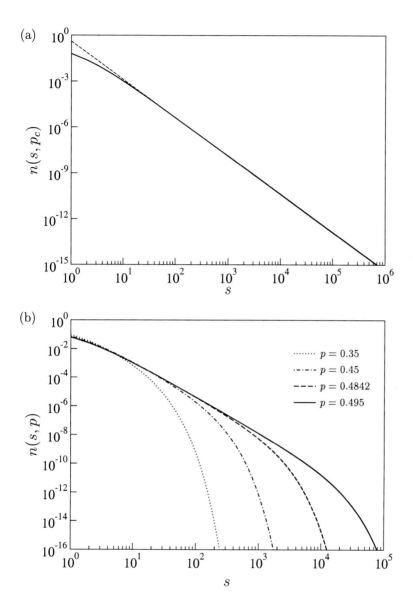

Fig. 1.13 Exact solution of the cluster number density for percolation on the Bethe lattice with $z = 3$ where $p_c = 1/2$. (a) The cluster number density, $n(s, p_c)$, versus the cluster size, s. For large cluster sizes, the cluster number density is well approximated by a power-law decay, $n(s, p_c) \propto s^{-\tau}$ for $s \gg 1$, with exponent $\tau = 5/2$. The dashed straight line has slope $-5/2$. (b) The cluster number density, $n(s, p)$, versus the cluster size, s. The four curves correspond to $p = 0.35, 0.45, 0.4842, 0.495$ with characteristic cluster sizes $s_\xi(p) \approx 10, 100, 1\,000, 10\,000$, respectively.

results together, the cluster number density $n(s,p)$ for the Bethe lattice is approximately of the form

$$n(s,p) \propto s^{-\tau} \exp\left(-s/s_\xi\right) \quad \text{for } s \gg 1, p \to p_c, \quad (1.46a)$$

$$s_\xi(p) \propto (p - p_c)^{-2} \quad \text{for } p \to p_c, \quad (1.46b)$$

where the characteristic cluster size diverges as a power law with exponent -2 in terms of $(p - p_c)$, when p tends to p_c.

In Figure 1.13(b) we display the exact solutions of the cluster number density versus the cluster size for four different values of occupation probabilities, $p = 0.35, 0.45, 0.4842, 0.495$. Each graph for the cluster number density has a characteristic cluster size, which is the typical size of the largest cluster in a realisation. From Equation (1.43b), we find that $s_\xi(p) \approx 10, 100, 1\,000, 10\,000$, respectively. For p approaching p_c and cluster sizes $s \ll s_\xi$, the cluster number density decays approximately as a power law in s, that is, $n(s,p) \propto s^{-\tau}$, while decaying faster than a power law for $s \gg s_\xi$. The graph corresponding to $p = p_c$ in Figure 1.13(a) is qualitatively different. There is no characteristic cluster size because $s_\xi(p_c) = \infty$. Thus, there are two distinct behaviours: for $p \neq p_c$, there exists a finite characteristic cluster size, which diverges for p approaching p_c, while at $p = p_c$, the characteristic cluster size is infinite.

As another test for the validity of the scaling form in Equation (1.46), we may calculate the average cluster size, $\chi(p)$, in terms of the cluster number density, by using Equation (1.44):

$$\chi(p) = \frac{\sum_{s=1}^\infty s^2 n(s,p)}{\sum_{s=1}^\infty s n(s,p)}$$

$$\propto \sum_{s=1}^\infty s^{2-\tau} \exp\left(-s/s_\xi\right)$$

$$\approx \int_1^\infty s^{2-\tau} \exp\left(-s/s_\xi\right) ds$$

$$= \int_{1/s_\xi}^\infty (us_\xi)^{2-\tau} \exp(-u) s_\xi \, du \quad \text{with } u = s/s_\xi$$

$$= s_\xi^{3-\tau} \int_{1/s_\xi}^\infty u^{2-\tau} \exp(-u) \, du, \quad (1.47)$$

where we have replaced the sum by an integral and made the substitution $u = s/s_\xi$. Letting p tend to p_c, the characteristic cluster size $s_\xi(p)$ diverges and the lower limit of the integral tends to zero. This integral is frequently encountered and is none other than the Gamma function

$$\Gamma(x) = \int_0^\infty u^{x-1} \exp(-u)\, du. \tag{1.48}$$

In our case, $x = 3 - \tau$ and the integral is just the number $\Gamma(3 - \tau)$. Thus the average cluster size $\chi(p)$ diverges like the characteristic cluster size $s_\xi(p)$ raised to the power $3 - \tau$. Equations (1.29) and (1.42) characterise, respectively, how $\chi(p)$ and $s_\xi(p)$ diverge when p tends to p_c^-. Therefore

$$\begin{aligned}\chi(p) &\propto (p_c - p)^{-1} \\ &\propto s_\xi^{3-\tau} \\ &\propto (p_c - p)^{2\tau - 6} \quad \text{for } p \to p_c^-,\end{aligned} \tag{1.49}$$

so we identify $\tau = 5/2$, which satisfies the constraint $2 < \tau \leq 3$ and is consistent with the power-law decay for large cluster sizes of the exact solution for the cluster number density displayed in Figure 1.13(a).

1.3.6 Correlation function

Given that a site i is occupied, the probability that a particular site j, l generations away, is occupied and belongs to the same finite cluster is the site-site correlation function, $g(i, j)$. Since there is a unique path between the sites i and j, $g(i, j) = p^l$. In the Bethe lattice, there is no Euclidean distance function between two general sites. However, treating site i as the parent in the centre of the Bethe lattice, the generation number of the site j provides a reasonable notion of distance. Accordingly, it is natural to calculate the average number of sites, $N(i; l)$, situated $l \geq 1$ generations away, that are occupied and belong to the same finite cluster as the parent site i in the centre. The expected number $N(i; l)$ is obtained by summing the site-site correlation function over all sites in the lth generation, that is,

$$N(i; l) = \sum_{j \in l} g(i, j) = n(l) p^l \quad \text{for } l \geq 1, \tag{1.50}$$

where $n(l)$ is the number of sites in the lth generation. In a Bethe lattice with coordination number z, each branch has $(z-1)^{l-1}$ sites in the lth generation. Since the parent in the centre provides z branches, there are in total $n(l) = z(z-1)^{l-1}$ sites in the lth generation. Therefore,

$$N(i; l) = z(z-1)^{l-1} p^l \quad \text{for } l \geq 1. \tag{1.51}$$

Restricting ourselves to $0 < p < p_c$, we can characterise the decay in the expected number of sites in the lth generation belonging to the same finite cluster as the parent in the centre in the time-honoured fashion

$$N(i;l) = \frac{z}{z-1}[(z-1)p]^l \quad \text{for } 0 < p < p_c$$
$$= \frac{z}{z-1}\exp(-l/l_\xi) \quad \text{for } 0 < p < p_c, l \geq 1, \quad (1.52)$$

where we have defined a characteristic generation

$$l_\xi(p) = -\frac{1}{\ln([z-1]p)} = -\frac{1}{\ln(p/p_c)} \quad \text{for } 0 < p < p_c, l \geq 1. \quad (1.53)$$

The characteristic generation $l_\xi(p)$ diverges for p approaching p_c from below, so that the exponential term in Equation (1.52) becomes one at $p = p_c$, leaving $N(i;l) = z/(z-1)$, independent of the generation l.

We can recover the average cluster size Equation (1.27) for $p < p_c$ by summing the site-site correlation function, $g(i,j)$, over all possible sites j,

$$\sum_j g(i,j) = g(i,i) + \sum_{j \neq i} g(i,j)$$
$$= 1 + \sum_{l=1}^\infty N(i;l)$$
$$= 1 + \sum_{l=1}^\infty \frac{z}{z-1}[(z-1)p]^l$$
$$= 1 + \frac{z}{(z-1)}\frac{(z-1)p}{1-(z-1)p} \quad \text{for } 0 < p < p_c$$
$$= \frac{1+p}{1-(z-1)p} \quad \text{for } 0 < p < p_c$$
$$= \chi(p) \quad \text{for } 0 < p < p_c. \quad (1.54)$$

The 'sum rule' in Equation (1.54) and the equivalent sum rule in one dimension in Equation (1.21) are special cases of a very general result known as the fluctuation-dissipation theorem.

1.4 Percolation in $d = 2$

The reader is now familiar with the techniques for extracting quantities of interest from the cluster number density such as the average cluster size and the characteristic cluster size. We have also been able to locate the transition to percolation by explicitly calculating the critical occupation probability. Furthermore, we have characterised the behaviour of the above quantities in the vicinity of the phase transition. We found that the location of the phase transition is dependent on the underlying lattice details, while the exponents characterising the behaviour of $\chi(p), s_\xi(p)$, and $P_\infty(p)$, near p_c, are not.

The exact results obtained thus far for the one-dimensional lattice and the Bethe lattice relied heavily on there being no loops. However, in general, clusters may form loops, making exact analytic results impossible but for a few special cases. Our programme now is to focus on percolation in two dimensions, which is nontrivial, motivating a more general approach to percolation [Stauffer and Aharony, 1994]. In doing so, we will lay the foundations for a general framework to encompass the phenomena encountered close to the continuous percolation phase transition. As we shall see later, this general framework can be applied to all systems exhibiting a continuous phase transition.

1.4.1 Transition to percolation

In one-dimensional percolation, the critical occupation probability $p_c = 1$, while for percolation on a Bethe lattice $p_c = 1/(z-1) < 1$ for coordination number $z > 2$. What is the critical occupation probability in two-dimensional site percolation on a square lattice? Looking back at Figure 1.2, a good guess would be p_c around 0.59. There is in fact no exact result for p_c on a square lattice, but numerics have closed in on $p_c = 0.59274621$ [Newman and Ziff, 2000]. For $p > p_c$ there is a single percolating cluster, while for $p \leq p_c$ there is no percolating cluster.

What is the probability that a site belongs to the percolating cluster? Figure 1.14(a) shows the numerically measured $P_\infty(p)$ for a finite lattice of size $L = 5\,000$. For finite lattice sizes, $P_\infty(p) > 0$ for $0 < p \leq p_c$. This is a finite-size effect. However, for increasing lattice size, the pick-up in $P_\infty(p)$ approaches p_c and for $L \to \infty$ we have $P_\infty(p) = 0$ for $p \leq p_c$.

Following the terminology of continuous phase transitions, we refer to $P_\infty(p)$ as the order parameter for percolation and to p_c as the critical point. The way that the order parameter picks up at the critical point $p = p_c$ is characterised by the critical exponent β, where

$$P_\infty(p) \propto (p - p_c)^\beta \quad \text{for } p \to p_c^+. \tag{1.55}$$

For the Bethe lattice with $z > 2$, the critical exponent $\beta = 1$, while for

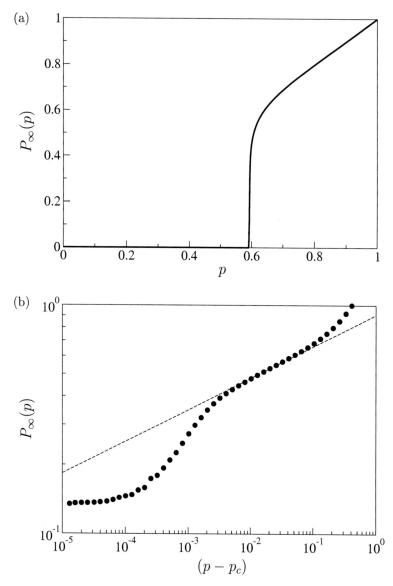

Fig. 1.14 Numerical results for the order parameter for two-dimensional site percolation on a square lattice of size $L = 5\,000$ where $p_c = 0.59274621$. (a) The order parameter, $P_\infty(p)$, versus the occupation probability, p. When $L \to \infty$, $P_\infty(p) = 0$ for $p \leq p_c$ but then picks up abruptly at $p = p_c$. For $p > p_c$, a finite fraction of the sites belong to the percolating cluster. (b) The order parameter, $P_\infty(p)$, versus $(p - p_c)$, the distance of p above p_c (solid circles). For $p \to p_c^+$, $P_\infty(p)$ picks up as a power law with exponent $\beta = 5/36$ in terms of $(p - p_c)$ when $L \to \infty$. The dashed straight line has slope $5/36$.

two-dimensional percolation, theory gives $\beta = 5/36$, independent of the underlying lattice (e.g. square, triangular or hexagonal), see for example [Grimmett, 1999; Francesco et al., 2001]. We will later justify the universality of β, that is, its insensitivity to the underlying lattice details.

In Figure 1.14(b) we attempt to confirm this theoretical value of β by plotting $\log P_\infty(p)$ against $\log(p - p_c)$ for $p > p_c$ measured in a finite lattice of size $L = 5\,000$. Without prior knowledge of the value of β, our attempt to measure it numerically would have failed, since the data do not follow an unambiguous straight line. It is, in fact, difficult to determine the critical exponent for the order parameter directly. In Section 1.8, we will develop a more powerful method for determining β, along with other critical exponents.

1.4.2 Average cluster size

Recalling Figure 1.2, we had the impression that the average cluster size increased with increasing $p < p_c$ before falling off for $p > p_c$. Bearing in mind that the average cluster size, by definition, only includes finite (non-percolating) clusters, the decrease for $p > p_c$ is due to the percolating cluster leaving less and less space for finite clusters.

Figure 1.15(a) shows the numerically measured average cluster size $\chi(p)$ for a square lattice of size $L = 5\,000$. For finite lattice sizes, the average cluster size cannot diverge at $p = p_c$. This is a finite-size effect. The divergence is capped by a maximum which increases with lattice size. For $L \to \infty$, the average cluster size $\chi(p)$ diverges for $p \to p_c$ as a power law with critical exponent $-\gamma$ in terms of the distance of p from p_c, that is,

$$\chi(p) \propto |p - p_c|^{-\gamma} \quad \text{for } p \to p_c. \tag{1.56}$$

For the Bethe lattice, the critical exponent $\gamma = 1$, while for two-dimensional percolation, theory gives $\gamma = 43/18$, independent of the lattice details.

In Figure 1.15(b) we attempt to confirm this theoretical value of γ by plotting $\log \chi(p)$ against $\log |p - p_c|$ for $p < p_c$ and $p > p_c$ measured in a finite lattice of size $L = 5\,000$. The data corresponding to $p < p_c$ apparently lie on a straight line for two orders of magnitude in $(p_c - p)$ before saturating because of the finiteness of the system. However, comparison with the theoretical dotted line shows that we have been misled into extracting an erroneous value for the critical exponent γ. In Section 1.8, we will describe the aforementioned method for determining critical exponents reliably.

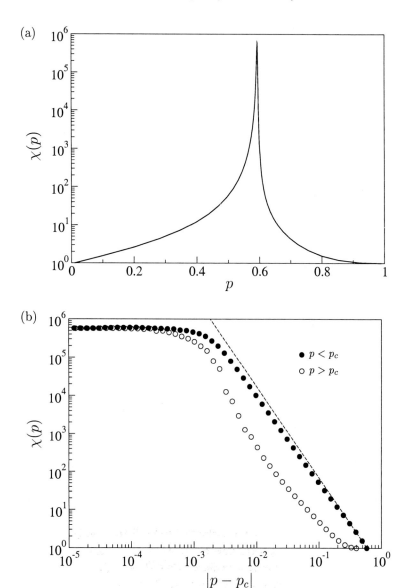

Fig. 1.15 Numerical results for the average cluster size for two-dimensional site percolation on a square lattice of size $L = 5\,000$ where $p_c = 0.59274621$. (a) The average cluster size, $\chi(p)$, versus the occupation probability, p. For $p \to p_c$, the average cluster size diverges but for finite lattices, the divergence is capped. For p tending to zero or one, $\chi(p)$ tends to one. (b) The average cluster size, $\chi(p)$, versus $|p - p_c|$, the distance of p from p_c. When $L \to \infty$, the average cluster size diverges as a power law $\chi(p) \propto |p - p_c|^{-\gamma}$ for $p \to p_c$ with $\gamma = 43/18$. The dashed straight line has slope $-43/18$.

1.4.3 Cluster number density – exact

Despite the above warnings, we will attempt to find the exact expression for the cluster number density in two-dimensional site percolation. Our starting point is the general form

$$n(s,p) = \sum_{t=1}^{\infty} g(s,t)\,(1-p)^t p^s, \qquad (1.57)$$

where $g(s,t)$ is the number of different clusters of size s with perimeter t. Unfortunately, unlike on the Bethe lattice, there is no unique relationship between the size of a cluster, s, and the perimeter t. Nevertheless, we can enumerate $g(s,t)$ by hand for small cluster sizes, see Figure 1.16. Thus for site percolation in a two-dimensional square lattice,

$$\begin{aligned}
n(1,p) &= (1-p)^4 p \\
n(2,p) &= 2(1-p)^6 p^2 \\
n(3,p) &= [2(1-p)^8 + 4(1-p)^7]\,p^3 \\
n(4,p) &= [2(1-p)^{10} + 8(1-p)^9 + 9(1-p)^8]\,p^4 \\
n(5,p) &= [2(1-p)^{12} + 12(1-p)^{11} + 28(1-p)^{10} + 20(1-p)^9 + (1-p)^8]\,p^5
\end{aligned} \qquad (1.58)$$

are the exact solutions of the cluster number density for clusters up to size $s=5$.

The number $\sum_t g(s,t)$ of different clusters of size s has been tabulated up to $s=53$, while $g(s,t)$ itself has been tabulated up to $s=40$ [Jensen, 2001]. The exact solutions of the cluster number density up to $s=40$ are shown in Figure 1.17. We cannot construct a theory of percolation based on clusters up to size $s=40$ – we must consider clusters of all sizes. Therefore, we must abandon our current approach and recognise that we cannot work out the exact solutions of the cluster number density for all cluster sizes.

1.4.4 Cluster number density – numerical

Since exact solutions are not available for large cluster sizes, we resort to a numerical approach to obtain the general form of the cluster number density $n(s,p)$ for any s and p.

Figure 1.18(a) shows numerical results for four different values of occupation probabilities, $p = 0.54, 0.55, 0.56, 0.57$. Each graph has a characteristic cluster size $s_\xi(p)$, which is the typical size of the largest cluster in a realisation, see Figure 1.2. For all cluster sizes $s \ll s_\xi$, the cluster number

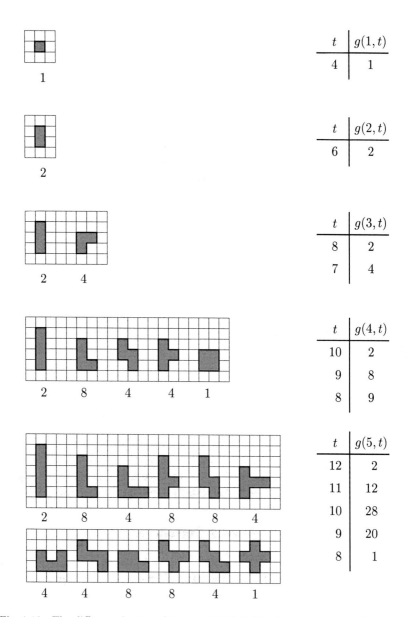

Fig. 1.16 The different clusters of size $s = 1, 2, 3, 4, 5$ for two-dimensional site percolation on a square lattice. Note that there are different perimeters, t, for a given cluster size when $s \geq 3$.

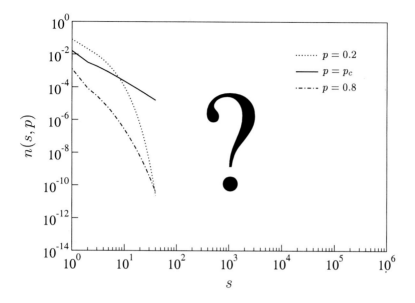

Fig. 1.17 Exact solution of the cluster number density, $n(s,p)$, versus the cluster size, s, up to size $s = 40$ for two-dimensional site percolation on a square lattice. The three curves correspond to occupation probabilities $p = 0.2, 0.59274621, 0.8$. No exact solutions are currently available above $s = 40$. Data courtesy of Dr Iwan Jensen, University of Melbourne, Australia.

density decays approximately as a power law in s, that is, $n(s,p) \propto s^{-\tau}$, while decaying faster than a power law for $s \gg s_\xi$. In the Bethe lattice, the critical exponent $\tau = 5/2$, while in two-dimensional percolation $\tau = 187/91$, independent of lattice details.

The graph corresponding to $p = p_c$ in Figure 1.18(b) is qualitatively different. There is apparently no characteristic cluster size. There are two distinct behaviours: for $p \neq p_c$, there exists a finite characteristic cluster size, which diverges for p approaching p_c, while at $p = p_c$, the characteristic cluster size is infinite. Just as in one-dimensional percolation and on the Bethe lattice, the characteristic cluster size $s_\xi(p)$ diverges as a power law with critical exponent $-1/\sigma$ in terms of the distance of p from p_c, that is,

$$s_\xi(p) \propto |p - p_c|^{-1/\sigma} \quad \text{for } p \to p_c. \tag{1.59}$$

For one-dimensional percolation, the critical exponent $\sigma = 1$, for the Bethe lattice with $z > 2$, $\sigma = 1/2$, while for two-dimensional percolation, theory gives $\sigma = 36/91$, independent of the lattice details.

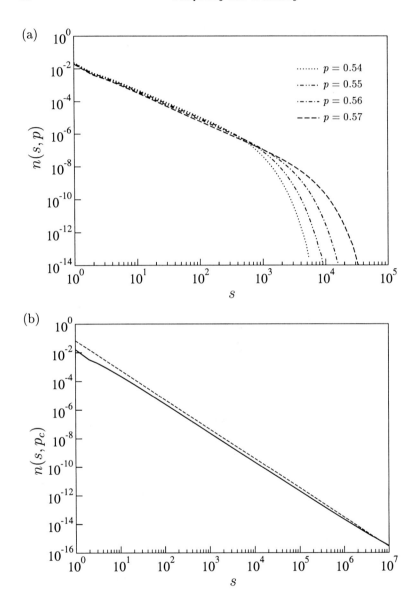

Fig. 1.18 Numerical results for the cluster number density, $n(s,p)$, versus the cluster size, s, for two-dimensional site percolation on a square lattice of size $L = 5\,000$ where $p_c = 0.59274621$. (a) For occupation probabilities $p = 0.54, 0.55, 0.56, 0.57$, the characteristic cluster size is finite and $n(s,p)$ decays as a power law for $1 \ll s \ll s_\xi$ and rapidly for $s \gg s_\xi$. (b) At $p = p_c$, the characteristic cluster size is infinite and for large cluster sizes, the cluster number density is well approximated by a power-law decay, $n(s,p_c) \propto s^{-\tau}$ for $s \gg 1$, with $\tau = 187/91$. The dashed straight line has slope $-187/91$.

1.5 Cluster Number Density – Scaling Ansatz

Since we are not able to calculate the cluster number density exactly for two-dimensional percolation for large clusters, we gather numerical evidence for the form of $n(s,p)$. There are two distinct behaviours of $n(s,p)$, depending on whether $p \neq p_c$ or $p = p_c$.

For $p \neq p_c$, there is a finite characteristic cluster size and, for a given occupation probability p, the characteristic cluster size $s_\xi(p)$ marks the crossover between a power-law decay and a rapid decay of $n(s,p)$, qualitatively,

$$n(s,p) \propto \begin{cases} s^{-\tau} & \text{for } 1 \ll s \ll s_\xi \\ \text{decays rapidly} & \text{for } s \gg s_\xi. \end{cases} \qquad (1.60)$$

When p approaches p_c, the critical exponent σ determines how the characteristic cluster size diverges, see Equation (1.59).

For $p = p_c$, the characteristic cluster size is infinite, so that for any finite cluster, $s \ll s_\xi$. Therefore, the cluster number density decays as a power law for large cluster sizes

$$n(s,p_c) \propto s^{-\tau} \quad \text{for } s \gg 1, \qquad (1.61)$$

except in one dimension where $n(s,p_c) = 0$.

With this qualitative picture in mind, and drawing on the results for the Bethe lattice, we propose the following scaling ansatz[1] for the cluster number density

$$n(s,p) \propto s^{-\tau} \exp(-s/s_\xi) \quad \text{for } p \to p_c, s \gg 1, \qquad (1.62a)$$

$$s_\xi(p) \propto |p - p_c|^{-1/\sigma} \quad \text{for } p \to p_c, \qquad (1.62b)$$

where the characteristic cluster size diverges like a power law in terms of the distance of p from p_c.

Evidently, the proposed ansatz encompasses the Bethe lattice. If our ansatz is to be general, it must also include the exact solutions of the trivial one-dimensional lattice and the numerical results for two-dimensional lattices. Recalling the results for the cluster number density and the characteristic cluster size in one dimension, we update Equation (1.3) and Equa-

---
[1] A 'scaling ansatz' is synonymous with a 'scaling hypothesis' and refers to a proposition that has not been derived analytically, but is consistent with known data. For distinction, we will reserve the term 'scaling form' for a result that has been derived analytically.

tion (1.5) by replacing $p = 1$ with p_c, thus

$$n(s,p) = (p_c - p)^2 \exp(-s/s_\xi), \tag{1.63a}$$
$$s_\xi(p) \propto (p_c - p)^{-1} \quad \text{for } p \to p_c^-. \tag{1.63b}$$

Clearly, the critical exponent $\sigma = 1$. Apparently, the critical exponent τ must be zero. However, this leads to a contradiction in a relation between the critical exponents τ, σ and γ that we now derive based on the scaling ansatz in Equation (1.62). Following the derivation of the average cluster size in Equation (1.47) and invoking Equation (1.56), we have

$$\begin{aligned}\chi(p) &= s_\xi^{3-\tau}\, \Gamma(3-\tau) &&\text{for } p \to p_c \\ &\propto |p - p_c|^{-(3-\tau)/\sigma} &&\text{for } p \to p_c \\ &\propto |p - p_c|^{-\gamma} &&\text{for } p \to p_c, \end{aligned} \tag{1.64}$$

implying the relation

$$\gamma = \frac{3-\tau}{\sigma}. \tag{1.65}$$

This relation is consistent for the values of the critical exponents on the Bethe lattice for $z > 2$ and on two-dimensional lattices. We will demonstrate shortly that this relation is general, and does not rely on the explicit form of the scaling ansatz in Equation (1.62). In one dimension, we have proved that $\sigma = 1$, see Equation (1.5), and $\gamma = 1$, see Equation (1.15), implying that $\tau = 2$. Assuming that the relation in Equation (1.65) is general, we attempt to rewrite the one-dimensional result for the cluster number density to be consistent with $\tau = 2$ rather than $\tau = 0$:

$$\begin{aligned}n(s,p) &= (p_c - p)^2 \exp(-s/s_\xi) \\ &= s^{-2}[s(p_c - p)]^2 \exp(-s/s_\xi) \\ &= s^{-2}(s/s_\xi)^2 \exp(-s/s_\xi) &&\text{for } p \to p_c^- \\ &= s^{-2}\mathcal{G}_{1\mathrm{d}}(s/s_\xi) &&\text{for } p \to p_c^-, \end{aligned} \tag{1.66}$$

where we have identified $(p_c - p)$ with $1/s_\xi$, which is valid when $p \to p_c^-$, and defined the function

$$\mathcal{G}_{1\mathrm{d}}(s/s_\xi) = (s/s_\xi)^2 \exp(-s/s_\xi), \tag{1.67}$$

where the argument is a rescaled cluster size. Thus we can satisfy the relation in Equation (1.65) since we can now identify $\tau = 2$. However, the expression for the cluster number density does not have a pure exponential decay as a function of s/s_ξ, but a quadratic term in s/s_ξ multiplied by an exponential decay. We therefore propose a yet more general scaling ansatz for the cluster number density, replacing the exponential decay with a general scaling function $\mathcal{G}(s/s_\xi)$ that also includes the one-dimensional result, that is,

$$n(s,p) \propto s^{-\tau}\mathcal{G}(s/s_\xi) \quad \text{for } p \to p_c, s \gg 1, \qquad (1.68a)$$

$$s_\xi(p) \propto |p - p_c|^{-1/\sigma} \quad \text{for } p \to p_c, \qquad (1.68b)$$

where the characteristic cluster size diverges like a power law in terms of the distance of p from p_c. The function $\mathcal{G}$ is known as the scaling function for the cluster number density.[2]

1.5.1 Scaling function and data collapse

An examination of the scaling ansatz for the cluster number density in Equation (1.68) reveals that the p-dependence of the cluster number density on the left-hand side has been taken up by the characteristic cluster size, $s_\xi(p)$, on the right-hand side. A given occupation probability p determines the typical size of the largest cluster. Clusters with size larger than this are extremely rare. Thus we would expect the scaling function $\mathcal{G}(s/s_\xi)$ to decay rapidly for large arguments $s/s_\xi \gg 1$. Note that at $p = p_c$, the characteristic cluster size is infinite, so that $n(s, p_c) \propto s^{-\tau}\mathcal{G}(0)$. Apart from in one dimension, $\mathcal{G}(0)$ is a non-zero constant, therefore at $p = p_c$, the cluster number density decays like a power law for $s \gg 1$. Assuming that the scaling function is well behaved for small arguments, we can Taylor expand it around zero,

$$\mathcal{G}(s/s_\xi) = \mathcal{G}(0) + \mathcal{G}'(0)(s/s_\xi) + \frac{1}{2}\mathcal{G}''(0)(s/s_\xi)^2 + \cdots . \qquad (1.69)$$

In one dimension, $\mathcal{G}_{1d}(0) = \mathcal{G}'_{1d}(0) = 0$ and $\mathcal{G}''_{1d}(0) = 2$ from Equation (1.67). In higher dimensions, however, $\mathcal{G}(0)$ is non-zero. Therefore, the leading order behaviour of the scaling function for $s/s_\xi \ll 1$:

$$\mathcal{G}(s/s_\xi) = \begin{cases} (s/s_\xi)^2 & \text{for } d = 1 \\ \mathcal{G}(0) & \text{for } d > 1. \end{cases} \qquad (1.70)$$

---

[2]Strictly speaking, not one but two scaling functions must be defined, $\mathcal{G}_\pm$, corresponding to $p > p_c$ and $p < p_c$ respectively. We do not discuss the technical reasons for this here, but refer the reader to Appendix C.

For small arguments, the scaling function is quadratic when $d = 1$ and constant when $d > 1$.

Explicitly, the scaling function in Equation (1.68a)

$$\mathcal{G}(s/s_\xi) \propto s^\tau n(s,p) \quad s \gg 1. \tag{1.71}$$

Although the right-hand side of Equation (1.71) is a function of two variables, s and p, or, s and $s_\xi(p)$, the left-hand side is only a function of one variable s/s_ξ, the relative cluster size. Therefore, the cluster number densities are described by a single function, when suitably transformed and viewed in the appropriate relative scale. For cluster sizes, the only relevant scale is the characteristic cluster size.

Assuming that we have a set $\{n(s,p_1), n(s,p_2), n(s,p_3), \ldots\}$ of cluster number densities as a function of s for occupation probabilities $p_1, p_2, p_3, \ldots$, the transformation and rescaling for the cluster number densities is performed in the following way:

(1) For all arguments s, multiply each cluster number density by s^τ to obtain $\{s^\tau n(s,p_1), s^\tau n(s,p_2), s^\tau n(s,p_3) \ldots\}$ as a function of s.
(2) For each transformed cluster number density $s^\tau n(s,p)$, rescale s to s/s_ξ, to obtain $s^\tau n(s,p)$ versus s/s_ξ.

According to Equation (1.71), all the cluster number densities should fall onto the same curve representing the graph of the scaling function $\mathcal{G}$. Such a procedure is known as a data collapse. We now attempt a data collapse for the one-dimensional lattice, the Bethe lattice and two-dimensional lattice, thereby exposing the graphs for the scaling functions $\mathcal{G}_{1d}, \mathcal{G}_{Bethe}, \mathcal{G}_{2d}$, respectively. Each example will provide some insight into the procedure of data collapse.

1.5.2 Scaling function and data collapse in $d = 1$

We take as our set of cluster number densities the graphs in Figure 1.4 with occupation probabilities $p = 0.4, 0.905, 0.99, 0.999, 0.9999$. By studying Figure 1.4, it is plausible that one operation on each degree of freedom should be performed to shift the graphs on top of each other. The first operation in the described procedure transforms each graph in the y-direction in a non-uniform manner by multiplying the cluster number density by s^τ. Figure 1.19(a) displays the transformed set, having multiplied each cluster number density by s^τ for all arguments s, where $\tau = 2$. The distinctive feature of each graph, namely the onset of the rapid decay, now all lie at

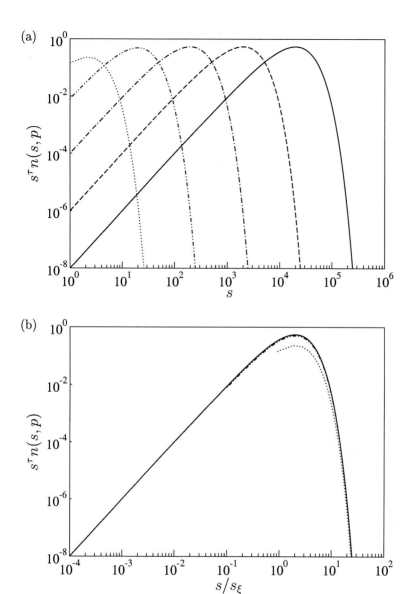

Fig. 1.19 Data collapse of the exact solution of the cluster number densities for one-dimensional percolation for $p = 0.4, 0.905, 0.99, 0.999, 0.9999$. (a) The transformed cluster number density, $s^\tau n(s,p)$, versus the cluster size, s, using the critical exponent $\tau = 2$. (b) The transformed cluster number density, $s^\tau n(s,p)$, versus the rescaled argument, s/s_ξ, where the characteristic cluster size $s_\xi(p) = -1/\ln p$. For $p \to p_c^-$, the curves collapse onto the graph for the scaling function $\mathcal{G}_{1\mathrm{d}}(s/s_\xi) = (s/s_\xi)^2 \exp(-s/s_\xi)$.

the same vertical position with only the horizontal position of this feature, given by the characteristic cluster size, distinguishing the graphs.

Thus, to make the graphs in Figure 1.19(a) collapse, we must rescale each argument by $s_\xi(p)$. Figure 1.19(b) displays the transformed and rescaled set, $s^\tau n(s,p)$ versus s/s_ξ. All graphs collapse onto one well-defined function, the scaling function

$$\mathcal{G}_{1\mathrm{d}}\left(s/s_\xi\right) = (s/s_\xi)^2 \exp(-s/s_\xi). \tag{1.72}$$

For $s/s_\xi \ll 1$, the function grows quadratically, and for $s/s_\xi \gg 1$ it decays rapidly. The crossover between these two behaviours is marked by $s/s_\xi \approx 1$.

The quality of the collapse improves with proximity to the critical point p_c. A scrutinising eye can detect that the graphs corresponding to $p = 0.4, 0.905$ deviate from the collapse. The deviation can be traced to the approximation of $(p_c - p)$ to $1/s_\xi$ in the derivation leading up to Equation (1.66). The validity of this approximation is visualised in Figure 1.5(b). For $p = 0.4$, the approximation is highly dubious, while for $p = 0.905$, the approximation is only slight.

1.5.3 Scaling function and data collapse on the Bethe lattice

We take as our set of cluster number densities the graphs in Figure 1.13(b) with occupation probabilities $p = 0.35, 0.45, 0.4842, 0.495$. We do not include the graph in Figure 1.13(a) corresponding to $p = p_c$, since in this case the characteristic cluster size is infinite, making the second operation in the data collapse meaningless.

The relative change in the cluster number density for the Bethe lattice is given by Equation (1.40). Figure 1.20(a) displays the relative change in the cluster number density multiplied by a p-dependent prefactor, versus the cluster size s. The onsets of the rapid decays all lie at the same vertical position, with only the horizontal positions of the onsets, given by the characteristic cluster sizes, distinguishing the graphs.

Accordingly, we rescale each argument by $s_\xi(p)$, see Figure 1.20(b), and obtain a perfect data collapse for all s and p onto the the scaling function

$$\mathcal{G}_{\mathrm{Bethe}}\left(s/s_\xi\right) = \exp\left(-s/s_\xi\right). \tag{1.73}$$

For $s/s_\xi \ll 1$, the scaling function is approximately constant, and for $s/s_\xi \gg 1$ it decays rapidly. The crossover between these two behaviours is marked by $s/s_\xi \approx 1$.

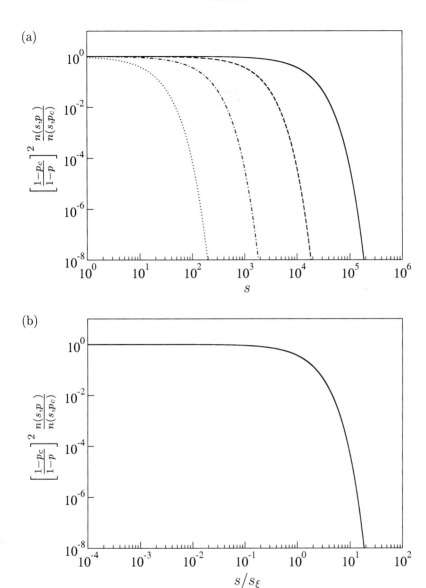

Fig. 1.20 Data collapse of the exact solution of the cluster number densities for percolation on the Bethe lattice for $p = 0.35, 0.45, 0.4842, 0.495$. (a) The relative change in cluster number densities, $[(1-p_c)/(1-p)]^2 \, n(s,p)/n(s,p_c)$, versus the cluster size, s. (b) The relative cluster number density, $[(1-p_c)/(1-p)]^2 \, n(s,p)/n(s,p_c)$, versus the rescaled argument, s/s_ξ, where the characteristic cluster size, $s_\xi(p) = -1/\ln(4p - 4p^2)$. The curves collapse onto the graph for the scaling function $\mathcal{G}_{\text{Bethe}}(s/s_\xi) = \exp\left(-s/s_\xi\right)$.

Only in special cases do we have the exact solution of the cluster number density at $p = p_c$ allowing us to calculate the relative change in the cluster number density applied in the data collapse procedure displayed in Figure 1.20. However, we have argued on general grounds that $n(s, p_c) \propto s^{-\tau}$ for $s \gg 1$ with exponent $\tau = 5/2$. Figure 1.21(a) displays the transformed cluster number density $s^\tau n(s, p)$ multiplied by the p-dependent prefactor, versus the cluster size s. The portions of the curves for $n(s, p)$ that were well approximated by $n(s,p) \propto s^{-\tau}$ are now flat. The onsets of the rapid decays all lie at the same vertical position, with only the horizontal positions of the onsets, given by the characteristic cluster sizes, distinguishing the graphs.

Accordingly, we rescale each argument by $s_\xi(p)$ and obtain the data collapse shown in Figure 1.21(b). When using the exact solution of the cluster number density at $p = p_c$, the data collapse is perfect. However, the approximation $n(s, p_c) \propto s^{-\tau}$ is only valid for large cluster sizes, see Figure 1.13. Although the departure from the decaying power law occurs in the range $1 \le s \lesssim 30$, upon rescaling with $s_\xi(p)$, the departure from the flat portion will be seen at smaller and smaller arguments s/s_ξ as p approaches p_c. Therefore, the curves do not collapse onto $\mathcal{G}_{\text{Bethe}}$ for small cluster sizes. For $s \gtrsim 30$, the collapse is perfect.

Similarly, only in special cases do we have knowledge of the p-dependent prefactor. This factor approaches one as $p \to p_c$ and will therefore not affect the cluster number densities for p close to p_c. Figure 1.22(a) displays the transformed set, having multiplied each cluster number density by s^τ for all arguments s, where $\tau = 5/2$. Again, the onsets of the rapid decays all lie at the same vertical position, with only the horizontal positions of the onsets, given by the characteristic cluster sizes, distinguishing the graphs.

After rescaling each argument by $s_\xi(p)$, we obtain the data collapse shown in Figure 1.22(b). The data collapse suffers in the small s behaviour of the cluster number density since the approximation $n(s, p_c) \propto s^{-\tau}$ is only valid in the limit of large cluster sizes. For $s \gtrsim 30$, the collapse is good, particularly for the graphs corresponding to p close to p_c, where the p-dependent prefactor in Equation (1.43a) approaches one.

In summary, for the Bethe lattice, the curves collapse onto the scaling function $\mathcal{G}_{\text{Bethe}} = \exp(-s/s_\xi)$ for $s \gg 1, p \to p_c$. For $s/s_\xi \ll 1$, the scaling function is approximately constant, and for $s/s_\xi \gg 1$ it decays rapidly. The crossover between these two behaviours is marked by $s/s_\xi \approx 1$.

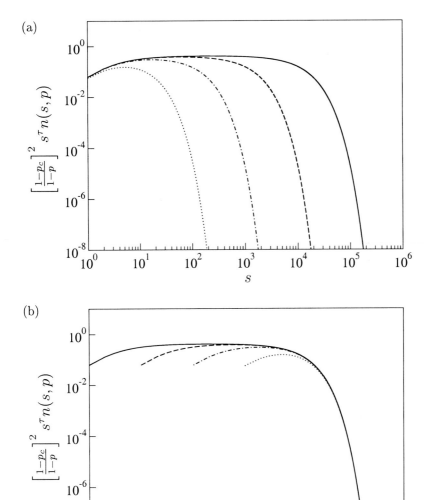

Fig. 1.21 Data collapse of the exact solution of the cluster number densities for percolation on the Bethe lattice for $p = 0.35, 0.45, 0.4842, 0.495$. (a) The transformed cluster number density, $[(1-p_c)/(1-p)]^2 s^\tau n(s,p)$, versus the cluster size, s, using the critical exponent $\tau = 5/2$. (b) The transformed cluster number density, $[(1-p_c)/(1-p)]^2 s^\tau n(s,p)$, versus the rescaled argument, s/s_ξ, where the characteristic cluster size, $s_\xi(p) = -1/\ln(4p-4p^2)$. For $s \gg 1$, the curves collapse onto the graph for the scaling function $\mathcal{G}_{\text{Bethe}}(s/s_\xi) = \exp\left(-s/s_\xi\right)$.

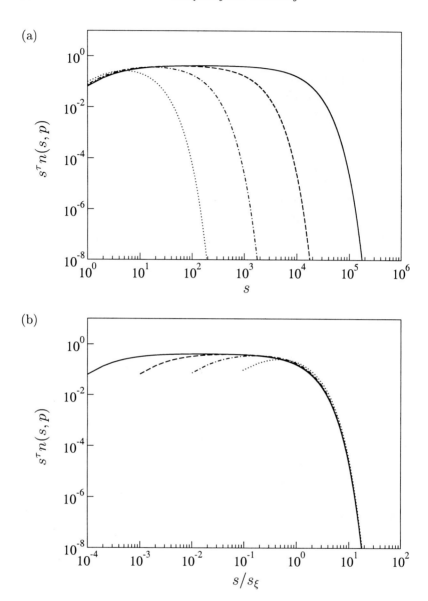

Fig. 1.22 Data collapse of the exact solution of the cluster number densities for percolation on the Bethe lattice for $p = 0.35, 0.45, 0.4842, 0.495$. (a) The transformed cluster number density, $s^\tau n(s,p)$, versus the cluster size, s, using the critical exponent $\tau = 5/2$. (b) The transformed cluster number density, $s^\tau n(s,p)$, versus the rescaled argument, s/s_ξ, where the characteristic cluster size, $s_\xi(p) = -1/\ln(4p - 4p^2)$. For $s \gg 1, p \to p_c$, the curves collapse onto the graph for the scaling function $\mathcal{G}_{\text{Bethe}}(s/s_\xi) = \exp\left(-s/s_\xi\right)$.

1.5.4 Scaling function and data collapse in $d = 2$

We consider percolation on a two-dimensional square lattice. We take as our set of cluster number densities the graphs in Figure 1.18(a) with occupation probabilities $p = 0.54, 0.55, 0.56, 0.57$ which are approaching the critical occupation probability p_c from below. We do not include the graph in Figure 1.18(b) corresponding to $p = p_c$ for the same reasons as above.

Figure 1.23(a) displays the transformed set, having multiplied each cluster number density by s^τ for all arguments s, where $\tau = 187/91$. The onsets of the rapid decays all lie at the same vertical position, with only the horizontal positions of the onsets, given by the characteristic cluster sizes $s_\xi(p) \propto |p - p_c|^{-1/\sigma}$, distinguishing the graphs.

Accordingly, we rescale each argument by $|p - p_c|^{-1/\sigma}$, using the critical exponent $\sigma = 36/91$. Figure 1.23(b) displays the transformed and rescaled set, $s^\tau n(s, p)$ versus $s|p - p_c|^{1/\sigma}$.

The scaling function $\mathcal{G}_{2d}$ for the two-dimensional lattice is given by the graph of the data collapse.[3] We have no explicit analytical form for the scaling function. However, for $s/s_\xi \ll 1$, we expect the scaling function to be constant, and for $s/s_\xi \gg 1$ to decay rapidly.

Note that the crossover between these two behaviours in Figure 1.23(b) is apparently displaced from $s/s_\xi = 1$. This is because the argument displayed, $s|p - p_c|^{1/\sigma}$, is only proportional to s/s_ξ and therefore the crossover location is displaced by some unknown factor.

For $s/s_\xi \gtrsim 10^{-2}$ the collapse is good. However, since the approximation $n(s, p_c) \propto s^{-\tau}$ is only valid for large cluster sizes, see Figure 1.18(b), the collapse suffers in the small s behaviour of the cluster number density. Moreover, it appears that the scaling function is not constant for small arguments but increases slightly with s for cluster sizes $1 \ll s \ll s_\xi$. The left-most portion corresponds to the peculiarities of small clusters, while the scaling ansatz only covers cluster sizes $s \gg 1$. Figure 1.18(b) reminds us that the critical exponent $\tau = 187/91$ is defined in the asymptotic limit $s \gg 1$. However, for finite s, there are corrections to the asymptotic behaviour which, in this case, is resulting in an effective (apparent) critical exponent. Therefore, when transforming the cluster number densities, the left-most portions of each graph will increase slightly rather than being constant. We will return to this issue later when discussing finite-size scaling in Section 1.8.

---

[3] Strictly speaking, since $p \to p_c^-$, the graph of the scaling function outlined in Figure 1.23(b) refers to the branch of $\mathcal{G}_{2d}$ associated with $p < p_c$.

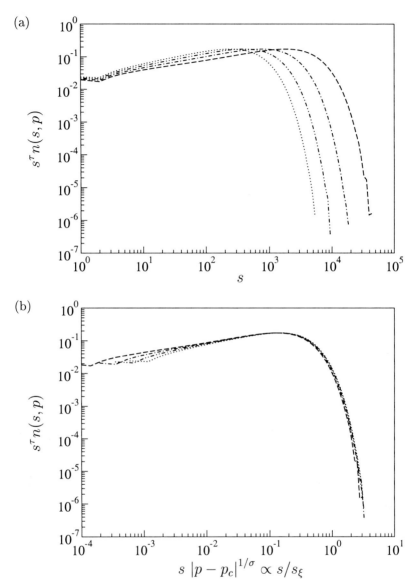

Fig. 1.23 Data collapse of the numerical results for the cluster number densities for two-dimensional site percolation on a square lattice of size $L = 1\,000$ for $p = 0.54, 0.55, 0.56, 0.57$. (a) The transformed cluster number density, $s^\tau n(s,p)$, versus the cluster size, s, using the critical exponent $\tau = 187/91$. (b) The transformed cluster number density, $s^\tau n(s,p)$, versus the rescaled argument, $s\,|p - p_c|^{1/\sigma}$, which is proportional to s/s_ξ, using $\sigma = 36/91$. For $s \gg 1, p \to p_c$, the curves collapse onto the graph for the scaling function $\mathcal{G}_{2\mathrm{d}}$.

1.6 Scaling Relations

We have introduced a variety of critical exponents characterising the behaviour of quantities of interest near the critical point. For example, the divergences of the average cluster size $\chi(p)$ and the characteristic cluster size $s_\xi(p)$ are described by the critical exponents γ and σ, respectively. The pick-up of the order parameter $P_\infty(p)$ at $p = p_c$, is described by the critical exponent β. Finally, the critical exponent τ relates the cluster number density $n(s,p)$ to the scaling function $\mathcal{G}(s/s_\xi)$ for $s \gg 1$.

Fortunately, these critical exponents are not independent. We have already derived a relation between the critical exponents γ, τ and σ in Equation (1.65). We will shortly derive an additional relation. Since we will then have two relations for four exponents, once any two exponents are specified, the remaining exponents are fixed. Such constraints among critical exponents are known as scaling relations.

The scaling relation, $\gamma = (3 - \tau)/\sigma$, was derived using the prototypical ansatz of the scaling function $\mathcal{G}(x) = \exp(-x)$, see Equation (1.62). We now put the general scaling ansatz for the cluster number density in Equation (1.68) to work by rederiving the above scaling relation.

The average cluster size is proportional to the second moment of the cluster number density. However, for the purpose of making the calculation more general, we will work out the kth moment of the cluster number density, that is,

$$M_k(p) = \sum_{s=1}^{\infty} s^k n(s,p) = \sum_{s=1}^{\infty} s^{k-\tau} \mathcal{G}(s/s_\xi), \quad (1.74)$$

assuming that the general scaling ansatz is valid for all cluster sizes. Replacing the sum by an integral and making the substitution $u = s/s_\xi$, we find

$$\begin{aligned} M_k(p) &\approx \int_1^\infty s^{k-\tau} \mathcal{G}(s/s_\xi) \, ds \\ &= \int_{1/s_\xi}^\infty (us_\xi)^{k-\tau} \mathcal{G}(u) s_\xi \, du \quad \text{with } u = s/s_\xi \\ &= s_\xi^{1+k-\tau} \int_{1/s_\xi}^\infty u^{k-\tau} \mathcal{G}(u) \, du \\ &\to s_\xi^{1+k-\tau} \int_0^\infty u^{k-\tau} \mathcal{G}(u) \, du \quad \text{for } p \to p_c. \end{aligned} \quad (1.75)$$

Letting p tend to p_c, the characteristic cluster size $s_\xi(p)$ diverges and the lower limit of the integral tends to zero. We will now determine when the definite integral converges. For dimensions $d > 1$, the scaling function is a non-zero constant for small arguments. In this case, for the integral to converge in the lower limit, we must impose the restriction $1 + k - \tau > 0$. For $d = 1$, the scaling function is proportional to u^2 for small arguments. The corresponding restriction $3 + k - \tau > 0$ is satisfied for $k \geq 0$, because in one dimension $\tau = 2$. In the upper limit, the scaling function falls off rapidly for $u \gg 1$, ensuring the convergence of the integral. Since the characteristic cluster size diverges, the contribution from small clusters is suppressed, thus the initial assumption that the general scaling ansatz for the cluster number density is valid for all cluster sizes is not crucial. The definite integral is just a number and thus the kth moment of the cluster number density

$$M_k(p) \propto s_\xi^{1+k-\tau} \qquad \text{for } p \to p_c$$
$$\propto |p - p_c|^{-(1+k-\tau)/\sigma} \qquad \text{for } p \to p_c, \qquad (1.76)$$

with the restriction $1 + k - \tau > 0$ for $d > 1$ and $k \geq 0$ for $d = 1$. The second moment corresponds to $k = 2$, and therefore we have rederived the scaling relation

$$\gamma = \frac{3 - \tau}{\sigma}, \qquad (1.77)$$

without assuming an explicit form for the scaling function $\mathcal{G}$. In one dimension, $\gamma = 1, \tau = 2, \sigma = 1$. In the Bethe lattice, $\gamma = 1, \tau = 5/2, \sigma = 1/2$. In two dimensions, $\gamma = 43/18, \tau = 187/91$, therefore $\sigma = 36/91$.

Equation (1.76) shows that it is not necessary to introduce new critical exponents for higher moments, because the separation between successive critical exponents is a constant.

An additional scaling relation results from considering the pick-up of the order parameter $P_\infty(p)$ for p approaching p_c from above. This is related to the first moment of the cluster number density by

$$P_\infty(p) = p - \sum_{s=1}^{\infty} s n(s, p), \qquad (1.78)$$

as we have already seen in Equation (1.17). To investigate the behaviour of $P_\infty(p)$ as $p \to p_c^+$, we focus on the vicinity of $p = p_c + (p - p_c)$ above p_c.

This is not possible in one dimension. In higher dimensions, however,

$$P_\infty(p) = p - \sum_{s=1}^{\infty} sn(s,p)$$

$$= p_c - \sum_{s=1}^{\infty} sn(s,p) + (p - p_c)$$

$$= \sum_{s=1}^{\infty} sn(s,p_c) - \sum_{s=1}^{\infty} sn(s,p) + (p - p_c), \quad (1.79)$$

where we have replaced p_c with just $\sum_{s=1}^{\infty} sn(s,p_c)$ since the order parameter $P_\infty(p_c) = 0$. Combining the sums and assuming once again that the general scaling ansatz for the cluster number density is valid for all s,

$$P_\infty(p) = \sum_{s=1}^{\infty}[sn(s,p_c) - sn(s,p)] + (p - p_c)$$

$$\propto \sum_{s=1}^{\infty} s^{1-\tau}[\mathcal{G}(0) - \mathcal{G}(s/s_\xi)] + (p - p_c). \quad (1.80)$$

Replacing the sum with an integral and making the substitution $u = s/s_\xi$ to reveal the scaling of the order parameter with $s_\xi(p)$, we find

$$P_\infty(p) \propto \int_1^{\infty} s^{1-\tau}[\mathcal{G}(0) - \mathcal{G}(s/s_\xi)]\, ds + (p - p_c)$$

$$\propto \int_{1/s_\xi}^{\infty} (us_\xi)^{1-\tau}[\mathcal{G}(0) - \mathcal{G}(u)]s_\xi\, du + (p - p_c) \quad \text{with } u = s/s_\xi$$

$$\propto s_\xi^{2-\tau} \int_{1/s_\xi}^{\infty} u^{1-\tau}[\mathcal{G}(0) - \mathcal{G}(u)]\, du + (p - p_c)$$

$$\propto s_\xi^{2-\tau} \int_0^{\infty} u^{1-\tau}[\mathcal{G}(0) - \mathcal{G}(u)]\, du + (p - p_c) \quad \text{for } p \to p_c^+, \quad (1.81)$$

because the lower limit of the integral tends to zero as p approaches p_c from above. To argue that the definite integral converges, we consider the lower and upper limit separately. For small arguments, $u \ll 1$, we Taylor expand the scaling function $\mathcal{G}(u) = \mathcal{G}(0) + \mathcal{G}'(0)u + \cdots$ to first order. Thus the integrand becomes $u^{2-\tau}\mathcal{G}'(0)$, so that the integral converges in the lower limit for $3 - \tau > 0$, in other words, $\tau < 3$. For large arguments, $u \gg 1$, the scaling function decays rapidly, thus the integrand becomes $u^{1-\tau}\mathcal{G}(0)$. Therefore the integral also converges in the upper limit for $2 - \tau < 0$, in other words, $\tau > 2$.

The pick-up of the order parameter approaching p_c from above is thus

$$P_\infty(p) \propto s_\xi^{2-\tau} \int_0^\infty u^{1-\tau}[\mathcal{G}(0) - \mathcal{G}(u)]\,du + (p - p_c) \quad \text{for } p \to p_c^+$$

$$\propto (p - p_c)^{(\tau-2)/\sigma} \int_0^\infty u^{1-\tau}[\mathcal{G}(0) - \mathcal{G}(u)]\,du + (p - p_c) \quad \text{for } p \to p_c^+.$$

There are two terms in $(p-p_c)$ with exponents $(\tau-2)/\sigma$ and 1, respectively. Since $(p - p_c)$ tends to zero, the dominating term will have the smaller of the two exponents. It is generally found that $(\tau - 2)/\sigma \leq 1$. Therefore,

$$P_\infty(p) \propto (p - p_c)^{(\tau-2)/\sigma} \propto (p - p_c)^\beta \quad \text{for } p \to p_c^+, \tag{1.82}$$

implying the scaling relation

$$\beta = \frac{\tau - 2}{\sigma}. \tag{1.83}$$

On the Bethe lattice, $\beta = 1, \tau = 5/2, \sigma = 1/2$ and in two dimensions, $\beta = 5/36, \tau = 187/91, \sigma = 36/91$. Although the above derivation is not applicable to one dimension, substituting the values of the critical exponents $\tau = 2, \sigma = 1$ regardless, implies that $\beta = 0$, which is nevertheless consistent, because $P_\infty(p) = (p - p_c)^0 = 1$ at $p = p_c$, see Figure 1.24. The scaling relation in Equation (1.83) is thus valid for all dimensions.

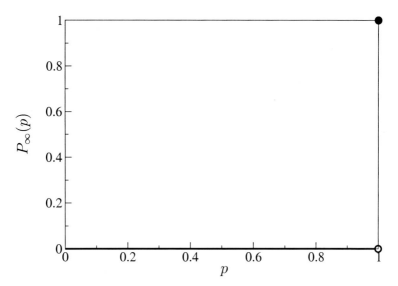

Fig. 1.24 Exact solution of the order parameter, $P_\infty(p)$, versus the occupation probability, p, for one-dimensional percolation. For $p < 1, P_\infty(p) = 0$ while at $p = 1, P_\infty(1) = 1$.

1.7 Geometric Properties of Clusters

At the outset, we asked whether the percolating cluster is a fractal. The cluster number density relates only to the statistics of cluster sizes, but holds no direct information about the geometric properties of clusters. We therefore introduce new quantities and ideas that are, in fact, widely applicable beyond percolation. In doing so, we will indeed show that the incipient infinite cluster is a fractal as the panel corresponding to $p = p_c$ in Figure 1.2 suggests. However, for $p > p_c$, there is a crossover from a fractal percolating cluster on length scales much smaller than a characteristic length scale to a uniform percolating cluster on length scales much larger that the characteristic length scale as the panels corresponding to $p > p_c$ in Figure 1.2 suggest.

1.7.1 *Self-similarity and fractal dimension*

By definition, $P_\infty(p)$ is the probability that a site belongs to the percolating cluster. Up until now, we have focused on the order parameter $P_\infty(p)$ as the quantity that signals the transition from the non-percolating phase with $P_\infty(p) = 0$ for $p \leq p_c$ into the percolating phase above p_c with $P_\infty(p) > 0$. For $p > p_c$, there exists a unique percolating cluster. At $p = p_c$, we will allow ourselves to refer to $P_\infty(p_c)$ as the density of the percolating cluster, even though, strictly speaking, only incipient infinite clusters exist. The following discussion therefore applies to $p \geq p_c$. In one dimension we cannot approach p_c from above, but for completeness we will comment on the trivial case of $d = 1$ when it requires special considerations.

Consider a percolating cluster in an infinite lattice, from which we mark out windows of size ℓ measured in units of the lattice spacing, see Figure 1.25(a).

Within a given window of size ℓ,

$$P_\infty(p;\ell) = \frac{M_\infty(p;\ell)}{\ell^2}, \qquad (1.84)$$

where $M_\infty(p;\ell)$ denotes the number of sites of the percolating cluster in that window. For convenience, we will refer to $M_\infty(p;\ell)$ as the 'mass' of the percolating cluster, although it is a dimensionless number. Accordingly, $P_\infty(p;\ell)$ can be alternatively interpreted as being the density (i.e., the number of sites per unit volume) of the percolating cluster within a window of size ℓ.

(a)

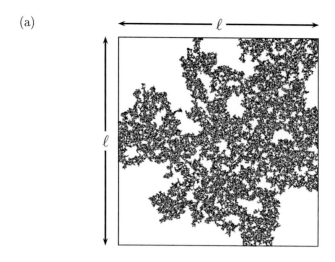

(b)

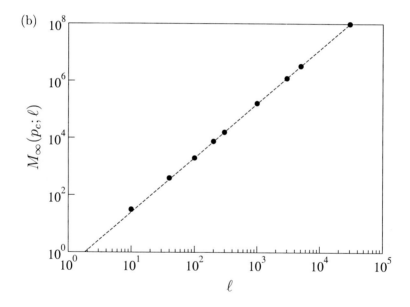

Fig. 1.25 Incipient infinite cluster in two-dimensional site percolation on a square lattice at occupation probability $p = p_c$. (a) Part of an incipient infinite cluster. Qualitatively, the incipient infinite cluster is an irregular, rough-edged structure, perforated by holes of all sizes limited only by the size of the window. The cluster extends into all the available space, however, its density decreases with distance. (b) Numerical results for the mass, $M_\infty(p_c; \ell)$, of an incipient infinite cluster versus the window size, ℓ (solid circles). For large ℓ, the mass increases as a power law in the window size, $M_\infty(p_c; \ell) \propto \ell^D$, where the exponent D is the fractal dimension. The dashed straight line has slope 91/48.

Since, by definition,

$$P_\infty(p) = \lim_{\ell \to \infty} P_\infty(p;\ell), \tag{1.85}$$

the density of the percolating cluster in the infinite lattice is zero at $p = p_c$, except in one dimension where $P_\infty(p_c) = 1$. How are we to reconcile the vanishing density with Equation (1.84)? To answer this, we study the mass of an incipient infinite cluster as a function of window size ℓ, see Figure 1.25(b).

The numerical data are consistent with

$$M_\infty(p_c;\ell) \propto \ell^D, \tag{1.86}$$

where D is known as the fractal dimension of the incipient infinite cluster. In two dimensions, $D = 91/48 \approx 1.90$. The power-law increase in the mass of an incipient infinite cluster with window size ℓ at $p = p_c$ is also seen in higher dimensions but with different fractal dimensions $D < d$, except in one dimension where $D = d = 1$. In the Bethe lattice, $D = 4$. Thus an incipient infinite cluster is indeed a fractal characterised by its fractal dimension D. The density of an incipient infinite cluster

$$P_\infty(p_c;\ell) = \frac{M_\infty(p_c;\ell)}{\ell^2} \propto \ell^{D-d} \tag{1.87}$$

decreases with increasing window size ℓ, for $d > 1$ and by Equation (1.85), $P_\infty(p_c) = 0$, as required.

This implies two unique properties of a fractal that may, at first, appear somewhat counterintuitive. First, the density of a fractal depends on the length scale ℓ on which it is measured. As many objects in Nature are fractal, such knowledge is crucial if one wants to estimate the density of an object on a length scale ℓ_2, based on the density measured on another length scale ℓ_1, see Appendix D. Second, the density of a fractal always decreases with increasing window size ℓ, irrespective of where on the fractal the window is placed.

Just by looking at the fractal in Figure 1.25(a), it would not be easy to conjecture the above properties. However, qualitatively, the decrease in the density of a fractal is caused by the holes inside it. If the holes were limited up to some characteristic size, then for sufficiently large windows, the density of a fractal would no longer decrease but become constant. If, on the other hand, there were no upper limit to the size of the holes, then larger windows would admit successively larger holes, maintaining the decrease in the density of a fractal.

Another important qualitative feature of fractals is self-similarity, which we will quantify later in Section 1.7.4. Figure 1.26 shows two incipient infinite clusters. Figure 1.26(a) (which is a reproduction of the incipient infinite cluster in Figure 1.25(a)) looks similar, at least statistically, to the incipient infinite cluster in Figure 1.26(b). The two incipient infinite clusters apparently have the same window size. However, a reader with keen eyesight will have spotted that the incipient infinite cluster in Figure 1.26(b) is in fact a rescaled version of the upper-right quarter of the incipient infinite cluster in Figure 1.26(a). This is the essence of self-similarity. A self-similar object looks alike on all length scales. For a further discussion of fractals, see Appendix D and for general reviews see [Feder, 1988; Falconer, 2003].

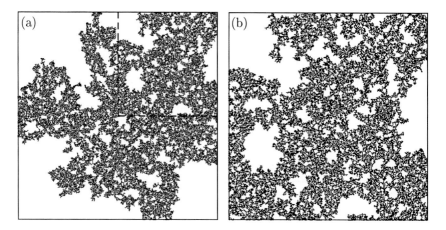

Fig. 1.26 Parts of two incipient infinite clusters in two-dimensional site percolation on a square lattice at occupation probability $p = p_c$. The incipient infinite cluster in (b) is a rescaled version of the upper right quarter of the incipient infinite cluster in (a).

1.7.2 Mass of a large but finite cluster at $p = p_c$

The relationship $M_\infty(p_c; \ell) \propto \ell^D$ applies to the mass of an incipient infinite cluster at $p = p_c$. How does the mass $M(s, p_c; \ell)$ of a large but finite cluster $s \gg 1$ behave at $p = p_c$? For small length scales ℓ, we do not know that the cluster we are considering is finite. Thus we would expect to find $M(s, p_c; \ell) \propto \ell^D$ where D is the fractal dimension of the incipient infinite cluster. However, for large length scales, contrary to the incipient infinite cluster, the finite cluster will be fully contained within a window of size ℓ with $M(s, p_c; \ell) = s$, see Figure 1.27.

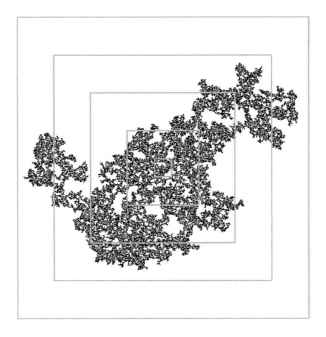

Fig. 1.27 A finite cluster of size $s = 42\,153$ in two-dimensional site percolation on a square lattice at occupation probability $p = p_c$. The origin of the windows is placed in the centre of mass of the s-cluster. For small window sizes, ℓ, the finite cluster appears fractal. When ℓ is large, the finite cluster will be fully contained within the windows, revealing that the cluster is, in fact, finite.

Thus the mass of a large s-cluster at $p = p_c$ satisfies

$$M(s, p_c; \ell) \propto \begin{cases} \ell^D & \text{for small } \ell \\ s & \text{for large } \ell. \end{cases} \quad (1.88)$$

In order to quantify what we mean by 'small' and 'large' ℓ, we define the radius of gyration, R_s, which is a measure of the linear scale of s-clusters in units of the lattice spacing. Assume that the positions of the occupied sites in a given s-cluster are denoted by $\mathbf{r}_i, i = 1, \ldots, s$. The centre of mass of the cluster is defined as

$$\mathbf{r}_{\text{cm}} = \frac{1}{s} \sum_{i=1}^{s} \mathbf{r}_i, \quad (1.89)$$

see Figure 1.28, while its radius of gyration squared is defined as the average

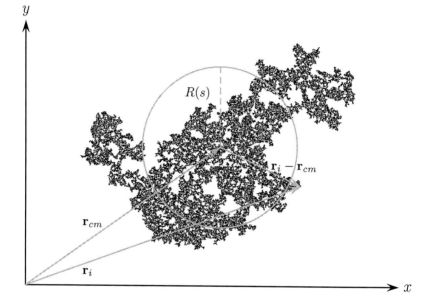

Fig. 1.28 A finite cluster of size $s = 42\,153$ in two-dimensional site percolation on a square lattice. The centre of mass is given by the position $\mathbf{r}_{\mathrm{cm}}$, and $|\mathbf{r}_i - \mathbf{r}_{\mathrm{cm}}|$ is the distance of occupied site $\mathbf{r}_i$ to the centre of mass $\mathbf{r}_{\mathrm{cm}}$. The radius of gyration, $R(s) \approx 153$ in units of the lattice spacing, is a measure of the linear size of the s-cluster (dashed line).

square distance to the centre of mass,

$$R^2(s) = \frac{1}{s} \sum_{i=1}^{s} |\mathbf{r}_i - \mathbf{r}_{\mathrm{cm}}|^2. \tag{1.90}$$

We denote by R_s^2 the average of $R^2(s)$ over the ensemble of all clusters of size s. Thus, more precisely,

$$M(s, p_c; \ell) \propto \begin{cases} \ell^D & \text{for } \ell \ll R_s \\ s & \text{for } \ell \gg R_s. \end{cases} \tag{1.91}$$

The average radius of gyration R_s is the sole length scale in the problem. For window sizes $\ell \lesssim R_s$, the cluster appears infinite to all intents and purposes. Therefore we would expect the relation $M(s, p_c; \ell) \propto \ell^D$ to hold all the way up to $\ell \approx R_s$, so that $s \propto R_s^D$. The same result is obtained by imposing continuity of $M(s, p_c; \ell)$ to ensure that the two limiting behaviours match up at $\ell \approx R_s$. From the right-hand side of Equation (1.91) it is apparent that only the ratio between the two length scales ℓ and R_s is relevant for characterising the behaviour of the mass of a large s-cluster at $p = p_c$ as a function of window size ℓ. We can summarise the crossover from the fractal behaviour for $\ell/R_s \ll 1$ to the trivial behaviour for $\ell/R_s \gg 1$ by introducing a crossover function, see Figure 1.29,

$$m(\ell/R_s) \propto \begin{cases} \text{constant} & \text{for } \ell/R_s \ll 1 \\ (\ell/R_s)^{-D} & \text{for } \ell/R_s \gg 1, \end{cases} \tag{1.92}$$

such that

$$M(s, p_c; \ell) = \ell^D m(\ell/R_s) \propto \begin{cases} \ell^D & \text{for } \ell/R_s \ll 1 \\ R_s^D & \text{for } \ell/R_s \gg 1. \end{cases} \tag{1.93}$$

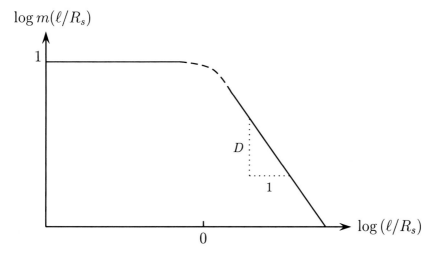

Fig. 1.29 A sketch of the crossover function, $m(\ell/R_s)$, for the mass of a large but finite cluster at $p = p_c$ versus the ratio, ℓ/R_s. The solid lines indicate the limiting behaviour of the crossover function, that is, constant for small arguments and decaying like a power law with exponent $-D$ for large arguments. The dashed line indicates that the crossover function is not known exactly in the region $\ell \approx R_s$.

On small length scales $\ell \ll R_s$, the crossover function is a constant and the cluster appears fractal. On large length scales $\ell \gg R_s$, the crossover function decays as a power law with exponent $-D$ and the cluster is finite with constant mass. The crossover takes place in the region $\ell \approx R_s$, where the exact form of the crossover function is not known.

1.7.3 Correlation length

For a particular characteristic cluster size $s_\xi(p)$ fixed by the occupation probability p, the associated radius of gyration defines a characteristic length scale that is proportional to the correlation length;[4] hence, we have

$$s_\xi \propto \xi^D. \tag{1.94}$$

Figure 1.30 is a sketch of the correlation length versus occupation probability. The vertical dotted line shows the position of the critical occupation probability p_c.

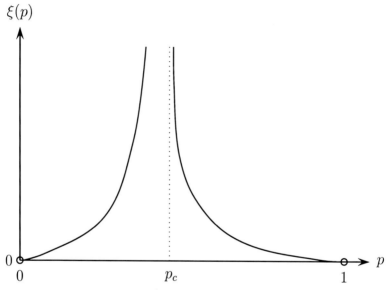

Fig. 1.30 A sketch of the correlation length, $\xi(p)$, versus the occupation probability, p. The vertical dotted line shows the position of p_c. For $p \to p_c$, the correlation length diverges as a power law with exponent $-\nu$ in terms of $|p - p_c|$, the distance of p from p_c, that is, $\xi(p) \propto |p - p_c|^{-\nu}$. For $p \to 0^+$ and $p \to 1^-$, the correlation length tends to zero.

The correlation length is the typical radius of the largest finite cluster, both below and above p_c. Since the correlation length is only defined for finite clusters, the percolating infinite cluster is excluded when $p > p_c$. Also, since finite clusters reside inside the holes of the percolating cluster when $p > p_c$, the correlation length can be identified with the typical radius of the largest holes in the percolating cluster, see Figure 1.2.

---
[4] For a discussion on the various different characteristic length scales and their interrelationship, see [Stauffer and Aharony, 1994].

Just as the characteristic cluster size diverges for p approaching p_c, so too does the correlation length. Thus at $p = p_c$, since there are finite clusters of all sizes, so too are there holes of all sizes in the incipient infinite cluster. The divergence is characterised by the critical exponent ν, where

$$\xi(p) \propto |p - p_c|^{-\nu} \quad \text{for } p \to p_c. \tag{1.95}$$

In one dimension, $\nu = 1$, see Equation (1.20). In two dimensions, $\nu = 4/3$. In the Bethe lattice, $\nu = 1/2$.

We have now introduced two 'new' critical exponents, the fractal dimension D of the incipient infinite cluster, and the exponent ν, characterising the divergence of the correlation length as p approaches p_c. To keep the number of independent critical exponents down to two, we must derive two additional scaling relations to accommodate D and ν.

The first of these follows immediately from Equation (1.94) and the definitions of the critical exponents σ and ν,

$$\begin{aligned} s_\xi &\propto \xi^D \\ &\propto |p - p_c|^{-D\nu} \quad \text{for } p \to p_c \\ &\propto |p - p_c|^{-1/\sigma} \quad \text{for } p \to p_c, \end{aligned} \tag{1.96}$$

implying the scaling relation

$$D = \frac{1}{\nu\sigma}. \tag{1.97}$$

In one dimension, $D = 1$, $\nu = 1$, and $\sigma = 1$. In two dimensions, $D = 91/48$, $\nu = 4/3$, and $\sigma = 36/91$. In the Bethe lattice, $D = 4$, $\nu = 1/2$, and $\sigma = 1/2$.

1.7.4 Mass of the percolating cluster for $p > p_c$

The second scaling relation comes about by considering the mass of the percolating cluster. As we have just seen, it is important to identify the relevant length scale in the problem. In the case of a large but finite cluster at $p = p_c$, where the correlation length $\xi(p)$ is infinite, the only relevant length scale is the average radius of gyration. The mass $M(s, p_c; \ell)$ of the finite cluster within a window of size ℓ therefore depends only on the ratio ℓ/R_s.

In the case of the percolating cluster for $p > p_c$, the only relevant length scale is the correlation length, $\xi(p)$. Thus we are led to consider the two limits, $\ell \ll \xi$ and $\ell \gg \xi$, in order to determine the mass $M_\infty(\xi; \ell)$ of the percolating cluster within a window of size ℓ at occupation probability p. Since a particular value of p fixes the correlation length, see Figure 1.30, we have replaced the p-dependence of the mass with a $\xi(p)$-dependence in order to make the comparison of length scales explicit.

For sufficiently small window sizes, $\ell \ll \xi$, it is not possible to detect whether the correlation length is finite. The percolating cluster appears fractal with holes of all sizes limited only by ℓ, thus

$$M_\infty(\xi; \ell) \propto \ell^D \quad \text{for } \ell \ll \xi. \tag{1.98}$$

Note that at $p = p_c$, the correlation length is infinite, guaranteeing that $\ell \ll \xi$ for all window sizes ℓ. The incipient infinite cluster is fractal on all length scales. When $p > p_c$, however, the opposite limit is accessible.

For sufficiently large window sizes, $\ell \gg \xi$, the holes are limited in size by $\xi(p)$ and are fully enclosed within the windows.

Given that we know how to calculate the mass of a fractal object, we divide the window into boxes of size $\xi(p)$ such that within each box the percolating cluster is a fractal with mass proportional to ξ^D, see Figure 1.31.

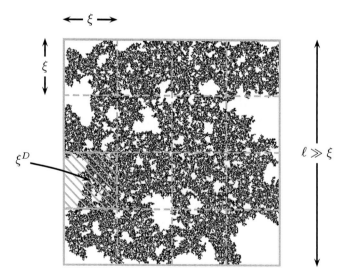

Fig. 1.31 A large window of size $\ell \gg \xi$ containing a percolating cluster at $p > p_c$ subdivided into $(\ell/\xi)^d$ boxes of size $\xi(p)$. On scales up to $\xi(p)$, the percolating cluster is a fractal. Therefore, the mass of the percolating cluster within each box is proportional to ξ^D.

In d-dimensions, the number of such boxes is $(\ell/\xi)^d$, therefore

$$\begin{aligned} M_\infty(\xi;\ell) &\propto (\ell/\xi)^d \xi^D \quad \text{for } \ell \gg \xi \\ &= \xi^{D-d} \ell^d \quad \text{for } \ell \gg \xi. \end{aligned} \quad (1.99)$$

Equation (1.99) is reminiscent of the formula mass = density × volume, identifying ξ^{D-d} as the density of the percolating cluster for length scales $\ell \gg \xi$, and ℓ^d as the volume of the window. Therefore, on scales $\ell \gg \xi$, the percolating cluster looks uniform with density ξ^{D-d}. But we also know that the density of the percolating cluster is simply given by the order parameter, that is,

$$\begin{aligned} M_\infty(\xi;\ell) &= P_\infty(p;\ell)\ell^d \\ &= P_\infty(p)\ell^d \quad \text{for } \ell \gg \xi \\ &\propto (p-p_c)^\beta \ell^d \quad \text{for } \ell \gg \xi, p \to p_c^+ \\ &\propto \xi^{-\beta/\nu} \ell^d \quad \text{for } \ell \gg \xi, p \to p_c^+, \end{aligned} \quad (1.100)$$

where, in the last step, we have exploited the behaviour of the correlation length near p_c. By comparing Equations (1.99) and (1.100), we read off the scaling relation

$$D - d = -\frac{\beta}{\nu}. \quad (1.101)$$

Such a scaling relation which involves the dimensionality d is commonly called a hyperscaling relation. Although the above derivation is not applicable to one dimension because we cannot approach p_c from above, substituting $D = 1$ and $d = 1$ regardless, implies $\beta = 0$, which is nevertheless consistent. In two dimensions, $D = 91/48$, $\beta = 5/36$, and $\nu = 4/3$. In the Bethe lattice, $D = 4$, $\beta = 1$, and $\nu = 1/2$ implies $d = 6$. We will be able to interpret this result shortly.

Collecting together the results for the two limits, the mass of the percolating cluster

$$M_\infty(\xi;\ell) \propto \begin{cases} \ell^D & \text{for } \ell \ll \xi \\ \ell^D (\ell/\xi)^{d-D} & \text{for } \ell \gg \xi, \end{cases} \quad (1.102)$$

where, for $\ell \gg \xi$, the right-hand side of Equation (1.99) has been reorganised.

Just as for a large finite cluster at $p = p_c$, we can summarise these two limiting behaviours by introducing a crossover function,

$$m_\infty(\ell/\xi) \propto \begin{cases} \text{constant} & \text{for } \ell \ll \xi \\ (\ell/\xi)^{d-D} & \text{for } \ell \gg \xi, \end{cases} \quad (1.103)$$

such that

$$M_\infty(\xi; \ell) = \ell^D m_\infty(\ell/\xi). \quad (1.104)$$

Equation (1.104) represents a crossover from fractal behaviour at length scales $\ell \ll \xi$ where the mass increases as ℓ^D, to non-fractal uniform behaviour at length scales $\ell \gg \xi$ where the mass increases as ℓ^d.

At $p = p_c$, the correlation length is infinite and only the fractal behaviour will be observed without any crossover to non-fractal uniform behaviour, that is,

$$M_\infty(\infty; \ell) = \ell^D m_\infty(0). \quad (1.105)$$

The left-hand side of Equation (1.104) is a function of the two length scales while the crossover function on the right-hand side is only a function of the ratio ℓ/ξ of the two length scales.

Since the crossover function, m_∞, is a function of the ratio ℓ/ξ, it is invariant under reduction of length scales $\xi \mapsto \xi/b, \ell \mapsto \ell/b$, where $b > 1$ is a dimensionless scale factor. After rescaling, the mass of the percolating cluster becomes

$$\begin{aligned} M_\infty(\xi/b; \ell/b) &= (\ell/b)^D m_\infty(\ell/\xi) \\ &= b^{-D} M_\infty(\xi; \ell); \end{aligned} \quad (1.106)$$

therefore, the mass of the percolating cluster satisfies

$$M_\infty(\xi; \ell) = b^D M_\infty(\xi/b; \ell/b) \quad \text{for } b > 1. \quad (1.107)$$

A function that obeys Equation (1.107) is known as a homogeneous function, see Appendix C.

We now show that we can recover the limiting forms for the mass of the percolating cluster in Equation (1.102). In doing so, we demonstrate that Equation (1.104), involving the crossover function, and Equation (1.107), expressing the homogeneity of the mass of the percolating cluster, are equivalent.

For both limits $\ell \ll \xi$ and $\ell \gg \xi$, our approach will be similar. We rescale all length scales such that the lesser of ξ and ℓ is equal to unity in terms of the lattice spacing. The lattice spacing constitutes a lower cutoff, below which it makes no sense to rescale.

Consider the limit $\ell \ll \xi$ and rescale ξ and ℓ by the factor $b = \ell$,

$$M_\infty(\xi; \ell) = \ell^D M_\infty(\xi/\ell; 1) \propto \ell^D \quad \text{for } \ell \ll \xi, \tag{1.108}$$

because the mass $M_\infty(\xi/\ell, 1)$ within a window of unit size and correlation length $\xi/\ell \gg 1$ is constant.

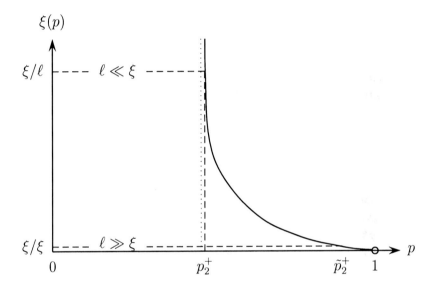

Fig. 1.32 A sketch of the correlation length, ξ, versus the occupation probabilities $p > p_c$. The vertical dotted line shows the position of p_c. For $p \to p_c^+$, the correlation length diverges as a power law with exponent $-\nu$ in terms of $(p - p_c)$, the distance of p above p_c, that is, $\xi(p) \propto (p - p_c)^{-\nu}$. For $p \to 1^-$, the correlation length tends to zero.

To justify that $M_\infty(\xi/\ell, 1)$ is a constant in the limit $\xi/\ell \gg 1$, we appeal to Figure 1.32, which is a sketch of the correlation length versus occupation probability for $p > p_c$. A particular value of p fixes the correlation length and vice versa. At $p = p_c$, the correlation length is infinite and holes of all sizes exist. For $p > p_c$ the correlation length is finite, so that there is a largest typical size of holes. With increasing p, the percolating cluster occupies more and more space. As the holes in the percolating cluster shrink, the correlation length decreases. For p approaching one, the largest typical size of a hole approaches one. At $p = 1$, there are no holes left, and the correlation length is zero.

At $p = p_c$, where ξ is infinite, and at $p = 1$, where ξ is zero, the correlation length remains invariant upon rescaling. However, reducing a finite correlation length by a factor $b > 1$, $\xi \mapsto \xi/b$, effectively corresponds to increasing the occupation probability p.

Consider an initial correlation length set by an initial occupation probability p_1^+ slightly above p_c, and $\ell \ll \xi$. Upon rescaling by the factor $b = \ell$, the correlation length reduces to ξ/ℓ, while the effective occupation probability increases slightly to p_2^+, as shown in Figure 1.32. Thus the mass $M_\infty(\xi/\ell, 1)$ within a window of unit size and correlation length $\xi/\ell \gg 1$ is the probability p_2^+ of occupying a single site.

Now, consider the limit $\ell \gg \xi$ and rescale ξ and ℓ by the factor $b = \xi$,

$$M_\infty(\xi; \ell) = \xi^D M_\infty(1; \ell/\xi)$$
$$\propto \xi^{D-d} \ell^d \qquad \text{for } \ell \gg \xi \qquad (1.109)$$

because the mass $M_\infty(1, \ell/\xi)$ within a window of size ℓ/ξ and unit correlation length equals the volume of the window $(\ell/\xi)^d$.

Consider once again an initial correlation length set by an initial occupation probability $\tilde{p}_1^+$ slightly above p_c, and $\ell \gg \xi$. Upon rescaling by the factor $b = \xi$, the correlation length reduces to 1, while the effective occupation probability increases markedly to $\tilde{p}_2^+ \approx 1$, as shown in Figure 1.32. Thus the mass $M_\infty(1, \ell/\xi)$ within a window of size ℓ/ξ and unit length is the probability $\tilde{p}_2^+ \approx 1$ of occupying a single site multiplied by the number of sites $(\ell/\xi)^d$.

By identification,

$$M_\infty(\xi; \ell) = \ell^D m_\infty(\ell/\xi)$$
$$= \begin{cases} \ell^D M_\infty(\xi/\ell; 1) & \text{for } \ell \ll \xi \\ \xi^D M_\infty(1; \ell/\xi) & \text{for } \ell \gg \xi. \end{cases} \qquad (1.110)$$

The limiting behaviours of the crossover function given in Equation (1.103) is therefore related to the mass of the percolating cluster in the manner

$$m_\infty(\ell/\xi) = \begin{cases} M_\infty(\xi/\ell; 1) & \text{for } \ell \ll \xi \\ (\ell/\xi)^{-D} M_\infty(1; \ell/\xi) & \text{for } \ell \gg \xi, \end{cases} \qquad (1.111)$$

where the right-hand side is only a function of the ratio ℓ/ξ.

It is also informative to examine the density of the percolating cluster in the two limits with the help of Equation (1.102),

$$P_\infty(\xi;\ell) = \frac{M_\infty(\xi;\ell)}{\ell^d} \propto \begin{cases} \ell^{D-d} & \text{for } \ell \ll \xi \\ \xi^{D-d} & \text{for } \ell \gg \xi; \end{cases} \quad (1.112)$$

see Figure 1.33. The length scale that will determine the density is the lesser of the two length scales ℓ and ξ. The density decreases as ℓ^{D-d} with increasing window size for length scales $\ell \ll \xi$ – the percolating cluster looks fractal. The density remains constant at value ξ^{D-d} for length scales $\ell \gg \xi$ – the percolating cluster looks uniform.

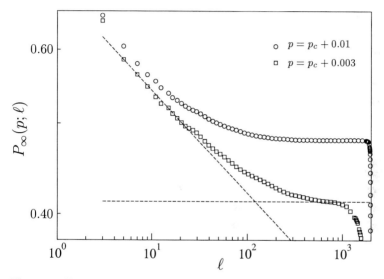

Fig. 1.33 Numerical results for the density of the percolating cluster, $P_\infty(p;\ell)$, versus the window size, ℓ, for two-dimensional site percolation on a square lattice of size $L = 5\,000$ at two different occupation probabilities $p > p_c$. Note that both axes are logarithmic. There is a crossover from fractal to uniform behaviour as ℓ increases.

1.8 Finite-Size Scaling

The problem of percolation is defined on an infinite lattice. In one dimension and on the Bethe lattice, exact results can be derived. For $d > 1$, we must resort to numerical results to verify theoretical predictions. Unfortunately, infinite lattices cannot be simulated. Fortunately, however, it is the size of the system relative to the correlation length that is important, rather than the absolute size of the system itself.

For $L \gg \xi$, all finite clusters up to ξ in linear size are contained within the lattice, ξ being the inherent cutoff radius. To all intents and purposes, such lattices appear infinite. All quantities, such as the average cluster size, remain the same for all lattices of size $L \gg \xi$, since they are determined by clusters up to size $s_\xi \propto \xi^D$. Because ξ increases as p tends to p_c, it becomes successively harder to maintain $L \gg \xi$. Ultimately, it is impossible to maintain $L \gg \xi$, because at $p = p_c$ the correlation length diverges, resulting in $L \ll \xi$ no matter what the lattice size.

For $L \ll \xi$, only finite clusters up to L in linear size are contained within the lattice, L being the imposed cutoff diameter. This is a finite-size effect. All quantities, such as the average cluster size, will now depend on the size L of the lattice, since they are determined by clusters up to size L^D by the finite size of the lattice.

At $p = p_c$, various quantities diverge, such as the correlation length, the characteristic cluster size, the average cluster size and all higher moments of the cluster number density. The divergences near p_c are characterised by critical exponents. For finite lattices, however, the divergences of all quantities are capped by a maximum which increases with lattice size. Therefore, the task of investigating critical behaviour numerically seems hopeless.

Far from being hopeless, we will now demonstrate how to exploit the inherent finiteness of the lattice at $p = p_c$, where necessarily $L \ll \xi$, to extract values for the critical exponents. This method, investigating the scaling of quantities as a function of lattice size at $p = p_c$, is known as finite-size scaling [Barber, 1983; Cardy, 1988; Privman, 1990]. For illustration, we consider the order parameter $P_\infty(p)$, the average cluster size $\chi(p)$, higher moments $M_k(p)$ of the cluster number density, and the cluster number density itself $n(s, p)$, although the method applies to all quantities associated with critical exponents.

1.8.1 Order parameter

Recall that in Figure 1.14 on page 32, we measured the order parameter as a function of occupation probability on a lattice of fixed size. We concluded that it is very difficult to identify a region with a straight line, let alone determine the critical exponent β. Only our prior knowledge of $\beta = 5/36$ guided us to the scaling region for the $L = 5\,000$ lattice. One approach to improve the situation would be to increase the lattice size in order to extend the scaling region over which the data are consistent with $P_\infty \propto (p - p_c)^\beta$. As a rule of thumb, the scaling region should persist for at least three orders of magnitude on both axes for a reliable estimate of the critical exponent.

However, a more powerful approach is to do the opposite in every sense. We measure the order parameter as a function of lattice size at the critical occupation probability. Furthermore, rather than increasing the lattice size above $L = 5\,000$, we decrease the lattice size!

To see why this procedure of investigating finite-size scaling works, we argue as follows. Consider an infinite lattice. Close to p_c, the correlation length diverges as $\xi \propto |p - p_c|^{-\nu}$, or in terms of the distance of p from the critical occupation probability p_c we have, $|p - p_c| \propto \xi^{-1/\nu}$. Hence, writing the order parameter as a function of correlation length,

$$P_\infty(p) \propto (p - p_c)^\beta \propto \xi^{-\beta/\nu} \quad \text{for } p \to p_c^+. \tag{1.113}$$

This result holds more generally for all lattice sizes $L \gg \xi$ since ξ is the lesser of the two length scales. Thus the measurement of the order parameter would be unchanged for different lattice sizes $L \gg \xi$. If, however, $L \ll \xi$, the lattice size is the lesser of the two length scales, and the measurement of the order parameter would depend on lattice size L. In summary,

$$P_\infty(p; L) \propto \begin{cases} \xi^{-\beta/\nu} & \text{for } L \gg \xi \\ L^{-\beta/\nu} & \text{for } 1 \ll L \ll \xi, \end{cases} \tag{1.114}$$

which is nothing more than Equation (1.112), using the scaling relation $D - d = -\beta/\nu$ and identifying the lattice size L with the window size ℓ. Figure 1.34(a) displays the order parameter versus occupation probability for different lattice sizes. For $p \lesssim 0.4$ and $p \gtrsim 0.8$ the order parameter is independent of lattice size, indicative of $L \gg \xi$.

At $p = p_c$, the correlation length is infinite and $L \ll \xi$ for all L. Thus the order parameter decays with lattice size as $P_\infty(p_c; L) \propto L^{-\beta/\nu}$, see inset in Figure 1.34(a), and only for $L \to \infty$ do we recover $P_\infty(p_c) = 0$. The way in which $P_\infty(p_c; L)$ decays to zero is characterised by the ratio β/ν, as revealed by Figure 1.34(b), where we have plotted $\log P_\infty(p_c; L)$ versus $\log L$. Figure 1.34(b) can be recovered from Figure 1.25(b) since $P_\infty(p_c; L) = M_\infty(p_c; L)/L^d$.

For the intermediate interval $0.4 \lesssim p \lesssim 0.8$, excluding $p = p_c$, one would observe a crossover around $L \approx \xi$ from $L \ll \xi$ behaviour to $L \gg \xi$ behaviour. Therefore extreme care must be exercised when interpreting data away from p_c in which a crossover may be operating. The three orders of magnitude rule of thumb for the scaling region will act as a safeguard in avoiding the region around $L \approx \xi$.

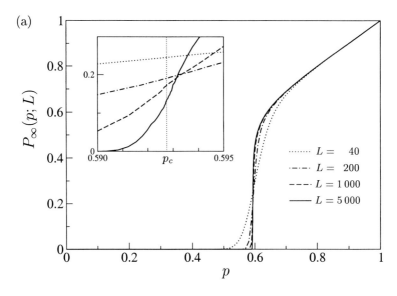

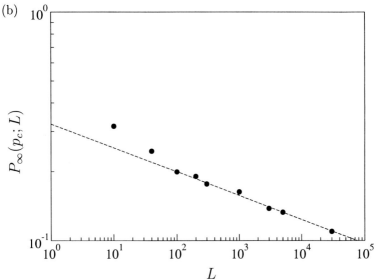

Fig. 1.34 Numerical results for the order parameter for two-dimensional site percolation on square lattices of size L. (a) The order parameter, $P_\infty(p;L)$, versus the occupation probability, p. When $L \to \infty$, $P_\infty(p) = 0$ for $p \leq p_c$ but then picks up abruptly at $p = p_c$. For $p \to p_c^+$, $P_\infty(p) \propto (p-p_c)^\beta$ with $\beta = 5/36$. Inset: At $p = p_c$ (vertical dotted line), the order parameter decays with increasing lattice size. (b) The order parameter at $p = p_c$, $P_\infty(p_c;L)$, versus the lattice size, L. The order parameter decays as a power law, $P_\infty(p_c;L) \propto L^{-\beta/\nu}$ for $L \gg 1$. The dashed straight line has slope $-5/48$.

1.8.2 Average cluster size and higher moments

The reasoning behind finite-size scaling for the average cluster size is the same as for the order parameter. Close to p_c, we can express the divergence of the average cluster size in terms of the correlation length,

$$\chi(p) \propto |p - p_c|^{-\gamma} \quad \text{for } p \to p_c$$
$$\propto \xi^{\gamma/\nu} \quad \text{for } p \to p_c. \quad (1.115)$$

This result holds generally for all lattice sizes $L \gg \xi$ since ξ is the lesser of the two length scales. However, when $L \ll \xi$, the lattice size is the lesser of the two length scales limiting the average cluster size. In summary,

$$\chi(p; L) \propto \begin{cases} \xi^{\gamma/\nu} & \text{for } L \gg \xi \\ L^{\gamma/\nu} & \text{for } 1 \ll L \ll \xi. \end{cases} \quad (1.116)$$

Figure 1.35(a) displays the average cluster size versus occupation probability for different lattice sizes. For $p \lesssim 0.4$ the average cluster size is independent of lattice size, indicative of $L \gg \xi$.

At $p = p_c$, the correlation length is infinite and $L \ll \xi$ for all L. Thus the average cluster size increases with lattice size as $\chi(p_c; L) \propto L^{\gamma/\nu}$ and only for $L \to \infty$ does the average cluster size diverge. The way in which $\chi(p_c; L)$ diverges is characterised by the ratio γ/ν, as revealed by Figure 1.35(b), where we have plotted $\log \chi(p_c; L)$ versus $\log L$.

The above derivation is readily extended to higher moments of the cluster number density, using Equation (1.76) on page 52. The kth moment of the cluster number density

$$M_k(p) \propto |p - p_c|^{-(1+k-\tau)/\sigma} \quad \text{for } p \to p_c$$
$$\propto \xi^{(1+k-\tau)/\sigma\nu} \quad \text{for } p \to p_c. \quad (1.117)$$

This result holds generally for all lattice sizes $L \gg \xi$ since ξ is the lesser of the two length scales. However, when $L \ll \xi$, the lattice size is the lesser of the two length scales limiting the average cluster size. In summary, using the scaling relation $D = 1/\sigma\nu$,

$$M_k(p; L) \propto \begin{cases} \xi^{D(1+k-\tau)} & \text{for } L \gg \xi \\ L^{D(1+k-\tau)} & \text{for } 1 \ll L \ll \xi, \end{cases} \quad (1.118)$$

which reduces to Equation (1.116) when $k = 2$.

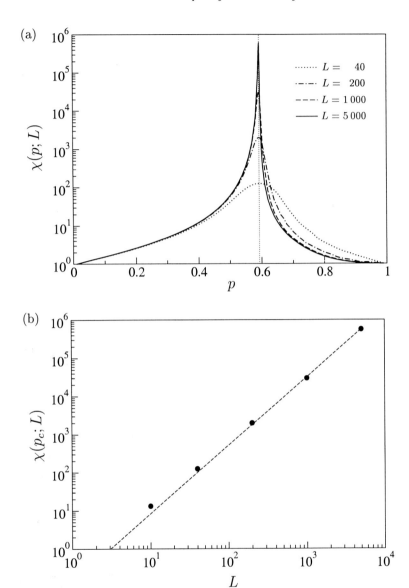

Fig. 1.35 Numerical results for the average cluster size for two-dimensional site percolation on square lattices of size L. (a) The average cluster size, $\chi(p;L)$, versus the occupation probability, p. When $L \to \infty$, $\chi(p) \propto |p - p_c|^{-\gamma}$ for $p \to p_c$ with $\gamma = 43/18$. For finite lattice sizes, the diverges at $p = p_c$ (vertical dotted line) is capped, but the average cluster size increases with lattice size. (b) The average cluster size at $p = p_c$, $\chi(p_c; L)$, versus the lattice size, L. The average cluster size increases as a power law, $\chi(p_c; L) \propto L^{\gamma/\nu}$ for $L \gg 1$. The dashed straight line has slope $43/24$.

1.8.3 Cluster number density

How does the cluster number density behave at $p = p_c$ as a function of lattice size? Figure 1.36 displays the cluster number density at $p = p_c$ versus cluster size for various lattice sizes. The characteristic cluster size increases with lattice size and only for $L \to \infty$ does the characteristic cluster size diverge.

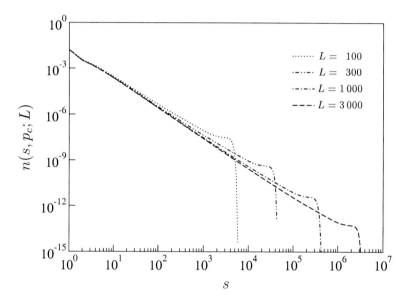

Fig. 1.36 Numerical results for the cluster number density, $n(s, p_c)$, versus the cluster size, s, for two-dimensional site percolation on square lattices of size $L = 100, 300, 1\,000, 3\,000$. At $p = p_c$, the correlation length is infinite so that $L \ll \xi$ for any lattice size L. The cluster number density is well approximated by a power-law decay, $n(s, p_c; L) \propto s^{-\tau}$ for cluster sizes $1 \ll s \ll L^D$, and the cluster number density decays rapidly for $s \gg L^D$.

Recall that from Equation (1.68)

$$n(s,p) \propto s^{-\tau} \mathcal{G}(s/s_\xi) \quad \text{for } p \to p_c, s \gg 1, \tag{1.119a}$$

where the characteristic cluster size, s_ξ, diverges as p tends to p_c like

$$\begin{aligned} s_\xi(p) &\propto |p - p_c|^{-1/\sigma} && \text{for } p \to p_c \\ &\propto \xi^{1/\sigma\nu} && \text{for } p \to p_c \\ &\propto \xi^D && \text{for } p \to p_c. \end{aligned} \tag{1.119b}$$

In an infinite system, the characteristic cluster size $s_\xi \propto \xi^D$. However, at $p = p_c$, the lattice size is always less than the correlation length and thus limits the characteristic cluster size, that is, for $1 \ll L \ll \xi$, the characteristic cluster size $s_\xi \propto L^D$. In summary,

$$n(s,p;L) \propto \begin{cases} s^{-\tau}\mathcal{G}(s/\xi^D) & \text{for } p \to p_c, s \gg 1, L \gg \xi \\ s^{-\tau}\tilde{\mathcal{G}}(s/L^D) & \text{for } p \to p_c, s \gg 1, 1 \ll L \ll \xi. \end{cases} \quad (1.120)$$

The finite-size scaling ansatz for the cluster number density

$$n(s,p_c;L) \propto s^{-\tau}\tilde{\mathcal{G}}(s/L^D) \quad \text{for } L \gg 1, s \gg 1 \quad (1.121)$$

implies that we can perform a data collapse by plotting the transformed cluster number density $s^\tau n(s,p_c;L)$ against the rescaled cluster size s/L^D. In doing so, we would not only lend support to the finite-size scaling ansatz, but also confirm the values of the critical exponents τ and D.

Figure 1.37(a) displays the result of the transformed cluster number density $s^\tau n(s,p_c;L)$ against cluster size s, with $\tau = 187/91$. The distinctive feature of each graph, namely the onset of the rapid decay, all lie at the same vertical position with only the horizontal position of this feature, given by L^D, distinguishing the graphs.

Figure 1.37(b) displays the result of rescaling the argument by L^D. Using $\tau = 187/91$ and $D = 91/48$, all the data fall on the same curve, which is the graph of the scaling function $\tilde{\mathcal{G}}$. Since the approximation $n(s,p_c;L) \propto s^{-\tau}$ is only valid for large cluster sizes, the curves do not collapse onto $\tilde{\mathcal{G}}$ for small cluster sizes. Moreover, we notice that the expected limiting behaviour of $\tilde{\mathcal{G}}$ being constant for small arguments is progressively revealed with increasing L.

In this example, we were fortunate enough to know the exact values of the critical exponents τ and D. Had we not known the exact values, performing a finite-size scaling data collapse would be a way to numerically estimate them. We remind the reader that when transforming the cluster number density, the value of τ must be chosen such that the distinctive feature of each graph, marking the onset of the rapid decay, aligns horizontally. Only in this way will the distinctive features collapse upon rescaling the cluster size by the characteristic cluster size L^D.

It is important to note that when transforming the cluster number density, one should not focus on collapsing or aligning the left-most portions of each graph by tuning the value of τ. The left-most portions correspond to the peculiarities of small clusters, while the scaling ansatz only covers

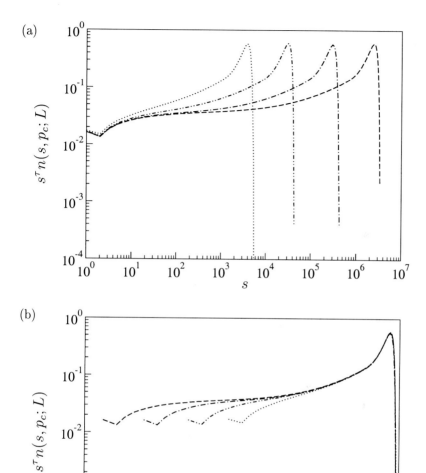

Fig. 1.37 Data collapse of the numerical results for the cluster number densities at $p = p_c$ for two-dimensional site percolation on square lattices of size $L = 100, 300, 1\,000, 3\,000$. (a) The transformed cluster number density, $s^\tau n(s, p_c; L)$, versus the cluster size, s, using the critical exponent $\tau = 187/91$. (b) The transformed cluster number density, $s^\tau n(s, p_c; L)$, versus the rescaled argument, s/L^D, which is proportional to s/s_ξ. For $s \gg 1, L \gg 1$, the curves collapse onto the graph for the scaling function $\tilde{\mathcal{G}}$.

cluster sizes $s \gg 1$. Therefore, the constant portion of the scaling function, in which the cluster number density decays like $s^{-\tau}$, would only be exposed for sufficiently large clusters $s \gg 1$, which at the same time must be much smaller than the characteristic cluster size set by the lattice size, that is, $s \ll L^D$. For most practical lattice sizes, this scaling region either does not exist or is so small that it cannot be identified reliably. We therefore urge the reader to focus on aligning the distinctive feature of each graph, related to the onset of the rapid decay, which is clearly identifiable if the data has been binned appropriately, see Appendix E.

Since all moments M_k are, by definition, derived from the cluster number density, we must be able to rederive the finite-size scaling result $M_k \propto L^{D(1+k-\tau)}$, valid for $L \gg 1$ at $p = p_c$. Using the finite-size scaling ansatz for the cluster number density, we find

$$\begin{aligned}
M_k(p_c; L) &= \sum_{s=1}^{\infty} s^k n(s, p_c; L) \\
&= \sum_{s=1}^{\infty} s^{k-\tau} \tilde{\mathcal{G}}(s/L^D) \qquad \text{for } L \gg 1 \\
&\approx \int_1^{\infty} s^{k-\tau} \tilde{\mathcal{G}}(s/L^D) \, ds \\
&= \int_{1/L^D}^{\infty} (uL^D)^{k-\tau} \tilde{\mathcal{G}}(u) L^D \, du \qquad \text{with } u = s/L^D \\
&= L^{(1+k-\tau)D} \int_{1/L^D}^{\infty} u^{k-\tau} \tilde{\mathcal{G}}(u) \, du \\
&\to L^{(1+k-\tau)D} \int_0^{\infty} u^{k-\tau} \tilde{\mathcal{G}}(u) \, du \qquad \text{for } L \gg 1, \qquad (1.122)
\end{aligned}$$

which is consistent with the limit $1 \ll L \ll \xi$ in Equation (1.118).

1.9 Non-Universal Critical Occupation Probabilities

We have concentrated entirely on site percolation, although so-called bond percolation is also widely studied. In order to define bond percolation, consider a two-dimensional square lattice composed of sites and bonds. In site percolation, all bonds are occupied with probability 1, while sites are occupied with probability p. In bond percolation, all sites are occupied with probability 1, while bonds are occupied with probability p. An example of each is shown in Figure 1.38.

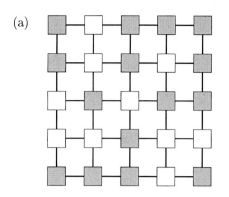

s	$N(s,p;L)$
1	2
2	1
4	1
7	1

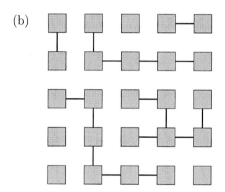

s	$N(s,p;L)$
1	4
2	2
5	1
6	2

Fig. 1.38 Two-dimensional percolation on a square lattice of size $L = 5$ made up of sites (squares) and bonds (links). Occupied sites are dark grey while unoccupied sites are white. Occupied bonds are shown as a link between sites while unoccupied bonds are not drawn. The tables to the right display the cluster size frequency, $N(s,p;L)$. (a) Site percolation (bonds always occupied). Sites are occupied with probability p. (b) Bond percolation (sites always occupied). Bonds are occupied with probability p.

For both site and bond percolation, a cluster is defined as a group of nearest-neighbouring occupied sites that are linked by occupied bonds. The cluster size is the number of sites in a cluster.[5]

Although we have focused on site percolation in one and two dimensions, and the Bethe lattice, for completeness we list the critical occupation probabilities for site and bond percolation in higher dimensions and various lattices in Table 1.1.

---

[5]In bond percolation, one may also define the cluster size as the number of bonds in a cluster.

Table 1.1 The critical occupation probabilities for various lattice types and dimensions in site and bond percolation. The current best estimates for the critical occupation probabilities, which are not known exactly, are given in decimal form. The second column lists the coordination number, z (the number of nearest neighbours) for a given lattice.

Lattice	z	Site percolation	Bond percolation
$d = 1$ line	2	1	1
$d = 2$ hexagonal	3	0.6971^a	$1 - 2\sin(\pi/18)^{b,c}$
square	4	0.59274621^d	$1/2^{b,e}$
triangular	6	$1/2^b$	$2\sin(\pi/18)^{b,c}$
$d = 3$ diamond	4	0.4301^f	0.3893^a
simple cubic	6	0.3116080^g	0.2488126^h
body-centred cubic	8	0.2459615^g	0.1802875^h
face-centred cubic	12	0.1992365^g	0.1201635^h
$d = 4$ hypercubic	8	0.196889^i	0.160130^i
$d = 5$ hypercubic	10	0.14081^i	0.118174^i
$d = 6$ hypercubic	12	0.1090^j	0.09420^k
$d = 7$ hypercubic	14	0.08893^l	0.078685^k
Bethe	z	$1/(z-1)^m$	$1/(z-1)^m$

[a][van der Marck, 1997].
[b][Sykes and Essam, 1964].
[c][Wierman, 1981].
[d][Newman and Ziff, 2000].
[e][Kesten, 1980].
[f][van der Marck, 1998].
[g][Lorenz and Ziff, 1998b].
[h][Lorenz and Ziff, 1998a].
[i][Paul et al., 2001].
[j][Hsu et al., 2005].
[k][Adler et al., 1990].
[l][Stauffer and Ziff, 2000].
[m][Fisher and Essam, 1961].

For the Bethe lattice, the critical occupation probability decreases with increasing coordination number. This trend is also observed for hypercubic lattices with coordination number $z = 2d$, where the critical occupation probability decreases with increasing dimension. Similarly, increasing the coordination number within a given dimension decreases the critical occupation probability. Clearly, the critical occupation probability depends heavily on the underlying lattice details. Therefore, the location of the phase transition, that is, the critical occupation probability, is non-universal.

1.10 Universal Critical Exponents

The critical exponents, by contrast, are universal. They only depend on dimensionality and are entirely insensitive to the underlying lattice details. Table 1.2 lists the critical exponents for various dimensions.

Table 1.2 The values of the critical exponents for percolation in dimensions $d = 1, 2, 3, 4, 5$, $d \geq 6$, and on the Bethe lattice. The critical exponents in $d = 3, 4, 5$ are not known exactly but the current best numerical results are listed with the uncertainty on the last digit(s) given by the figure(s) in the brackets. Two of the critical exponents have been measured numerically and the remaining critical exponents are evaluated from scaling relations.

Exponent: Quantity	$d=1$	$d=2$	$d=3$	$d=4$	$d=5$	$d \geq 6$	Bethe
$\beta : P_\infty(p) \propto (p-p_c)^\beta$	0 (dis)	5/36	0.4181(8)	0.657(9)	0.830(10)	1	1
$\gamma : \chi(p) \propto \|p-p_c\|^{-\gamma}$	1	43/18	1.793(3)	1.442(16)	1.185(5)[a]	1	1
$\nu : \xi(p) \propto \|p-p_c\|^{-\nu}$	1	4/3[b]	0.8765(16)	0.689(10)[c]	0.569(5)	1/2	1/2
$\sigma : s_\xi(p) \propto \|p-p_c\|^{-1/\sigma}$	1	36/91	0.4522(8)[d]	0.476(5)	0.496(4)	1/2	1/2
$\tau : n(s,p) \propto s^{-\tau} \mathcal{G}(s/s_\xi)$	2	187/91	2.18906(6)[d]	2.313(3)[e]	2.412(4)[e]	5/2	5/2
$D : s_\xi \propto \xi^D$	1	91/48[b]	2.523(6)	3.05(5)	3.54(4)	4	4

[a][Adler et al., 1990].
[b][den Nijs, 1979; Nienhuis, 1982].
[c][Ballesteros et al., 1997].
[d][Ballesteros et al., 1999].
[e][Paul et al., 2001].

Exact values are known in one and two dimensions,[6] in six or more dimensions and on the Bethe lattice. The critical exponents for $d \geq 6$ remain unchanged. The dimension above which the critical exponents remain unchanged is known as the upper critical dimension, d_u. For percolation, $d_u = 6$. Note that the critical exponents for $d \geq d_u$ and the Bethe lattice are identical. In some sense, the Bethe lattice corresponds to an infinite-dimensional lattice, because there are no loops of occupied sites. Likewise, the probability of forming a loop of occupied sites on a d-dimensional hypercubic lattice decreases with increasing dimensionality, and vanishes on an infinite-dimensional lattice. Another similarity is that on the Bethe lattice and on an infinite-dimensional lattice, the number of surface sites is proportional to the number of bulk sites.

---

[6]Derived from conformal field theory, see for example [Grimmett, 1999; Francesco et al., 2001].

Using the values of the critical exponents $D = 4, \beta = 1$, and $\nu = 1/2$ for the Bethe lattice, the hyperscaling relation $d = D + \beta/\nu$, derived previously in Equation (1.101), already indicated the significance of $d = 6$ as an upper critical dimension.

1.11 Real-Space Renormalisation

A classic approach to solving a problem is to identify its characteristic scale in order to break it down into smaller uncorrelated pieces. In this way, one can work with a smaller problem within the characteristic scale, safe in the knowledge that the solution to the original problem can be constructed from the solution of the smaller problem. We have already exploited such an approach in the derivation of Equation (1.99), see Figure 1.31. However, at a critical point, there is no characteristic scale, rendering such an approach impossible. But since there is no characteristic scale, critical systems are self-similar. It is precisely the self-similarity, emerging at $p = p_c$, that we will take advantage of in order to generalise our approach to scale invariance [Wilson, 1971a; Cardy, 1996; Fisher, 1998].

1.11.1 *Self-similarity and the correlation length*

Previously, we have discussed the self-similarity of incipient infinite clusters with reference to Figure 1.26. We found that, statistically, an incipient infinite cluster looks the same on all length scales. At $p = p_c$, the correlation length $\xi = \infty$. There is no characteristic length scale. Mathematically, this is expressed in the power-law relation

$$M_\infty(\infty; \ell) \propto \ell^D, \qquad (1.123)$$

which says that the mass of an incipient infinite cluster within a window of size ℓ scales like a fractal. This is made possible by an incipient infinite cluster having holes of all sizes, which in turn contain clusters of all sizes.

There are, of course, also two trivially self-similar configurations. They are given by the empty lattice and the fully occupied lattice which look the same on all length scales. At $p = 0$ and $p = 1$, the correlation length $\xi = 0$. There is no characteristic length scale.

Hitherto, we have interpreted the correlation length ξ as the typical radius of the largest cluster or, equivalently, for $p > p_c$, the typical radius of the largest hole in the percolating cluster. In the present setting, it is instructive to think of the correlation length in terms of 'fluctuations' away from the two trivially self-similar configurations.

At $p = 0$, there are no fluctuations away from the empty lattice and $\xi = 0$. For p close to zero, the correlation length sets the upper scale of the fluctuations away from the empty lattice, see the panel corresponding to $p = 0.10$ in Figure 1.2 on page 4. The fluctuations increase with p and diverge as p tends to p_c^-. Only at $p = p_c$ are there fluctuations of all sizes away from the empty lattice and $\xi = \infty$.

At $p = 1$, there are no fluctuations away from the fully occupied lattice and $\xi = 0$. For p close to one, the correlation length sets the upper scale of the fluctuations away from the fully occupied lattice, see the panel corresponding to $p = 0.90$ in Figure 1.2. The fluctuations increase with decreasing p and diverge as p tends to p_c^+. Only at $p = p_c$ are there fluctuations of all sizes away from the fully occupied lattice and $\xi = \infty$.

Hence, at $p = p_c$, the system is delicately poised in a non-trivial self-similar state between two trivial self-similar configurations corresponding to $p = 0$ and $p = 1$.

1.11.2 Self-similarity and fixed points

Self-similarity can be identified with the fixed points of a rescaling transformation. Qualitatively, we imagine applying a rescaling transformation to a system in which all length scales are reduced by a factor $b > 1$. In particular, the correlation length $\xi \mapsto \xi/b$. Under repeated rescaling, all finite correlation lengths $0 < \xi < \infty$ would eventually be reduced to zero. The correlation length $\xi = 0$ is a fixed point of the rescaling transformation, because the transformation leaves the correlation length unchanged. In fact, the rescaling transformation has only two values of ξ that solve the fixed-point equation,

$$\xi = \frac{\xi}{b} \quad \Leftrightarrow \quad \xi = \begin{cases} 0 & \text{'trivial'} \\ \infty & \text{'non-trivial'}. \end{cases} \quad (1.124)$$

The fixed point $\xi = 0$ is associated with the trivially self-similar configurations at $p = 0$ and $p = 1$, while the fixed point $\xi = \infty$ is associated with the self-similar state at the critical point $p = p_c$. Therefore, the fixed-point equation $\xi = \xi/b$ implies scale invariance, which is indeed the requirement for self-similarity.

To investigate the behaviour of the rescaling transformation away from the fixed points, consider a large but finite value of ξ corresponding to an oc-

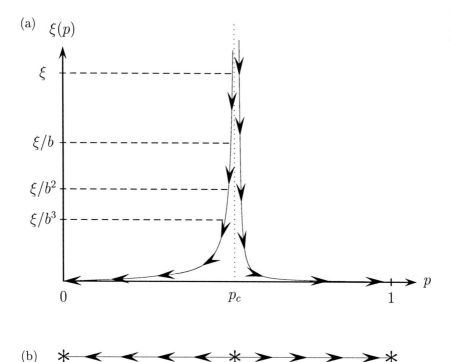

Fig. 1.39 (a) A sketch of the correlation length, ξ, versus the occupation probability, p. The vertical dotted line shows the position of p_c. For $p \to p_c$, the correlation length diverges as a power law with exponent $-\nu$ in terms of $|p - p_c|$, the distance of p from p_c, that is, $\xi(p) \propto |p_c - p|^{-\nu}$. When applying the rescaling transformation, the correlation length is reduced by a factor b. (b) The corresponding flow in parameter space. The fixed point $p = p_c$ is unstable. For $p < p_c$, applying the rescaling transformation will induce a flow towards the stable fixed point $p = 0$. For $p > p_c$, applying the rescaling transformation will induce a flow towards the stable fixed point $p = 1$.

cupation probability p_1^- slightly below p_c, see Figure 1.39(a). After applying a rescaling transformation once, the correlation length is reduced from ξ to ξ/b corresponding to an occupation probability $p_2^- < p_1^-$. Successive applications of the rescaling transformation reduce the original correlation length by a factor b each time, that is, $\xi \mapsto \xi/b \mapsto \xi/b^2 \mapsto \cdots$ with corresponding occupation probabilities $p_1^- > p_2^- > p_3^- > \cdots$. This process can be visualised as a flow from ξ towards the fixed point $\lim_{n \to \infty} \xi/b^n = 0$ or, equivalently, as a flow in p-space from p_1^- towards the fixed point $\lim_{n \to \infty} p_n^- = 0$, see Figure 1.39(b).

Similarly, consider a large but finite value of ξ corresponding to an occupation probability p_1^+ slightly above p_c. Successive applications of the rescaling transformation reduce the original correlation length by a factor b each time, that is, $\xi \mapsto \xi/b \mapsto \xi/b^2 \mapsto \cdots$ with corresponding occupation probabilities $p_1^+ < p_2^+ < p_3^+ < \cdots$. This process can be visualised as a flow from ξ towards the fixed point $\lim_{n\to\infty} \xi/b^n = 0$ or, equivalently, as a flow in p-space from p_1^+ towards the fixed point $\lim_{n\to\infty} p_n^+ = 1$.

Thus the flows will be directed away from $p = p_c$ either towards $p = 0$ (for all initial $p < p_c$) or $p = 1$ (for all initial $p > p_c$). The fixed points $p = 0$ and $p = 1$ are therefore known as stable fixed points, while $p = p_c$ is known as an unstable fixed point.

No flow takes place when the original correlation length $\xi = \infty$, corresponding to the critical occupation probability p_c, nor when $\xi = 0$, corresponding to either $p = 0$ or $p = 1$. Thus there is no flow at the fixed points. These three fixed points of the rescaling transformation are therefore associated with self-similar states.

1.11.3 Coarse graining and rescaling

Recall that the divergence of the correlation length close to p_c is characterised by the critical exponent ν. The critical exponent ν is intimately related to the rescaling transformation since it determines the speed of the flows away from the unstable fixed point p_c, as can be seen by inspecting the graph in Figure 1.39(a).

Let

$$\xi(p) = \text{constant} \, |p - p_c|^{-\nu} \quad \text{for } p \to p_c \tag{1.125}$$

denote the correlation length in the original system. After applying a rescaling transformation, the correlation length is reduced by a factor b, so that $\xi \mapsto \xi/b$. Therefore

$$\frac{\xi}{b} = \frac{\text{constant}|p-p_c|^{-\nu}}{b} = \text{constant}\,|T_b(p) - p_c|^{-\nu} \quad \text{for } p \to p_c, \tag{1.126}$$

where $T_b(p)$ is the rescaling transformation giving the new occupation probability in the rescaled system with the smaller correlation length ξ/b. After rearrangement,

$$\left(\frac{|T_b(p) - p_c|}{|p - p_c|}\right)^{-\nu} = \frac{1}{b}, \tag{1.127}$$

so that the critical exponent ν is given by

$$\nu = \frac{\log b}{\log\left(\frac{|T_b(p)-p_c|}{|p-p_c|}\right)} \quad \text{for } p \to p_c, \tag{1.128}$$

which explicitly relates ν to the speed of the flows away from the unstable fixed point at p_c. However, since p_c is a fixed point for the rescaling transformation, $T_b(p_c) = p_c$, Equation (1.128) can be rewritten as

$$\nu = \frac{\log b}{\log\left(\frac{|T_b(p)-T_b(p_c)|}{|p-p_c|}\right)} \quad \text{for } p \to p_c$$

$$= \frac{\log b}{\log\left(\left.\frac{dT_b}{dp}\right|_{p_c}\right)} \quad \text{for } p \to p_c. \tag{1.129}$$

The rescaling transformation T_b, sketched in Figure 1.39, is only known if we have full information about the system, and only if we know the rescaling transformation can we extract the critical exponent ν. Therefore, Equation (1.129) is not a practical prescription for calculating ν.

However, ν is defined in the vicinity of the critical point p_c where the correlation length is diverging. Therefore, ν is determined by the large length scale behaviour and fluctuations on scales smaller than ξ are irrelevant. This affords a crucial simplification and suggests a simpler transformation, R_b, that incorporates coarsening over a length scale b with rescaling. The coarsening step eliminates fluctuations on scales less than b. The transformation R_b is known as the real-space renormalisation group transformation.

Note that coarsening an empty or fully occupied lattice has no effect since $\xi = 0$, implying that there are no fluctuations present in these lattices. Therefore the real-space renormalisation group transformation $R_b(p)$ has the same two trivial fixed points as the rescaling transformation T_b, namely $p = 0$ and $p = 1$. It is not clear, a priori, whether R_b and T_b have the same non-trivial fixed point. For now, let $p^\star$ denote the fixed points of the real-space renormalisation group transformation R_b, that is,

$$R_b(p^\star) = p^\star. \tag{1.130}$$

In Equation (1.129), we now replace the rescaling transformation T_b and its non-trivial fixed point p_c with the simpler real-space renormalisation group transformation R_b and its non-trivial fixed point $p^\star$, to find that

$$\nu \approx \frac{\log b}{\log\left(\frac{|R_b(p) - R_b(p^\star)|}{|p - p^\star|}\right)} \quad \text{for } p \to p^\star$$

$$= \frac{\log b}{\log\left(\left.\frac{dR_b}{dp}\right|_{p^\star}\right)} \quad \text{for } p \to p^\star. \quad (1.131)$$

Equation (1.131) is a more practical prescription for estimating ν. Given the real-space renormalisation group transformation $R_b(p)$ as a function of occupation probability p, we locate its non-trivial fixed point $p^\star$ as one of the solutions to Equation (1.130). We then evaluate the derivative of R_b at its non-trivial fixed point $p^\star$ and substitute the result into the denominator of Equation (1.131). The numerator is given by the logarithm of the length scale b over which the coarsening takes place.

1.11.4 Real-space renormalisation group procedure

We now outline a three-step procedure to perform a real-space renormalisation group transformation [Reynolds et al., 1977].

(1) Divide the lattice into blocks of linear size b (in terms of the lattice constant) with each block containing at least a few sites, see Figure 1.40(a).
(2) Replace each block of sites by a single block site of size b which is occupied with probability $R_b(p)$ according to the renormalisation group transformation, see Figure 1.40(b).
(3) Rescale all lengths by the factor b to restore the original lattice spacing, see Figure 1.40(c).

We schematically perform the three steps of the renormalisation group transformation in Figure 1.40. In the first step, see Figure 1.40(a), the lattice is divided into blocks of size ba, maintaining the symmetry of the lattice such that the renormalisation group transformation may be carried out indefinitely. The size of the sites is given by the lattice spacing a. The sites are occupied with probability p and the correlation length is ξ. In the second step, see Figure 1.40(b), all the b^2 sites in each block are replaced by a single block site of size ba. The block sites are occupied with probability $R_b(p)$ and the correlation length is ξ/b. The block sites are of size ba, but the original lattice spacing is a. Therefore, in step three, see Figure 1.40(c), all length scales are reduced by a factor b to make the block size identical

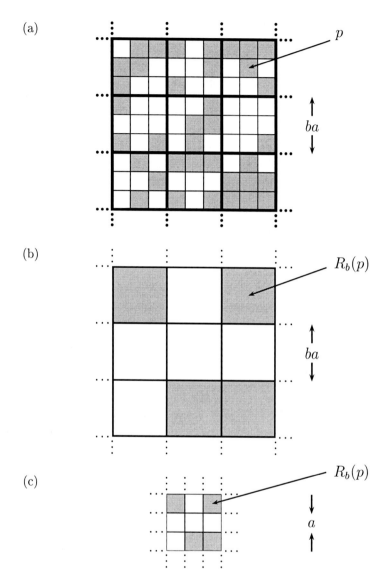

Fig. 1.40 Real-space renormalisation group transformation of two-dimensional site percolation on a square lattice with sites of size a. Sites are occupied with probability p. (a) The lattice is divided into blocks of size ba, each containing b^2 sites. (b) Each block is coarse grained and all of its sites are replaced by a single block site of size ba occupied with probability $R_b(p)$. (c) All length scales are reduced by the factor b. The block size is thereby reduced to a, the same size as the sites in the original lattice: block sites become sites. We therefore obtain a rescaled version of the original lattice where, because of coarsening, sites are occupied with probability $R_b(p)$.

to the lattice spacing and block sites become sites. These sites are occupied with probability $R_b(p)$ and the correlation length is ξ/b.

It is important to note that there is no unique way to coarse grain. However, all coarse graining procedures should eliminate fluctuations on scales less than size b. One possibility would be to employ the spanning-cluster rule, which occupies a block site if a cluster spans the block in specified directions. Another possibility would be to employ a majority rule, which occupies a block site if the majority of the original sites inside the block are occupied. The majority rule has been applied in Figure 1.40.

As well as smearing out fluctuations on scales less than b, the coarse graining reduces the number of degrees of freedom in the system. For example, going from Figure 1.40(a) to Figure 1.40(b), the number of degrees of freedom in each block is reduced from 3×3 to 1, thereby reducing the number of possible microstates from $2^{3 \times 3} = 512$ to $2^1 = 2$. In d dimensions, upon renormalisation the number of degrees of freedom in the lattice $N \mapsto N/b^d$, thereby reducing the number of possible microstates from 2^N to $2^{N/b^d}$, when using hypercubic blocks of size ba. Therefore, we lose information upon coarse graining. The coarse graining procedure is not invertible and cannot be undone uniquely; hence the nomenclature of renormalisation group transformation is, strictly speaking, unfortunate since a group must contain the inverse of all its elements.

In Figure 1.41, we put the renormalisation group transformation to work with the vertically spanning-cluster rule. The renormalisation group transformation is applied from top to bottom on three lattices with occupation probabilities $p_1^- < p^\star, p_1 = p^\star, p_1^+ > p^\star$. For $p_1^- < p^\star$ in the left-hand column, the renormalisation group transformation induces a flow towards the trivial fixed point $p = 0$. For $p_1^+ > p^\star$ in the right-hand column, the renormalisation group transformation induces a flow towards the trivial fixed point $p = 1$. For $p_1 = p^\star$ in the middle column, there is no flow associated with the renormalisation group transformation since the initial occupation probability is at the non-trivial fixed point.

The renormalisation group transformation applied at $p^\star$ reproduces systems with occupation probability $R_b(p^\star) = p^\star$. These systems are invariant with respect to the renormalisation group transformation. In particular, the largest cluster in each panel of the middle column is statistically invariant under the renormalisation group transformation and looks the same. Comparing these largest clusters with those of Figure 1.2, we see that the fixed point for this particular transformation appears to lie somewhere in the range $p_c < p^\star < 0.65$. In fact, $p^\star = 0.619\ldots$ for the renormalisation

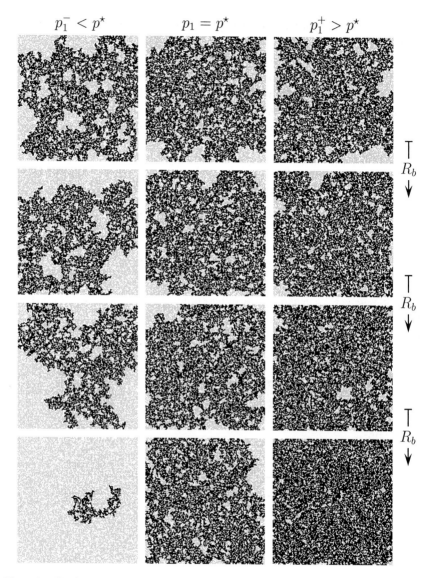

Fig. 1.41 Real-space renormalisation group transformation of two-dimensional site percolation on a square lattice. The panels are windows of size $\ell = 150$ inside larger lattices. The largest cluster in each panel has been shaded black. The three panels in the top row correspond to lattices with occupation probabilities $p_1^- < p^\star, p_1 = p^\star, p_1^+ > p^\star$ from left to right, where $p^\star$ is the unstable fixed point for the renormalisation group transformation. In each of the three columns the renormalisation group transformation R_b is carried out three times, from top to bottom, revealing large scale behaviour. Coarsening is achieved by employing the spanning-cluster rule in the vertical direction with $b = 3$.

group transformation applied in Figure 1.41. Therefore, it is not generally true that the fixed point $p^\star$ for the renormalisation group transformation is identical to the critical occupation probability p_c.

We now explicitly apply the real-space renormalisation group procedure on a few simple examples [Reynolds et al., 1977], starting with one-dimensional percolation where the fixed point of the transformation coincides with the critical occupation probability and the critical exponent ν is correctly predicted.

1.11.5 *Renormalisation in $d = 1$*

Consider a one-dimensional lattice in which sites are occupied with probability p. First, we divide the lattice into blocks of size ba, see Figure 1.42(a). Second, all b sites within a block are replaced by a single block site of size ba. If we employ the spanning-cluster rule as the coarsening procedure, the block sites are occupied with probability $R_b(p) = p^b$, since all b sites within the block must be occupied for the cluster to span, see Figure 1.42(b). Third, all length scales are reduced by the factor b to make the block size identical to the original lattice spacing, see Figure 1.42(c).

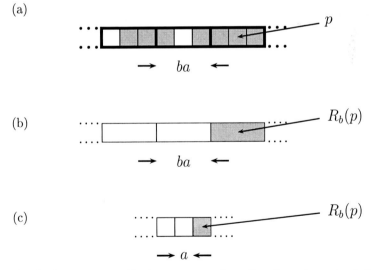

Fig. 1.42 Real-space renormalisation group transformation of one-dimensional site percolation. (a) The lattice is divided into blocks, each containing b sites. (b) Each block is coarse grained and all of its sites are replaced by a single block site occupied with probability $R_b(p) = p^b$. (c) All length scales are reduced by the factor b to obtain a renormalised version of the original lattice.

Figure 1.43(a) displays the graph for the real-space renormalisation group transformation $R_b(p) = p^b$ with $b = 3$. We solve the fixed point equation $R_b(p^\star) = p^{\star b} = p^\star$ to find that

$$p^\star = \begin{cases} 0 \\ 1. \end{cases} \quad (1.132)$$

Note that the fixed point equation can be solved graphically: the fixed points $p^\star$ will lie at the intersections between the graph for the renormalisation group transformation $R_b(p)$ and the identity transformation.

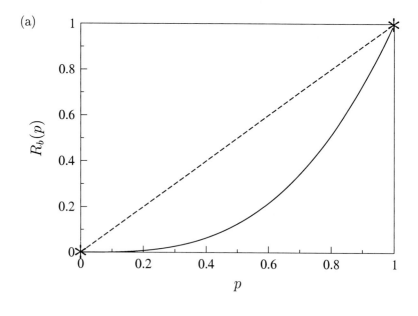

Fig. 1.43 (a) The renormalisation group transformation, $R_b(p)$, versus occupation probability, p, for one-dimensional site percolation using the spanning-cluster rule with $b = 3$. The fixed points (∗) lie at the intersections between the graph for the renormalisation group transformation and the dashed line $R_b(p) = p$. (b) The associated renormalisation group transformation flow in p-space. In the renormalised lattice, sites are occupied with probability $R_b(p) = p^b$. The fixed point $p^\star = 1$ is unstable. For $0 < p < 1$, applying the renormalisation group transformation will induce a flow towards the stable fixed point $p^\star = 0$.

The fixed point $p^* = 0$ corresponds to the empty lattice. The fixed point $p^* = 1$ corresponds to the fully occupied lattice. If the initial occupation probability $0 < p < 1$, the renormalisation group transformation induces a flow towards the stable fixed point $p^* = 0$ because $R_b(p) = p^b < p$, see Figure 1.43(b). Thus $p^* = 1$ is an unstable fixed point and is identified as the non-trivial fixed point in this example.

Figure 1.44(a) visualises the flow in configuration space induced by applying the renormalisation group transformation with the spanning-cluster rule repeatedly on a one-dimensional lattice with initial occupation probability $p < 1$. For $p < 1$, the renormalisation group transformation induces a flow towards the trivial stable fixed point $p^* = 0$ associated with the empty configuration. Figure 1.44(b) displays the flow in the correlation length. For one-dimensional percolation, we can demonstrate explicitly, that the correlation length in the renormalised lattice is reduced by the factor b:

$$\begin{aligned}\xi\left(R_b(p)\right) &= -\frac{1}{\ln R_b(p)} \\ &= -\frac{1}{\ln p^b} \\ &= \frac{\xi(p)}{b}.\end{aligned} \qquad (1.133)$$

In order to determine ν from Equation (1.131), we evaluate the derivative of the renormalisation group transformation at the non-trivial fixed point $p^* = 1$, that is,

$$\left.\frac{dR_b}{dp}\right|_{p^*=1} = \left.bp^{b-1}\right|_{p^*=1} = b, \qquad (1.134)$$

implying

$$\nu = \frac{\log b}{\log\left(\left.\frac{dR_b}{dp}\right|_{p^*=1}\right)} = 1. \qquad (1.135)$$

In one dimension we thus find that the real-space renormalisation group transformation correctly predicts the critical exponent $\nu = 1$, and the value of the non-trivial fixed point coincides with the critical occupation probability $p_c = 1$, see Tables 1.1 and 1.2.

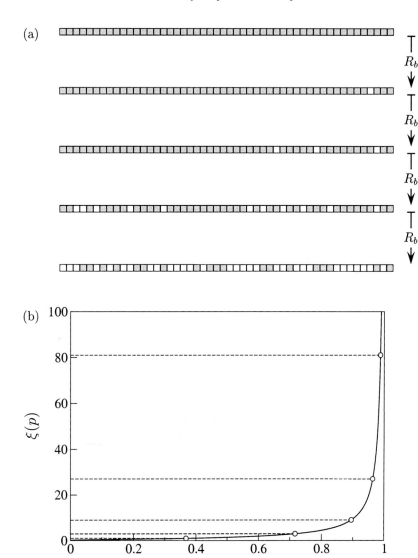

Fig. 1.44 Real-space renormalisation group transformation of one-dimensional site percolation. (a) The panels are windows of size $\ell = 50$ inside a larger lattice. The top panel has $p \approx 0.9877$ with $\xi(p) \approx 81$. The renormalisation group transformation R_b is carried out four times, from top to bottom, revealing the large scale behaviour. Coarsening is achieved by employing the spanning-cluster rule with $b = 3$. (b) Exact solution of the correlation length, $\xi = -1/\ln p$, versus the occupation probability, p. When applying the renormalisation group transformation, the occupation probability p is renormalised into $R_b(p) = p^b$ and the correlation length is reduced by the factor b.

1.11.6 Renormalisation in $d = 2$ on a triangular lattice

Consider a triangular lattice in which sites are occupied with probability p. First, we divide the lattice into blocks of size ba containing three sites each, see Figure 1.45(a). Second, all sites within a block are replaced by a single block site of size ba. Third, all length scales are reduced by the factor $b = \sqrt{3}$ to restore the original lattice spacing, see Figure 1.45(b).

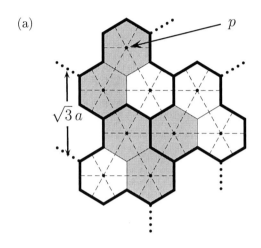

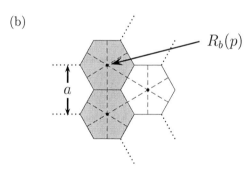

Fig. 1.45 Real-space renormalisation group transformation of two-dimensional site percolation on a triangular lattice. (a) The lattice is divided into blocks each containing three sites. (b) Each block is coarse grained and all of its sites are replaced by a single block site occupied with probability $R_b(p) = 3p^2 - 2p^3$ and all length scales are reduced by the factor $b = \sqrt{3}$ to obtain a renormalised version of the original lattice. Note that the renormalised lattice is rotated by 30° counterclockwise relative to the original lattice.

We employ the spanning-cluster rule without any specified direction as the coarsening procedure. The block sites are occupied with probability

$R_b(p) = p^3 + 3p^2(1-p) = 3p^2 - 2p^3$, since either all three or any two sites must be occupied for a cluster to span, see Figure 1.46. In this example, the spanning-cluster rule is identical to the majority rule, which occupies a block site if the majority of the original sites inside the block are occupied.

$$p^3 + 3p^2(1-p)$$

Fig. 1.46 The $2^3 = 8$ different configurations of a block containing 3 sites for two-dimensional site percolation on a triangular lattice. The four left-most configurations have a spanning cluster and by adding the probabilities for each configuration, we find $R_b(p) = p^3 + 3p^2(1-p) = 3p^2 - 2p^3$.

We rearrange the fixed point equation $R_b(p^\star) = 3p^{\star 2} - 2p^{\star 3} = p^\star$ to find that

$$p^\star(p^\star - 1)(2p^\star - 1) = 0 \quad \Leftrightarrow \quad p^\star = \begin{cases} 0 \\ 1/2 \\ 1. \end{cases} \quad (1.136)$$

The fixed point $p^\star = 0$ corresponds to the empty lattice. The fixed point $p^\star = 1$ corresponds to the fully occupied lattice. If the initial occupation probability $0 < p < 1/2$, the renormalised occupation probability is less than p and the renormalisation group transformation induces a flow from the unstable fixed point $p^\star = 1/2$ towards the stable fixed point $p^\star = 0$, see Figure 1.47. Similarly, if the initial occupation probability $1/2 < p < 1$, the renormalised occupation probability is greater than p and the renormalisation group transformation induces a flow from the unstable fixed point $p^\star = 1/2$ towards the stable fixed point $p^\star = 1$. Thus $p^\star = 1/2$ is an unstable fixed point and is identified as the non-trivial fixed point.

In order to determine ν from Equation (1.131), we evaluate the derivative of the renormalisation group transformation at the non-trivial fixed point $p^\star = 1/2$, that is,

$$\left.\frac{dR_b}{dp}\right|_{p^\star = \frac{1}{2}} = \left.(6p^\star - 6p^{\star 2})\right|_{p^\star = \frac{1}{2}} = \frac{3}{2}, \quad (1.137)$$

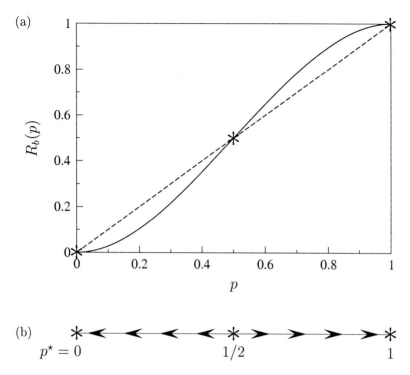

Fig. 1.47 (a) The renormalisation group transformation, $R_b(p) = 3p^2 - 2p^3$, versus the occupation probability, p, for two-dimensional site percolation on a triangular lattice using the spanning-cluster rule with $b = \sqrt{3}$. The fixed points ($*$) lie at the intersections between the graph for the renormalisation group transformation and the dashed line $R_b(p) = p$. (b) The associated renormalisation group transformation flow in p-space. The fixed point $p^\star = 1/2$ is unstable. For $0 < p < 1/2$, applying the renormalisation group transformation will induce a flow towards the stable fixed point $p^\star = 0$. For $1/2 < p < 1$, applying the renormalisation group transformation will induce a flow towards the stable fixed point $p^\star = 1$.

implying

$$\nu = \frac{\log b}{\log\left(\left.\frac{dR_b}{dp}\right|_{p^\star=\frac{1}{2}}\right)} = \frac{\log\sqrt{3}}{\log(3/2)} \approx 1.355. \qquad (1.138)$$

In two-dimensional site percolation on a triangular lattice we thus find that the real-space renormalisation group transformation predicts the critical exponent $\nu \approx 1.355$, and the value of the non-trivial fixed point coincides with the critical occupation probability $p_c = 1/2$, see Table 1.1. In two dimensions, the exact value of the critical exponent $\nu = 4/3$, see Table 1.2.

1.11.7 Renormalisation in $d = 2$ on a square lattice

Consider a square lattice in which sites are occupied with probability p. First, we divide the lattice into blocks of size ba containing b^2 sites each. Second, all sites within a block are replaced by a single block site of size ba. We employ the spanning-cluster rule in the vertical direction as the coarsening procedure. For simplicity, consider $b = 2$. There is always a vertically spanning cluster if all four sites or three sites are occupied. Of the six configurations with two sites occupied, only two are vertically spanning, see Figure 1.48. Therefore the block sites are occupied with probability $R_b(p) = p^4 + 4p^3(1-p) + 2p^2(1-p)^2 = 2p^2 - p^4$. Third, all length scales are reduced by the factor $b = 2$ to make the block size identical to the original lattice spacing.

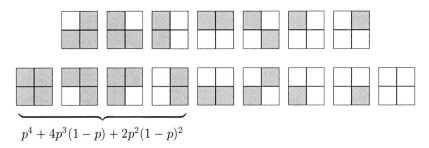

$$p^4 + 4p^3(1-p) + 2p^2(1-p)^2$$

Fig. 1.48 The $2^4 = 16$ different configurations of a $b = 2$ block containing 4 sites for two-dimensional site percolation on a square lattice. The seven left-most configurations have a vertically spanning cluster and by adding the probabilities for each configuration, we find $R_b(p) = p^4 + 4p^3(1-p) + 2p^2(1-p)^2 = 2p^2 - p^4$.

We rearrange the fixed point equation $R_b(p^\star) = 2p^{\star 2} - p^{\star 4} = p^\star$ to find that

$$p^\star(p^\star - 1)(p^{\star 2} + p^\star - 1) = 0 \quad \Leftrightarrow \quad p^\star = \begin{cases} 0 \\ (-1 \pm \sqrt{5})/2 \\ 1. \end{cases} \quad (1.139)$$

The negative solution $(-1-\sqrt{5})/2$ to the fixed point equation is unphysical and we discount it. The fixed point $p^\star = 0$ corresponds to the empty lattice. The fixed point $p^\star = 1$ corresponds to the fully occupied lattice. If the initial occupation probability $0 < p < (-1+\sqrt{5})/2$, the renormalised occupation probability is less than p and the renormalisation group transformation induces a flow from the unstable fixed point $p^\star = (-1+\sqrt{5})/2$

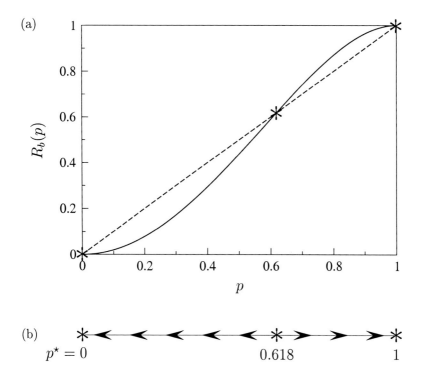

Fig. 1.49 (a) The renormalisation group transformation, $R_b(p)2p^2 - p^4$, versus the occupation probability, p, for two-dimensional site percolation on a square lattice using the spanning-cluster rule in the vertical direction with $b = 2$. The fixed points ($*$) lie at the intersections between the graph for the renormalisation group transformation and the dashed line $R_b(p) = p$. (b) The associated renormalisation group transformation flow in p-space. The fixed point $p^\star = 0.618$ is unstable. For $0 < p < 0.618$, applying the renormalisation group transformation will induce a flow towards the stable fixed point $p^\star = 0$. For $0.618 < p < 1$, applying the renormalisation group transformation will induce a flow towards the stable fixed point $p^\star = 1$.

towards the stable fixed point $p^\star = 0$, see Figure 1.49. Similarly, if the initial occupation probability $(-1 + \sqrt{5})/2 < p < 1$, the renormalised occupation probability is greater than p and the renormalisation group transformation induces a flow from the unstable fixed point $p^\star = (-1 + \sqrt{5})/2$ towards the stable fixed point $p^\star = 1$. Thus $p^\star = (-1 + \sqrt{5})/2$ is an unstable fixed point and is identified as the non-trivial fixed point.

In order to determine ν from Equation (1.131), we evaluate the derivative of the renormalisation group transformation at the non-trivial fixed point $p^\star = (-1+\sqrt{5})/2$, that is,

$$\left.\frac{dR_b}{dp}\right|_{p^\star} = \left.(4p^\star - 4p^{\star 3})\right|_{p^\star} = (\sqrt{5}-1)^2, \qquad (1.140)$$

implying

$$\nu = \frac{\log b}{\log\left(\left.\frac{dR_b}{dp}\right)\right|_{p^\star}} = \frac{\log 2}{2\log(\sqrt{5}-1)} \approx 1.635. \qquad (1.141)$$

In two-dimensional site percolation on a square lattice we thus find that the real-space renormalisation group transformation predicts the critical exponent $\nu \approx 1.635$, while the value of the non-trivial fixed point estimates the critical occupation probability $p_c \approx 0.618$. In two dimensions, the exact value of the critical exponent $\nu = 4/3$, see Table 1.2, while $p_c = 0.59274621$, see Table 1.1.

1.11.8 Approximation via the truncation of parameter space

The renormalisation group transformation $R_b(p)$ is exact in one dimension. When $d > 1$, the one parameter renormalisation group transformation only gives approximate results for the critical exponent ν and the critical occupation probability p_c. For the renormalisation group transformation to remain exact in $d > 1$, we must generalise the one-dimensional parameter space p to an infinite-dimensional parameter space $\{p\}$ that includes, for example, connections between sites on the renormalised lattice that are not nearest neighbours.

Consider the renormalisation group transformation for two-dimensional site percolation on a square lattice displayed in Figure 1.50, using the spanning cluster rule in the vertical direction with $b = 2$. In the renormalised lattice, sites A and D are occupied but they do not belong to the same cluster if only nearest neighbours connections are applied. However, in the original lattice, blocks A and D are connected via block C. Therefore, the identity of the cluster is not preserved in the renormalised lattice; to preserve the cluster identity, we have to introduce next-nearest-neighbour connections between sites in the renormalised lattice.

Hence, repeatedly applying the renormalisation group transformation will introduce connections over further and further distant sites. Moreover, sites in the renormalised lattice may not be occupied randomly and independently. Therefore, in an exact renormalisation group transformation,

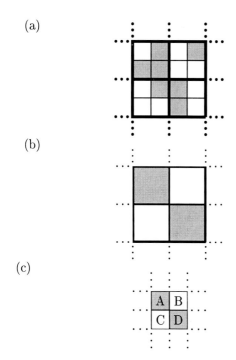

Fig. 1.50 The renormalisation group transformation for two-dimensional site percolation on a square lattice using the spanning-cluster rule in the vertical direction with $b = 2$. (a) The lattice is divided into blocks of size $2a$ each containing 4 sites. (b) Each block is coarse grained and replaced by a single block site that is occupied if there is a vertically spanning cluster. (c) All length scales are reduced by the factor $b = 2$: block sites become sites in the renormalised lattice. In the renormalised lattice, sites A and D are occupied while sites B and C are empty.

we would have to replace the simple one-dimensional parameter space p with a generalised infinite-dimensional parameter space $\{p\}$ that includes all possible connections and correlations in addition to the occupation probability p [Young and Stinchcombe, 1975]. Indeed, after repeatedly applying the renormalisation group transformation, we would end up with an infinite number of parameters $\{p\}$, so some approximation scheme must be devised.

In our examples of the renormalisation group transformation in $d = 2$ on square and triangular lattices, we have tacitly truncated the infinite-dimensional parameter space to a single parameter, that is, replaced the exact renormalisation group transformation by an approximation:

$$R_b(\{p\}) \approx R_b(p). \qquad (1.142)$$

The approximation via the truncation of the infinite-dimensional parameter space is reflected in the values we have obtained for ν and p_c, which are only in very approximate agreement with the theoretically known values – except in $d = 1$, where no truncation is necessary.

1.12 Summary

In this chapter we introduced percolation, one of the simplest models undergoing a continuous phase transition. Our goal was to give a statistical description of finite cluster sizes in terms of the cluster number density. We found that this was possible in one dimension and in the mean field, but not in two dimensions, because we were unable to enumerate the cluster degeneracy factor exactly. We therefore appealed to numerical simulations which indicated that the cluster number density decays like a power law for large cluster sizes, up until some characteristic cluster size, and then rapidly thereafter.

We were able to summarise these observations in a general scaling ansatz for the cluster number density which encompasses the scaling forms derived for the cluster number densities in one dimension and on the Bethe lattice.

At the critical point we argued that the average and characteristic cluster size both diverge, based on our experience with one-dimensional and mean-field percolation, and that the decay of the cluster number density was a pure power law. To characterise these power law divergences and decays we introduced critical exponents. We found that these critical exponents were related through scaling relations and were not all independent.

While the scaling ansatz would not allow us to write down the cluster number density exactly, we nevertheless found that we were able to summarise one-dimensional, mean-field and two-dimensional percolation in a unifying scaling framework. Such a framework is particularly suitable in the vicinity of a phase transition, where it naturally incorporates the idea of scale invariance.

Our intuition of scale invariance was enhanced by considering the geometrical nature of clusters. We discovered that clusters are examples of statistical fractals, characterised by a fractal dimension, and that scale invariance at the critical occupation probability is accompanied by the emergence of an incipient infinite cluster.

We associated the density of the percolating cluster with a so-called order parameter, that was non-zero above the critical occupation probability and zero elsewhere. We assigned yet another critical exponent to characterise the abrupt manner in which the order parameter picks up just above the critical occupation probability, and we were able to relate this critical exponent to geometrical properties of clusters through scaling relations.

In finite lattices, we showed how to adapt our conclusions when the correlation length of the clusters exceeds the size of the lattice. Investigating the scaling of quantities as a function of system size at $p = p_c$ where $\xi = \infty$ is a powerful procedure to extract numerical estimates of critical exponent.

Finally, by invoking the so-called renormalisation group, we developed a general framework for exploiting scale invariance. In this approach, we associate self-similar states with the fixed point of a renormalisation group transformation, the effect of which is to preserve large-distance behaviour and average out fluctuations over small scales. Repeated application of this transformation induces a flow, which we track by using the renormalised correlation length. By studying the action of the renormalisation group transformation in the vicinity of a critical point, we are able to express critical exponents in terms of the derivative of the transformation. To illustrate the approach, we considered some real-space renormalisation schemes in one- and two-dimensional percolation.

The renormalisation group transformation is exact in one dimension. Applying the renormalisation group transformation in $d > 1$ generates connections over further and further distant sites, and this is accompanied by a growth in the parameter space $\{p\}$. When truncating the parameter space to only the renormalised occupation probability $R_b(p)$, we are left with an approximation to the exact renormalisation group transformation $R_b(\{p\})$. Therefore, the critical exponents are only determined approximately.

Exercises

1.1 *Moments and moment ratio of the cluster number density in $d = 1$.*

Consider one-dimensional site percolation on an infinite lattice.

(i) Using the exact solution for the cluster number density in one-dimensional percolation, show that the kth moment with $k \geq 2$ of the cluster number density

$$M_k(p) = \sum_{s=1}^{\infty} s^k n(s,p) \to \Gamma_k (p_c - p)^{-\gamma_k} \quad \text{for } p \to p_c^-, \quad (1.1.1)$$

and identify the critical exponent γ_k, and the critical amplitude Γ_k.

(ii) Express the moment ratio

$$g_k = \frac{M_k M_1^{k-2}}{M_2^{k-1}} \quad \text{for } p \to p_c^-, k \geq 2, \quad (1.1.2)$$

in terms of the critical amplitudes and hence find the value of g_k.

1.2 *Site percolation and site-bond percolation in $d = 1$.*

Consider one-dimensional site percolation on an infinite lattice.

(i) (a) What is the critical occupation probability p_c?

(b) Determine the cluster number density $n(s,p)$, that is, the number of s-clusters per lattice site at occupation probability p.

(c) Explain why

$$\sum_{s=1}^{\infty} sn(s,p) = p \quad \text{for } p < 1. \quad (1.2.1)$$

Why is this identity not valid for $p = 1$?

(d) Derive the identity

$$\sum_{s=1}^{\infty} s^2 p^s = p\frac{1+p}{(1-p)^3} \quad \text{for } p < 1. \quad (1.2.2)$$

(e) Calculate the average cluster size

$$\chi(p) = \frac{\sum_{s=1}^{\infty} s^2 n(s,p)}{\sum_{s=1}^{\infty} sn(s,p)}. \quad (1.2.3)$$

(f) Determine the critical amplitude Γ, and the critical exponent γ, such that

$$\chi(p) = \Gamma(p_c - p)^{-\gamma} \quad \text{for } p \to p_c^-. \tag{1.2.4}$$

Consider one-dimensional site-bond percolation on an infinite lattice. Sites are occupied with probability p while bonds are occupied with probability q, see Figure 1.2.1. A cluster of size s is defined as having s consecutive occupied sites with $s-1$ intermediate occupied bonds. For example, the left-most cluster in Figure 1.2.1 has size $s=3$. The cluster terminates to the right because a bond is empty and to the left because a site is empty.

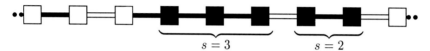

Fig. 1.2.1 Part of an infinite one-dimensional lattice where occupied sites and bonds are black while empty sites and bonds are white. Two clusters are present with $s=3$ and $s=2$, respectively.

(ii) (a) What is the critical point (p_c, q_c) for site-bond percolation?
(b) Show that the cluster number density

$$n(s, p, q) = p^s q^{s-1} (1 - pq)^2. \tag{1.2.5}$$

(c) Calculate the average cluster size

$$\chi(p, q) = \frac{\sum_{s=1}^{\infty} s^2 n(s, p, q)}{\sum_{s=1}^{\infty} s n(s, p, q)}. \tag{1.2.6}$$

Comment on the result.

1.3 *Percolation in $d=1$ on a lattice with periodic boundary conditions.*

Consider one-dimensional percolation on a finite circular ring of L sites. The geometrical periodicity of the ring will ensure that all sites in the finite lattice have two neighbours. This is the method of periodic boundary conditions and can be extended to any dimension d.

(i) Find $n(s, p)$.
(ii) Explain why the probability that a site belongs to a finite cluster is $\sum_{s=1}^{L-1} sn(s, p)$ rather than $\sum_{s=1}^{L} sn(s, p)$.

(iii) Show that the probability of a site belonging to the percolating, 'infinite' cluster $P_\infty(p; L) = p^L$, by calculating the right-hand side of the identity

$$P_\infty(p; L) = p - \sum_{s=1}^{L-1} sn(s, p). \qquad (1.3.1)$$

(iv) (a) Using the identity

$$\xi(p) = -\frac{1}{\ln p} \qquad (1.3.2)$$

express the order parameter $P_\infty(p; L)$ as a function of the system size L and the correlation length ξ.

(b) Write the order parameter using the scaling form

$$P_\infty(\xi; L) = \xi^{-\beta/\nu} \mathcal{P}(L/\xi), \qquad (1.3.3)$$

and identify the ratio of critical exponents β/ν and the associated scaling function $\mathcal{P}$. How does $\mathcal{P}(x)$ behave for $x \ll 1$ and $x \gg 1$?

1.4 *Cluster number density scaling functions in $d = 1$ and the Bethe lattice.*

The general scaling ansatz for the cluster number density is

$$n(s, p) = a\, s^{-\tau} \mathcal{G}(s/s_\xi) \quad \text{for } p \to p_c, s \gg 1, \qquad (1.4.1\text{a})$$

$$s_\xi = b\,|p - p_c|^{-1/\sigma} \quad \text{for } p \to p_c, \qquad (1.4.1\text{b})$$

where the critical exponents τ and σ and the scaling function $\mathcal{G}$ are universal, that is, independent of the underlying microscopic details and dependent only on the dimensionality of the problem. Meanwhile, the proportionality constants a and b and the critical occupation probability p_c depend on the lattice details, and are therefore non-universal.

(i) (a) Show that the cluster number density for $d = 1$ percolation is a special case of Equation (1.4.1) and identify the constants a, b, the critical exponents τ, σ, and the scaling function $\mathcal{G}_{1d}$.

(b) Plot the graph of the scaling function $\mathcal{G}_{1d}(s/s_\xi)$ versus the rescaled cluster size s/s_ξ in a double-logarithmic plot. Identify the limiting behaviour of the scaling function for $s/s_\xi \ll 1$ and $s/s_\xi \gg 1$, respectively.

(c) Taylor expand the scaling function $\mathcal{G}_{1d}(s/s_\xi)$ around zero to leading order.

(ii) (a) Show that the cluster number density for percolation on a Bethe lattice with $z = 3$ is a special case of Equation (1.4.1) and identify the constant b, the critical exponents τ and σ, and the scaling function $\mathcal{G}_{\text{Bethe}}$. Assuming Equation (1.4.1) was valid for all s, how would you determine the constant a?

(b) Plot the graph of the scaling function $\mathcal{G}_{\text{Bethe}}(s/s_\xi)$ versus the rescaled cluster size s/s_ξ in a double-logarithmic plot. Identify the limiting behaviour of the scaling function for $s/s_\xi \ll 1$ and $s/s_\xi \gg 1$, respectively.

(c) Taylor expand the scaling function $\mathcal{G}_{\text{Bethe}}(s/s_\xi)$ around zero to leading order.

1.5 *Moments of the cluster number density.*

Assume the general ansatz for the cluster number density

$$n(s, p) = a\, s^{-\tau} \mathcal{G}(s/s_\xi) \quad \text{for } p \to p_c, s \gg 1, \tag{1.5.1a}$$

$$s_\xi = b\, |p - p_c|^{-1/\sigma} \quad \text{for } p \to p_c \tag{1.5.1b}$$

is valid for all cluster sizes s.

(i) Show that the kth moment with $k \geq 2$ of the cluster number density

$$M_k(p) = \sum_{s=1}^{\infty} s^k n(s, p) \to \Gamma_k\, |p - p_c|^{-\gamma_k} \quad \text{for } p \to p_c \tag{1.5.2}$$

and identify the critical exponent γ_k and the critical amplitude Γ_k.

(ii) Express the moment ratio

$$g_k = \frac{M_k M_1^{k-2}}{M_2^{k-1}} \quad \text{for } p \to p_c, k \geq 2, \tag{1.5.3}$$

in terms of the critical amplitudes and hence find the value of g_k.

(iii) Combine the results of Exercise 1.5(i) and Exercise 1.4(i) to recover the result derived in Exercise 1.1(i).

1.6 *Universality of the ratio of amplitudes for the average cluster size.*

Assume that the general scaling ansatz for the cluster number density

$$n(s, p) = a\, s^{-\tau} \mathcal{G}_\pm(s/s_\xi) \quad \text{for } p \to p_c^\pm, s \gg 1, \tag{1.6.1a}$$

$$s_\xi = b\, |p - p_c|^{-1/\sigma} \quad \text{for } p \to p_c \tag{1.6.1b}$$

is valid for all s where the scaling functions $\mathcal{G}_\pm$ correspond to $p > p_c$ and $p < p_c$, respectively.

(i) Calculate explicitly the average cluster size $\chi(p) = \Gamma^-(p_c - p)^{-\gamma^-}$ for $p < p_c$, keeping all the proportionality factors, and identify the critical exponent γ^- and the critical amplitude Γ^-.

(ii) Calculate explicitly the average cluster size $\chi(p) = \Gamma^+(p - p_c)^{-\gamma^+}$ for $p > p_c$, keeping all the proportionality factors, and identify the critical exponent γ^+ and the critical amplitude Γ^+.

(iii) (a) Show that $\gamma^- = \gamma^+$.
  (b) Prove that the ratio of critical amplitudes Γ^+/Γ^- is universal, that is, independent of the constants a and b.
  (c) How would you identify the ratio of the critical amplitudes Γ^+/Γ^- for two-dimensional percolation using the numerical results displayed in Figure 1.15 on page 34?

1.7 *The order parameter on a Bethe lattice with coordination number z.*

Consider a Bethe lattice with coordination number z. Let $Q_\infty(p)$ denote the probability that a given branch does not connect to the percolating cluster.

(i) Argue why

$$P_\infty(p) = p\left[1 - Q_\infty^z(p)\right], \tag{1.7.1}$$

where $Q_\infty(p)$ must obey the equation

$$Q_\infty(p) = 1 - p + pQ_\infty^{z-1}(p). \tag{1.7.2}$$

(ii) Taylor expand the right-hand side of the identity

$$Q_\infty^{z-1}(p) = (1 - [1 - Q_\infty(p)])^{z-1} \tag{1.7.3}$$

up to second order in $[1 - Q_\infty(p)]$. Use this Taylor expansion in Equation (1.7.2) and solve the corresponding quadratic equation to show that

$$Q_\infty(p) = \begin{cases} 1 & \text{for } p \leq p_c \\ 1 - \frac{2p(z-1)-2}{p(z-1)(z-2)} & \text{for } p > p_c. \end{cases} \tag{1.7.4}$$

(iii) Hence, by substituting the solution Equation (1.7.4) into Equation (1.7.1) and using another Taylor expansion, deduce the approximate solution

$$P_\infty(p) = A(p - p_c) \quad \text{for } p \to p_c^+ \qquad (1.7.5)$$

and find p_c and the proportionality constant A.

(iv) Check that the general results in Equation (1.7.4) and Equation (1.7.5) are consistent with the solutions for percolation on the Bethe lattice with coordination number $z = 3$ derived in Section 1.3.4.

1.8 *Finite-size scaling and scaling function for the average cluster size.*

For percolation on a lattice of infinite size $L = \infty$, the average cluster size at occupation probability p diverges like

$$\chi(p; L = \infty) \propto |p - p_c|^{-\gamma} \quad \text{for } p \to p_c. \qquad (1.8.1)$$

(i) Assume that the general scaling ansatz for the cluster number density

$$n(s, p) = a s^{-\tau} \mathcal{G}(s/s_\xi) \quad \text{for } p \to p_c, s \gg 1, \qquad (1.8.2)$$

is valid for all s. The scaling function $\mathcal{G}$ is constant for small arguments and decays rapidly for large arguments. Derive a scaling relation between the critical exponents γ, τ and σ.

(ii) Let $\xi(p)$ denote the correlation length at occupation probability p. Show that for p close to p_c,

$$\chi(\xi; L = \infty) \propto \xi^{\gamma/\nu} \quad \text{for } p \to p_c. \qquad (1.8.3)$$

(iii) (a) Argue why, for finite lattices with $L \ll \xi$, one would expect

$$\chi(\xi; L) \propto L^{\gamma/\nu} \quad \text{for } p \to p_c, L \ll \xi. \qquad (1.8.4)$$

(b) For finite lattices, $L < \infty$, show that the average cluster size $\chi(\xi; L)$ obeys the scaling law

$$\chi(\xi; L) = \xi^{\gamma/\nu} \mathcal{X}(L/\xi). \qquad (1.8.5)$$

Identify the behaviour of the scaling function $\mathcal{X}$ for $L \gg \xi$ and $L \ll \xi$ and sketch its graph on a double-logarithmic plot.

(c) Numerically, how would you determine the ratio γ/ν?

(iv) In finite lattices, $L < \infty$, you may assume the cluster number density at $p = p_c$ obeys the scaling law

$$n(s, p_c; L) = s^{-\tau} \tilde{\mathcal{G}}\left(s/L^D\right) \quad \text{for } s \gg 1, L \gg 1, \qquad (1.8.6)$$

where D is the fractal dimension and the scaling function $\tilde{\mathcal{G}}(x)$ is constant for $x \ll 1$ and decays rapidly for $x \gg 1$.

Combining Equations (1.8.5) and (1.8.6), derive the scaling relation

$$\gamma/\nu = D(3 - \tau). \tag{1.8.7}$$

1.9 *Finite-size scaling and scaling function for the order parameter.*

Consider percolation on a lattice of infinite size $L = \infty$ with critical occupation probability p_c.

(i) (a) Define the order parameter $P_\infty(p; L = \infty)$ for the geometrical phase transition.
  (b) Describe the behaviour of the order parameter as a function of p. Illustrate your explanation with a sketch.
  (c) Let $n(s, p; L = \infty)$ denote the number of s-clusters per lattice site. Justify the relation

$$P_\infty(p; L = \infty) = p - \sum_{s=1}^{\infty} sn(s, p; L = \infty). \tag{1.9.1}$$

In the following we consider $p > p_c$. Let $\xi(p)$ denote the correlation length and assume that the order parameter becoming non-zero for p approaching p_c from above is characterised by the critical exponent β, that is,

$$P_\infty(p; L = \infty) \propto (p - p_c)^\beta \quad \text{for } p \to p_c^+. \tag{1.9.2}$$

(ii) (a) Show that the order parameter $P_\infty(\xi; L = \infty) \propto \xi^{-\beta/\nu}$ for $p \to p_c^+$, where ν is the critical exponent characterising the divergence of the correlation length as $p \to p_c$.
  (b) Argue why, for finite lattices with $L \ll \xi$, one would expect

$$P_\infty(\xi; L) \propto L^{-\beta/\nu} \quad \text{for } p \to p_c^+, L \ll \xi. \tag{1.9.3}$$

  (c) Numerically, how would you determine the ratio $-\beta/\nu$?

(iii) In finite lattices, $L < \infty$, you may assume the cluster number density at $p = p_c$ obeys the scaling law for all cluster sizes

$$n(s, p_c; L) = s^{-\tau} \tilde{\mathcal{G}}(s/L^D) \quad \text{for } L \gg 1, s \gg 1, \tag{1.9.4}$$

where τ is the cluster number exponent, D is the fractal dimension and the scaling function $\tilde{\mathcal{G}}(x)$ is constant for $x \ll 1$ and decays rapidly for $x \gg 1$. In addition, you may assume that

$$P_\infty(p_c; L) = \sum_{s=1}^\infty s \, s^{-\tau} \tilde{\mathcal{G}}(0) - \sum_{s=1}^\infty s n(s, p_c; L). \qquad (1.9.5)$$

(a) Show that Equation (1.9.5) is correct in the limit of $L \to \infty$.
(b) Combining Equations (1.9.3), (1.9.4) and (1.9.5), derive the scaling relation

$$-\beta/\nu = D(2-\tau). \qquad (1.9.6)$$

1.10 *Probability of having a percolating cluster on a lattice of size L.*

(i) Let $\Pi_\infty(p; L = \infty)$ denote the probability of having a percolating cluster at occupation probability p in an infinite lattice. Describe the behaviour of the function $\Pi_\infty(p; L = \infty)$ as a function of p. Illustrate your explanation with a sketch of the graph.

(ii) Consider one-dimensional site percolation on a finite lattice.

(a) Let $\Pi_\infty(p; L)$ denote the probability of having a percolating cluster in a lattice of size L. What is the function $\Pi_\infty(p; L)$? Explain your answer and sketch the graph of the function.

(b) If ξ is the correlation length at occupation probability p, show that

$$\Pi_\infty(\xi; L) = \exp(-L/\xi). \qquad (1.10.1)$$

(c) Hence, show that

$$\Pi_\infty(p; L) = \mathcal{F}_{1d}\left[(p_c - p)L\right] \quad \text{for } p \to p_c^-, \qquad (1.10.2)$$

and identify the scaling function $\mathcal{F}_{1d}$. How does $\mathcal{F}_{1d}(x)$ behave for $x \ll 1$ and $x \gg 1$?

(iii) In higher dimensions there exists a scaling function Π such that

$$\Pi_\infty(\xi; L) = \Pi(L/\xi) \quad \text{for } p \to p_c. \qquad (1.10.3)$$

(a) Hence, show that

$$\Pi_\infty(p; L) = \tilde{\Pi}\left[|p_c - p|L^{1/\nu}\right] \text{ for } p \to p_c, \qquad (1.10.4)$$

and relate the scaling function $\tilde{\Pi}$ to Π.

(b) What is the limiting function $d\Pi_\infty/dp$ when $L \to \infty$?

1.11 *Real-space renormalisation group transformation on a square lattice.*

  (i) Define and outline the procedure of real-space renormalisation transformation applied to the percolation theory problem.

  (ii) Consider site percolation on a square lattice in two dimensions. Using blocks of size 2×2 and adapting the spanning cluster rule (in any direction) to define the real-space renormalisation group transformation, show that

$$R_b(p) = p^4 - 4p^3 + 4p^2. \qquad (1.11.1)$$

  (iii) Find the fixed points for the real-space renormalisation group transformation in Equation (1.11.1) and comment on their nature. What are the correlation lengths ξ associated with the respective fixed points? Discuss the concept of flow in p-space associated with the real-space renormalisation group transformation R_b.

  (iv) Identify the critical occupation probability p_c, derive the equation used to determine the correlation length exponent ν predicted by the real-space renormalisation group transformation, and evaluate ν. Compare the findings to the analytic results and comment on the discrepancies.

  (v) Discuss the concept of universality in the theory of percolation. Give examples of quantities which are universal and non-universal, respectively.

1.12 *Real-space renormalisation group transformation on a square lattice.*

In bond percolation, each bond between neighbouring lattice sites is occupied with probability p and empty with probability $(1-p)$. The bond percolation threshold for a square lattice $p_c = 0.5$. In a real-space renormalisation group transformation on the square lattice with unit lattice spacing, the lattice is replaced by a new renormalised lattice, with super-bonds of length $b = 2$ occupied with probability $R_b(p)$, following the procedure shown in Figure 1.12.1

  (i) Assuming that the super-bond between **A** and **B** in the renormalised lattice is occupied if there exists a connected path from **A** to **B** along the four bonds in lattice (c), show that

$$R_b(p) = p^4 - 4p^3 + 4p^2. \qquad (1.12.1)$$

  (ii) (a) Solve graphically the fixed point equation for the renormalisation group transformation in Equation (1.12.1).

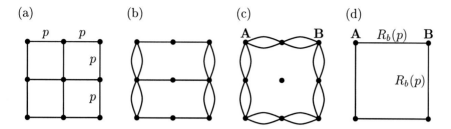

Fig. 1.12.1 (a) Original lattice with unit lattice spacing where each bond is occupied with probability p. (b) Lattice where every second column in the original lattice is moved one lattice unit to the left. (c) Lattice where, in addition, every second row in the original lattice is moved one lattice unit upwards. In this lattice, there are two bonds between each site. (d) Renormalised lattice with lattice spacing $b = 2$ where each super-bond is occupied with probability $R_b(p)$.

 (b) Describe the flow in p-space and the renormalisation of the correlation length when applying the renormalisation group transformation repeatedly.
 (c) Identify clearly the correlation lengths associated with the fixed points $p^\star$ of the renormalisation group transformation and hence explain why fixed points are associated with scale invariance.
(iii) (a) Derive a form for the critical exponent ν in terms of the renormalisation group transformation.
 (b) Hence, identify the critical occupation probability p_c, and determine the correlation length exponent ν predicted by the renormalisation group transformation in Equation (1.12.1).

1.13 *Renormalisation and finite-size scaling of the cluster number density.*

Consider the following percolation problem involving a cluster of a very large but finite size s at the critical occupation probability $p = p_c$. The part of the cluster contained within a window of linear size ℓ has mass $M(\ell, R_s)$, where R_s denotes the radius of gyration.

 (i) Define R_s and discuss the dependence of M on ℓ and R_s. Writing

$$M(\ell, R_s) = \ell^{D_1} m(\ell/R_s^{D_2}), \qquad (1.13.1)$$

identify the critical exponents D_1 and D_2, and find the behaviour of the crossover function $m(x)$ for $x \ll 1$ and $x \gg 1$.
 (ii) Consider a d-dimensional lattice of linear size L at the critical occupation probability $p = p_c$ and a real-space renormalisation

group transformation of the lattice by a factor b. Explain why a cluster containing s sites in the lattice L^d is mapped into a cluster having $s' = s/b^D$ sites in the new lattice $(L/b)^d$, where D denotes the fractal dimension.

(iii) Let $n(s, p)$ denote the cluster number density, that is, the average number of clusters containing s sites per lattice site. Justify the relation

$$sn(s, p_c; L) = b^{-d} s' n(s', p_c; L/b), \qquad (1.13.2)$$

with $s' = s/b^D$.

(iv) Assuming the general scaling ansatz for the cluster number density

$$n(s, p) = s^{-\tau} \mathcal{G}(s/s_\xi) \quad \text{for } p \to p_c, s \gg 1, \qquad (1.13.3)$$

show that in a d-dimensional lattice of size L one would expect

$$n(s, p_c; L) = s^{-\tau} \tilde{\mathcal{G}}(s/L^D) \quad \text{for } s \gg 1, L \gg 1. \qquad (1.13.4)$$

How does the scaling function $\tilde{\mathcal{G}}(x)$ behave for $x \ll 1$ and $x \gg 1$?

(v) Combine Equations (1.13.2) and (1.13.4) to show the hyper-scaling relation

$$\tau = \frac{d + D}{D}. \qquad (1.13.5)$$

Chapter 2
Ising Model

2.1 Introduction

Criticality refers to the behaviour of extended systems at a phase transition where no characteristic scale exists. Thermodynamically, a phase transition occurs when there is a singularity in the free energy. The liquid-gas, conductor-superconductor, fluid-superfluid, or paramagnetic-ferromagnetic phase transitions are common examples. The Ising model of a ferromagnet is one of the simplest models displaying the paramagnetic-ferromagnetic phase transition, that is, the spontaneous emergence of magnetisation in zero external field as the temperature is lowered below a certain critical temperature. At the critical point of critical temperature and zero external field there is no characteristic scale. As in the case of percolation, the scale invariance is intimately related to fixed points of a rescaling transformation.

The Ising model has had an enormous impact on modern physics in general and statistical physics in particular, but also on other areas of science, including biology and neuroscience [Hopfield, 1982; Amit, 1989; Majewski *et al.*, 2001], economics [Sornette, 2003] and sociology [Weidlich, 2001] among others. The importance of the Ising model cannot be overstated. At present, hundreds of papers in these research areas are published each year on models inspired by the Ising model.

There are many excellent introductions to critical phenomena, with extensive discussions of the Ising model. Among our favourites, but by no means exhaustive, are [Goldenfeld, 1992; Cardy, 1996; Pathria, 1996].

Our programme is similar to that of percolation. In percolation, once the cluster number density is known, all other quantities follow. Likewise, in the Ising model the task is to calculate the partition function. Then a standard recipe can be followed to find all other quantities of interest.

Before considering the Ising model, we solve a model of non-interacting spins to gain practice in statistical mechanical calculations. The one-dimensional Ising model can also be readily solved analytically by calculating the partition function, and we find that there is no phase transition in zero external field at finite temperature.

Next, we make use of a so-called 'mean-field' approach where the fluctuations of spins are neglected. This simplification aids the calculation of the partition function, and a phase transition in zero external field is predicted at a finite temperature for the mean-field Ising model. However, the insight that the mean-field Ising model provides is more qualitative than quantitative. For example, the mean-field Ising model predicts that the values of the critical exponents are independent of dimension, which is not the case in reality. We account for the success of the mean-field approach in general terms by invoking the Landau theory of continuous phase transitions.

The Ising model is non-trivial in two dimensions. We state the analytic solution of Onsager,[1] the derivation of which lies outside the scope of this book, and present numerical results for known analytical results.

Just as for the phase transition in percolation, we summarise our knowledge of the behaviour of quantities in the vicinity of the critical point by introducing scaling functions, from which we are able to derive scaling relations among critical exponents and describe finite-size scaling.

Finally, a heuristic scaling picture based on a block spin transformation of the Ising model provides us with a deeper understanding of its critical behaviour at a phase transition, and this in turn leads to a discussion of the renormalisation group transformation, universality and scaling in general.

2.1.1 Definition of the Ising model

Consider once again a two-dimensional square lattice composed of $N = L \times L$ sites. Every site i is occupied by a so-called spin, s_i. For a magnetic material, we may think of the spins as the magnetic dipoles positioned on the crystal structure lattice. In uniaxial magnetic materials, the magnetic dipole interactions constrain the spins to point parallel or anti-parallel along a given direction. Therefore, for simplicity, we assume that the spins can only be in one of two states, either spin-up, $s_i = +1$, or spin-down, $s_i = -1$, see Figure 2.1. By convention, the upwards direction is taken to be positive.

---

[1] Onsager was awarded the Nobel prize in chemistry 1968, not for solving the Ising model but 'for the discovery of the reciprocal relations bearing his name, which are fundamental for the thermodynamics of irreversible processes.'

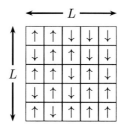

Fig. 2.1 A microstate of the two-dimensional Ising model on a square lattice of size $L = 5$. Every site is occupied by a spin that can be in one of two states: spin-up or spin-down. Of the $N = 25$ spins, 14 are spin-up and 11 are spin-down.

The spins at positions i and j interact with one another. For a pair of parallel spins we assign an interaction energy of $-J_{ij}$, while for a pair of anti-parallel spins we assign an interaction energy of $+J_{ij}$. The four possible configurations of a pair of spins and their corresponding interaction energies are given in Table 2.1.

Table 2.1 The four possible configurations of a pair of spins s_i and s_j at positions i and j and their interaction energies in terms of the coupling constants J_{ij}.

i	j	s_i	s_j	interaction energy
↑	↑	+1	+1	$-J_{ij}$
↓	↓	−1	−1	$-J_{ij}$
↑	↓	+1	−1	$+J_{ij}$
↓	↑	−1	+1	$+J_{ij}$

Thus we can compactly summarise the total internal interaction energy

$$E_{\text{int}} = -\sum_{ij} J_{ij} s_i s_j, \qquad (2.1)$$

where J_{ij} are known as the coupling constants between spins s_i and s_j and the sum runs over all distinct pairs of spins.

In addition to the internal spin-spin interaction, we can impose a uniform external field, H, which acts upon every spin. A spin aligned parallel with the external field has energy $-|H|$ associated with the spin-external field interaction, while a spin aligned anti-parallel with the external field has energy $+|H|$. The external energy for each spin is thus $-Hs_i$ and the total external energy for the system is

$$E_{\text{ext}} = -H \sum_{i=1}^{N} s_i, \qquad (2.2)$$

where the sum runs over all N spins.

Combining the internal and the external contributions, we find that the total energy of a system of N spins $s_i = \pm 1$ with coupling constants J_{ij} placed in a uniform external field H is

$$E_{\{s_i\}} = E_{\text{int}} + E_{\text{ext}} = -\sum_{ij} J_{ij} s_i s_j - H \sum_{i=1}^{N} s_i, \qquad (2.3)$$

where the subscript $\{s_i\} = \{s_1, s_2, \ldots, s_N\}$ is a shorthand notation for the full microscopic description of a particular microstate. To simplify the model even further, we only consider a constant interaction strength J between nearest-neighbour spins, such that the coupling constants

$$J_{ij} = \begin{cases} J & \text{for nearest neighbours } ij \\ 0 & \text{otherwise.} \end{cases} \qquad (2.4)$$

For $J > 0$, the system can lower its internal energy by aligning the spins parallel. Conversely, for $J < 0$, the system can lower its internal energy by aligning the spins anti-parallel. Henceforth, we will only consider $J > 0$.

The total energy for a system of N spins $s_i = \pm 1$ with constant positive nearest-neighbour interactions J placed in a uniform external field H is

$$E_{\{s_i\}} = -J \sum_{\langle ij \rangle} s_i s_j - H \sum_{i=1}^{N} s_i, \qquad (2.5)$$

where the notation $\langle ij \rangle$ restricts the first sum to run over all distinct nearest-neighbour pairs. This defines the celebrated Ising model for a ferromagnet [Ising, 1925].

We wish to consider the behaviour of the Ising model as a function of the temperature, T, which defines an energy scale $k_B T$, where k_B is Boltzmann's constant. Assume for the moment a zero external field $H = 0$. For temperatures $k_B T \ll J$, the spin-spin interactions are relatively strong, so that the spins tend to align with one another, reducing the system's internal energy. For temperatures $k_B T \gg J$, the spin-spin interactions are relatively weak, so that the spins are effectively non-interacting and point up and down randomly.

Qualitatively, we therefore anticipate a transition in the Ising model in zero external field from a high-temperature disordered phase, where the spins are pointing up and down randomly, to a low-temperature ordered phase, where the spins are aligned with one another. In fact, we will find that this order-disorder phase transition is continuous but abrupt, like the phase transition in percolation.

2.1.2 *Review of equilibrium statistical mechanics*

Statistical mechanics attempts to derive the thermodynamic laws of macroscopic quantities from a microscopic description of a system. One can only measure the temporal average of a macroscopic observable. Microscopically, the temporal average of an observable is identified as a suitably weighted ensemble average, $\langle A \rangle$, over all possible microstates.[2] Therefore, if $p_{\{s_i\}}$ is the probability of the system being in a particular microstate $\{s_i\}$ with observable $A_{\{s_i\}}$, its ensemble average

$$\langle A \rangle = \sum_{\{s_i\}} p_{\{s_i\}} A_{\{s_i\}}, \qquad (2.6)$$

where the sum runs over all possible 2^N microstates $\{s_i\}$, that is,

$$\sum_{\{s_i\}} = \sum_{s_1=\pm 1} \sum_{s_2=\pm 1} \cdots \sum_{s_N=\pm 1}. \qquad (2.7)$$

In the canonical ensemble the temperature and volume of the system is kept fixed. The probability $p_{\{s_i\}}$ to find the system in a microstate $\{s_i\}$ with energy $E_{\{s_i\}}$ is given by the Boltzmann distribution

$$p_{\{s_i\}} = \frac{\exp(-\beta E_{\{s_i\}})}{\sum_{\{s_i\}} \exp(-\beta E_{\{s_i\}})}, \qquad (2.8)$$

see Appendix F. Therefore, the ensemble average of the observable $A_{\{s_i\}}$ is given by

$$\langle A \rangle = \frac{1}{Z} \sum_{\{s_i\}} \exp(-\beta E_{\{s_i\}}) A_{\{s_i\}}, \qquad (2.9)$$

---

[2] Under certain conditions, temporal and ensemble averages are equal. This is a statement of the ergodic theorem which lies at the foundations of equilibrium statistical mechanics.

where the denominator is the partition function

$$Z(T, H, N) = \sum_{\{s_i\}} \exp(-\beta E_{\{s_i\}}), \tag{2.10}$$

and $\beta = 1/k_B T$ is known as the inverse temperature.

The partition function represents a suitably weighted average over all the possible microstates and provides the link between the microscopic and macroscopic descriptions of a system. The partition function depends on the temperature T, the external field H, and the number of spins N. Therefore, all ensemble averages of observables also depend on T, H, and N. For ease of notation, however, we will only indicate the functional dependencies when necessary.

An example of an observable in the Ising model is the total magnetisation

$$M_{\{s_i\}} = \sum_{i=1}^{N} s_i \tag{2.11}$$

which is the difference in the number of spins pointing up and down. The average total magnetisation

$$\langle M \rangle = \frac{1}{Z} \sum_{\{s_i\}} \exp(-\beta E_{\{s_i\}}) M_{\{s_i\}}. \tag{2.12}$$

Similarly, the average total energy

$$\langle E \rangle = \frac{1}{Z} \sum_{\{s_i\}} \exp(-\beta E_{\{s_i\}}) E_{\{s_i\}}. \tag{2.13}$$

In the canonical ensemble, it is not the average total energy that is minimised at equilibrium, but rather the free energy[3]

$$F = \langle E \rangle - TS, \tag{2.14}$$

where S is the total entropy of the system. The average total energy and the total entropy are not independent and both play a role in the minimisation of the free energy. For high temperatures, the entropic term generally dominates the free energy, while for low temperatures, the energy term generally dominates the free energy.

---

[3] Throughout, 'free energy' is equivalent to 'Helmholtz free energy'.

In the canonical ensemble, the total free energy is defined by the bridge equation

$$F(T, H) = -k_B T \ln Z. \qquad (2.15)$$

For a derivation of the equivalence of Equations (2.14) and (2.15) we refer the reader to Exercise 2.1.

The total free energy is proportional to the number of spins and is therefore an extensive quantity. As in percolation, we normalise quantities of interest by the number of degrees of freedom in the system. We therefore work with the free energy per spin, $f = F/N$, the average magnetisation per spin, $m = \langle M \rangle / N$, and the average energy per spin, $\varepsilon = \langle E \rangle / N$.

All thermodynamic quantities can be extracted by taking suitable partial derivatives of the partition function or the free energy. By way of illustration, we show that the average magnetisation per spin is

$$m(T, H) = -\left(\frac{\partial f}{\partial H}\right)_T, \qquad (2.16)$$

where the temperature T is kept fixed in the derivative of the free energy per spin f with respect to the external field H. The statistical mechanical definition of the free energy yields

$$\begin{aligned}
-N \left(\frac{\partial f}{\partial H}\right)_T &= k_B T \frac{\partial}{\partial H} \ln Z \\
&= k_B T \frac{1}{Z} \frac{\partial}{\partial H} Z \\
&= k_B T \frac{1}{Z} \frac{\partial}{\partial H} \sum_{\{s_i\}} \exp(-\beta E_{\{s_i\}}) \\
&= \frac{1}{Z} \sum_{\{s_i\}} \exp(-\beta E_{\{s_i\}}) M_{\{s_i\}} \qquad (2.17)
\end{aligned}$$

which is indeed the average total magnetisation according to Equation (2.12).

The sensitivity of the average magnetisation per spin to changes in the external field at a fixed temperature is given by the susceptibility per spin

$$\chi(T, H) = \left(\frac{\partial m}{\partial H}\right)_T, \qquad (2.18)$$

which is an example of a response function. Graphically, the susceptibility per spin is the slope of the surface of the average magnetisation per spin as a function of the external field for a given temperature. The variance of the total magnetisation is related to the susceptibility through

$$\begin{aligned}
N\chi &= \left(\frac{\partial \langle M \rangle}{\partial H}\right)_T \\
&= \frac{\partial}{\partial H}\left(\frac{1}{Z}\sum_{\{s_i\}}\exp(-\beta E_{\{s_i\}})M_{\{s_i\}}\right) \\
&= \frac{1}{Z}\sum_{\{s_i\}}\exp(-\beta E_{\{s_i\}})\beta M_{\{s_i\}}^2 - \frac{1}{Z^2}\frac{\partial Z}{\partial H}\sum_{\{s_i\}}\exp(-\beta E_{\{s_i\}})M_{\{s_i\}} \\
&= \beta\frac{1}{Z}\sum_{\{s_i\}}\exp(-\beta E_{\{s_i\}})M_{\{s_i\}}^2 - \beta\left(\frac{1}{Z}\sum_{\{s_i\}}\exp(-\beta E_{\{s_i\}})M_{\{s_i\}}\right)^2 \\
&= \beta\left(\langle M^2 \rangle - \langle M \rangle^2\right). \quad (2.19)
\end{aligned}$$

Thus, the susceptibility per spin is related to the variance of the total magnetisation by

$$k_B T \chi = \frac{1}{N}\left(\langle M^2 \rangle - \langle M \rangle^2\right). \quad (2.20)$$

The sensitivity of the average energy per spin to changes in the temperature at fixed external field is given by the heat capacity per spin, referred to as the specific heat

$$c(T, H) = \left(\frac{\partial \varepsilon}{\partial T}\right)_H, \quad (2.21)$$

which is another example of a response function. Graphically, the specific heat is the slope of the surface of the average energy per spin as a function of the temperature for a given external field. The variance of the total energy is related to the specific heat through

$$k_B T^2 c = \frac{1}{N}\left(\langle E^2 \rangle - \langle E \rangle^2\right). \quad (2.22)$$

The relationships in Equations (2.20) and (2.22) between the response functions $\chi(T, H)$ and $c(T, H)$ and the variances of the total magnetisation and energy, respectively, are examples of a general fluctuation-dissipation theorem [Callen and Welton, 1951].

For the reader's convenience, we collect the relevant thermodynamic relations for a magnetic system in Table 2.2.

Table 2.2 Relations for common thermodynamic quantities. The free energy per spin, f, is related to the partition function via the bridge Equation (2.15). The average magnetisation per spin, m, the energy per spin, ε, and the entropy per spin, S/N, are related to the first partial derivatives of the free energy per spin. The response functions, that is, the susceptibility per spin, χ, and the specific heat, c, are related to the second partial derivatives of the free energy per spin.

Quantity	Relation	Response function
Partition function:	$Z = \sum_{\{s_i\}} \exp\left(-\beta E_{\{s_i\}}\right)$	
Free energy per spin:	$f = -\frac{1}{N} k_B T \ln Z$	
Magnetisation per spin:	$m = -\left(\frac{\partial f}{\partial H}\right)_T$	$\chi = \left(\frac{\partial m}{\partial H}\right)_T = -\left(\frac{\partial^2 f}{\partial H^2}\right)_T$
Energy per spin:	$\varepsilon = -\frac{1}{N}\left(\frac{\partial \ln Z}{\partial \beta}\right)_H = f - T\left(\frac{\partial f}{\partial T}\right)_H$	$c = \left(\frac{\partial \varepsilon}{\partial T}\right)_H = -T\left(\frac{\partial^2 f}{\partial T^2}\right)_H$
Entropy per spin:	$S/N = -\left(\frac{\partial f}{\partial T}\right)_H = \frac{1}{T}(\varepsilon - f)$	

2.1.3 Thermodynamic limit

In percolation, we neglected the effect of boundary sites by letting the system size tend to infinity. We will make use of this so-called thermodynamic limit in the Ising model in order to work with bulk thermodynamic quantities which do not depend on the shape of the system and the boundary conditions imposed thereon.

Specifically, in d dimensions, the total free energy F for a finite system of $N = L^d$ spins can be separated into a bulk contribution, F_{bulk}, and a boundary contribution, F_{boundary}, which are proportional to L^d and L^{d-1}, respectively. When considering the free energy per spin, boundary effects decrease with increasing system size and disappear altogether in the thermodynamic limit. Therefore, in the thermodynamic limit, the free energy of the system per spin reduces to the bulk free energy per spin,

$$\lim_{N\to\infty} \frac{F}{N} = \lim_{N\to\infty} \frac{F_{\text{bulk}} + F_{\text{boundary}}}{N} = \lim_{N\to\infty} \frac{F_{\text{bulk}}}{N}. \quad (2.23)$$

For a system of N spins, there is a finite number of 2^N terms in the partition function, see Equation (2.10). Hence, the free energy is guaranteed to be an analytic function with all derivatives well-defined everywhere. In particular, left and right partial derivatives with respect to temperature and external field are equal and finite for all values of temperature and external field. However, in the thermodynamic limit, there is an infinite number of terms in the partition function. In this case, the free energy is no longer guaranteed to be analytic, and there is at least a possibility that it is not.

Should the free energy become non-analytic in the thermodynamic limit, thermodynamic quantities may be ill-defined, that is, left and right partial derivatives are unequal or diverge for some particular values of temperature and external field. In fact, it is exactly at such values that the system will undergo a phase transition.

In the following we assume that the thermodynamic limit has already been taken prior to limits involving temperature and external field.

2.2 System of Non-Interacting Spins

To gain experience in statistical mechanical calculations, we consider a system of N spins $s_i = \pm 1$ with no interactions $J = 0$, placed in a uniform external field H. Since the spins are non-interacting, the results we obtain are independent of the underlying lattice and the dimensionality. We later compare the results with those of the Ising model ($J \neq 0$), thereby highlighting the role of spin-spin interactions. In addition, the results serve as a cross reference for the behaviour of the Ising model in the weak-coupling limit $J/k_B T \to 0$.

The total energy of a system of non-interacting spins

$$E_{\{s_i\}} = -H \sum_{i=1}^{N} s_i. \quad (2.24)$$

The ratio of the average energy and the entropic contribution to the free energy in Equation (2.14) is $\langle E \rangle / (TS) \propto H/(k_B T)$, so that the behaviour of the thermodynamic quantities is determined by the relative strength of the external field energy compared to the thermal energy.

In the limit $H/(k_B T) \to \pm 0$, the free energy is minimised by maximising the entropy. The microstates consist of spins pointing up and down randomly, corresponding to maximum disorder, see Figure 2.2(a). There is no average total magnetisation, $\langle M \rangle = 0$, and the average total energy is zero, $\langle E \rangle = 0$.

In the limit $H/(k_B T) \to \pm \infty$, the free energy is minimised by minimising the energy. There is one ground state corresponding to $H > 0$, and another corresponding to $H < 0$. These two microstates consist of all spins aligned with the external field, corresponding to minimum disorder, see Figure 2.2(b). The average total magnetisation $\langle M \rangle = \pm N$ and average total energy $\langle E \rangle = -N|H|$ are extensive quantities since they are proportional to the number of spins in the system. Therefore, the average magnetisation per spin $m = \pm 1$ and the energy per spin $\varepsilon = -|H|$ are intensive quantities since they are independent of the size of the system.

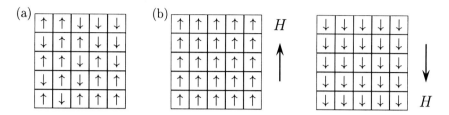

Fig. 2.2 Microstates of a system with $N = 25$ non-interacting spins. (a) A typical microstate in the limit of high temperature and weak field, $H/(k_B T) \to 0$. The spins point up or down randomly and the microstates have maximum disorder with zero total average magnetisation and energy. (b) The two microstates in the limit of low temperature and strong field, $H/(k_B T) \to \pm\infty$. All the spins align with the external field, H, and the microstates have minimum disorder with average total magnetisation $\langle M \rangle = \pm N$ and average total energy $\langle E \rangle = -N|H|$.

2.2.1 Partition function and free energy

The analytical solution of the model starts with a calculation of the partition function,

$$\begin{aligned}
Z(T,H) &= \sum_{\{s_i\}} \exp\left(-\beta E_{\{s_i\}}\right) \\
&= \sum_{\{s_i\}} \exp\left(\beta H \sum_{i=1}^{N} s_i\right) \\
&= \sum_{\{s_i\}} \exp\left(\beta H s_1\right) \exp\left(\beta H s_2\right) \cdots \exp\left(\beta H s_N\right) \\
&= \sum_{s_1=\pm 1} \sum_{s_2=\pm 1} \cdots \sum_{s_N=\pm 1} \exp\left(\beta H s_1\right) \exp\left(\beta H s_2\right) \cdots \exp\left(\beta H s_N\right) \\
&= \sum_{s_1=\pm 1} \exp\left(\beta H s_1\right) \sum_{s_2=\pm 1} \exp\left(\beta H s_2\right) \cdots \sum_{s_N=\pm 1} \exp\left(\beta H s_N\right) \\
&= \left(\exp\left(\beta H\right) + \exp\left(-\beta H\right)\right)^N \\
&= \left(2\cosh \beta H\right)^N.
\end{aligned} \quad (2.25)$$

The calculation is relatively simple because the spins are non-interacting and the product of sums can be factorised into sums over individual spins.

The total free energy is evaluated using the bridge Equation (2.15):

$$F(T,H) = -k_B T \ln\left(2\cosh\beta H\right)^N$$
$$= -Nk_B T \ln\left(2\cosh\beta H\right). \quad (2.26)$$

Therefore, the free energy per spin

$$f(T,H) = -k_B T \ln\left(2\cosh\beta H\right). \quad (2.27)$$

Figure 2.3 displays the free energy per spin versus the temperature and the external field. The free energy per spin is analytic everywhere implying that the surface is smooth everywhere without any singularities. Hence there is no phase transition in a system of non-interacting spins. Indeed, a phase transition is a cooperative phenomenon for which an interaction among spins is essential.

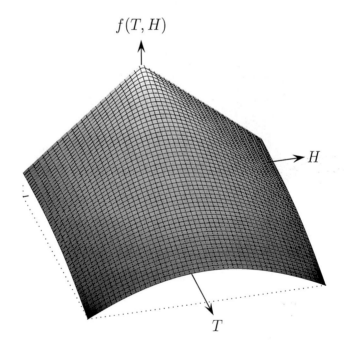

Fig. 2.3 Exact solution of the free energy per spin, $f(T,H)$, versus the temperature, T, and the external field, H, for a system of non-interacting spins. The free energy is analytic everywhere; hence there is no phase transition in a system of non-interacting spins.

2.2.2 Magnetisation and susceptibility

Graphically, the average magnetisation per spin is minus the slope of the surface of the free energy per spin as a function of the external field for a given temperature. Mathematically, the average magnetisation per spin

$$m(T, H) = -\left(\frac{\partial f}{\partial H}\right)_T = k_B T \frac{2 \sinh \beta H}{2 \cosh \beta H} \beta = \tanh \beta H. \qquad (2.28)$$

Figure 2.4 displays the magnetisation per spin versus the temperature and the external field.

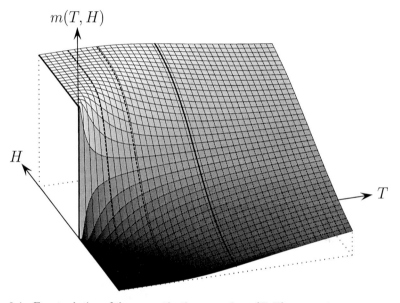

Fig. 2.4 Exact solution of the magnetisation per spin, $m(T, H)$, versus the temperature, T, and the external field, H, for a system of non-interacting spins. In zero external field, the magnetisation is zero for any temperature.

Since the magnetisation is related to the derivative of the free energy with respect to the external field at fixed temperature, it is natural to plot $m(T, H)$ for fixed T as a function of H, see Figure 2.5(a). For fixed $T > 0$, the magnetisation is zero at $H = 0$, but its graph becomes progressively steeper in the vicinity of $H = 0$ as T approaches zero. In the limit $T \to 0$, the graph for the magnetisation is a step function with $\lim_{H \to 0^{\pm}} m(T, H) = \pm 1$. However, there is no spontaneous magnetisation in zero external field for any temperature in a system of non-interacting spins.

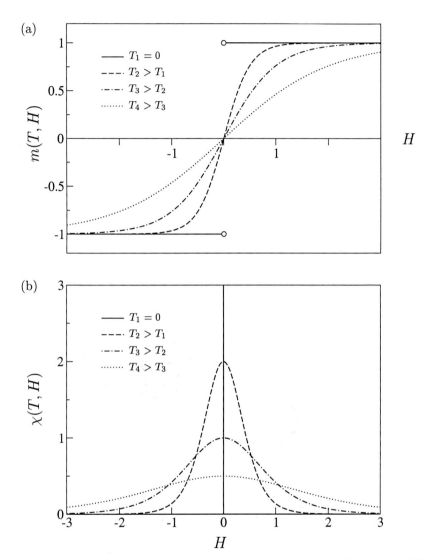

Fig. 2.5 Exact solution of the average magnetisation per spin and the susceptibility per spin for a system of non-interacting spins. (a) The average magnetisation per spin, $m(T, H)$, versus the external field, H, for temperatures $T_4 > T_3 > T_2 > T_1 = 0$. At high temperatures, a strong external field must be applied to align the spins (dotted line), but as the temperature is lowered the spins align more easily (dashed lines). At zero temperature, all spins align with the external field (solid line). In zero external field, $H = 0$, there is no spontaneous magnetisation at any temperature. (b) The susceptibility per spin, $\chi(T, H)$, versus the external field, H, for temperatures $T_4 > T_3 > T_2 > T_1 = 0$.

The susceptibility per spin

$$\chi(T,H) = \left(\frac{\partial m}{\partial H}\right)_T = \beta \operatorname{sech}^2 \beta H \qquad (2.29)$$

is the slope of the graph of the magnetisation per spin as a function of external field for a given temperature, see Figure 2.5(b). In the limit of relatively small external fields $|H| \ll k_B T$, we use the expansion $\operatorname{sech} x = 1 - x^2/2 + \cdots$, see Appendix A, so that

$$\chi(T,0) = \frac{1}{k_B T}. \qquad (2.30)$$

Thus, in zero external field, the susceptibility per spin increases with decreasing temperature and diverges for T approaching zero, as can be seen by inspecting the slopes of the graphs at $H = 0$ in Figure 2.5(a).

According to the fluctuation-dissipation theorem in Equation (2.20), the variance of the total magnetisation per spin

$$k_B T \chi = \operatorname{sech}^2 \beta H = \frac{1}{N} \left(\langle M^2 \rangle - \langle M \rangle^2 \right), \qquad (2.31)$$

implying that for $H \neq 0$, the relative fluctuations of the magnetisation

$$\frac{\sqrt{\langle M^2 \rangle - \langle M \rangle^2}}{\langle M \rangle} = \frac{1}{\sqrt{N}} \operatorname{csch} \beta H \propto \frac{1}{\sqrt{N}} \quad \text{for } H \neq 0. \qquad (2.32)$$

Therefore, the distribution of the total magnetisation $M_{\{s_i\}}$ becomes more and more sharply peaked around its average value $\langle M \rangle$ with increasing system size. This is the standard behaviour for a system of non-interacting spins.

2.2.3 Energy and specific heat

The average energy per spin

$$\varepsilon(T,H) = -\frac{1}{N}\left(\frac{\partial \ln Z}{\partial \beta}\right)_H = -\frac{2\sinh \beta H}{2\cosh \beta H} H = -H \tanh \beta H; \qquad (2.33)$$

see Figure 2.6(a). The slope of the graph in Figure 2.6(a) as a function of temperature is the specific heat

$$c(T,H) = \left(\frac{\partial \varepsilon}{\partial T}\right)_H = -H \operatorname{sech}^2(\beta H) \frac{-H}{k_B T^2} = \frac{H^2}{k_B T^2} \operatorname{sech}^2 \beta H; \qquad (2.34)$$

see Figure 2.6(b).

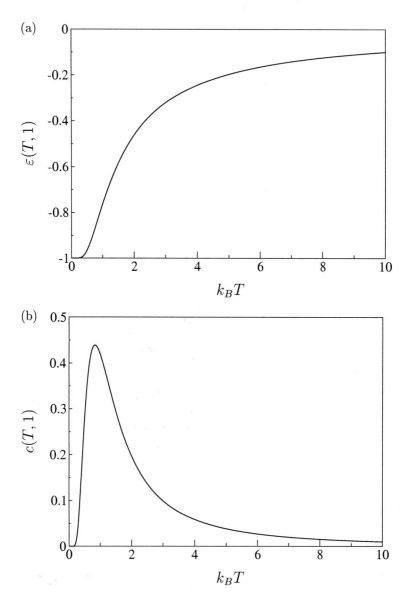

Fig. 2.6 Exact solution of the average energy per spin and the specific heat for a system of non-interacting spins. (a) The average energy per spin at external field $H = 1$, $\varepsilon(T,1)$, versus the temperature, k_BT. (b) The specific heat at external field $H = 1$, $c(T,1)$, versus the temperature, k_BT. The specific heat has a maximum around $k_BT = H$ and tends to zero for $k_BT \gg 1$, where the thermal energy is much greater than the external field, and $k_BT \ll 1$, where the external field is much greater than the thermal energy.

According to the fluctuation-dissipation theorem in Equation (2.22), the variance of the total energy per spin

$$k_B T^2 c = H^2 \text{sech}^2 \beta H = \frac{1}{N}\left(\langle E^2 \rangle - \langle E \rangle^2\right), \qquad (2.35)$$

implying that for $H \neq 0$, the relative fluctuation of the energy

$$\frac{\sqrt{\langle E^2 \rangle - \langle E \rangle^2}}{\langle E \rangle} = -\frac{\text{sign}(H)}{\sqrt{N}} \text{csch}\, \beta H \propto \frac{1}{\sqrt{N}} \quad \text{for } H \neq 0. \qquad (2.36)$$

Therefore, the distribution of the total energy $E_{\{s_i\}}$ becomes more and more sharply peaked around its average value $\langle E \rangle$ with increasing system size.

2.3 Quantities of Interest

Having considered the simpler case of a system of non-interacting spins, we turn to the Ising model proper, where the total energy

$$E_{\{s_i\}} = -J \sum_{\langle ij \rangle} s_i s_j - H \sum_{i=1}^{N} s_i. \qquad (2.37)$$

2.3.1 *Magnetisation*

Identifying the magnetisation as a quantity of interest, we investigate whether the short-ranged nearest-neighbour interaction among the spins in the Ising model gives rise to spontaneous magnetisation not induced by an external field. For now, we set the external field $H = 0$. A competition exists between the thermal energy $k_B T$, tending to randomise the orientation of spins, and the interaction energy J, tending to align spins. This is encapsulated in the minimisation of free energy

$$F = \langle E \rangle - TS. \qquad (2.38)$$

Note that the entropic contribution to F is proportional to temperature and that the ratio of the two terms $\langle E \rangle/(TS) \propto J/(k_B T)$. Thus, at relatively high temperatures $J/(k_B T) \ll 1$, the second term dominates so that the free energy is minimised by maximising the entropy. This is achieved by randomising the orientation of spins. Meanwhile, at relatively low temperatures $J/(k_B T) \gg 1$, the first term dominates so that the free energy is minimised by minimising the energy. This is achieved by aligning spins.

The above considerations suggest that there may be a phase transition from a disordered (paramagnetic) high-temperature phase to an ordered (ferromagnetic) low-temperature phase, which takes place when the thermal energy and the interaction energy are comparable. The average magnetisation serves as an order parameter. From what we have learned about the continuous phase transitions in percolation, we anticipate a continuous but abrupt pick-up of the average magnetisation per spin at a critical temperature T_c,

$$m_0(T) = \lim_{H \to 0^{\pm}} m(T, H) \propto \begin{cases} 0 & \text{for } T \geq T_c \\ \pm(T_c - T)^{\beta} & \text{for } T \to T_c^-, \end{cases} \quad (2.39)$$

see the solid curve in Figure 2.7.

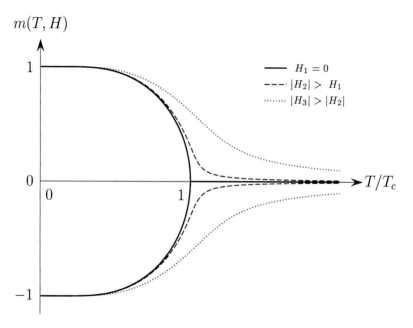

Fig. 2.7 A sketch of the magnetisation per spin, $m(T, H)$, versus the relative temperature, T/T_c, for the Ising model. The solid curve refers to the spontaneous magnetisation in zero external field $m_0(T)$. For $T \geq T_c$, the spins are equally likely to be pointing up and down on average, and the magnetisation is zero. The magnetisation picks up abruptly at $T = T_c$ and for $T < T_c$ a finite fraction of the spins are aligned. At $T = 0$, all spins point in the same direction. The dashed and dotted curves refer to the magnetisation in small external fields $|H_3| > |H_2| > H_1 = 0$. The critical exponent, δ, describes the increase in the magnitude of the magnetisation as a function of external field, H, at the critical temperature $T = T_c$.

The critical exponent β for the Ising model characterises the pick-up of the magnetisation at $T = T_c$ and takes a different numerical value to that of percolation.[4] In zero external field, the graph for $m_0(T)$ is not differentiable at $T = T_c$ and has a vertical tangent line at $T = T_c$ if $\beta < 1$.

In addition to studying the behaviour of the magnetisation as a function of temperature in zero external field, one might also examine its behaviour as a function of external field at the critical temperature. The magnetisation only vanishes in the limit of zero external field at the critical temperature and increases in magnitude in small external fields, see Figure 2.7. The critical exponent δ characterises how the magnetisation vanishes as the external field tends to zero at the critical temperature,

$$m(T_c, H) \propto \text{sign}(H)|H|^{1/\delta} \quad \text{for } |H| \to 0, T = T_c, \qquad (2.40)$$

where $\text{sign}(H)$ is the sign of the external field H.

2.3.2 Response functions

The susceptibility is related to the variance in the total magnetisation, see Equation (2.20). In contrast to a system of non-interacting spins, the variance per spin of the magnetisation diverges at the critical temperature in the Ising model. At relatively high temperatures $J/(k_BT) \ll 1$, the Ising model is effectively a system of non-interacting spins, and we can borrow the result that the variance per spin decays as $1/(k_BT)$ in zero external field, see Equation (2.30). As the temperature is lowered, the fluctuations away from the average magnetisation per spin $m = 0$ increase as the spins become increasingly more interactive, so that they can no longer be considered independent of each other, and at the critical temperature the fluctuations diverge. Therefore, at the critical point of critical temperature and zero external field, the spins are correlated throughout the system.

At relatively low temperatures $J/(k_BT) \gg 1$, we can expect that the fluctuations away from the average magnetisation per spin $m = \pm 1$ are negligible since the thermal noise is relatively weak. The thermal noise increases with temperature, driving larger and larger fluctuations away from the average magnetisation. At the critical temperature, the fluctuations diverge and wash out any net magnetisation.

In fact, in zero external field, the susceptibility per spin diverges in the vicinity of the critical temperature as a power law with exponent γ in terms

---
[4] Note that the critical exponent β is not to be confused with the inverse temperature $\beta = 1/k_BT$, that appears, for example, in the Boltzmann distribution.

of $|T - T_c|$, the distance of T from T_c,

$$\chi(T,0) \propto |T - T_c|^{-\gamma} \quad \text{for } T \to T_c. \tag{2.41}$$

Similarly, in zero external field, the specific heat diverges in the vicinity of the critical temperature as a power law with exponent α, in terms of $|T - T_c|$, the distance of T from T_c,

$$c(T,0) \propto |T - T_c|^{-\alpha} \quad \text{for } T \to T_c. \tag{2.42}$$

2.3.3 Correlation length and spin-spin correlation function

Just as for percolation, we can introduce a correlation length, $\xi(T, H)$, which, loosely speaking, sets the typical linear scale of the largest cluster of aligned spins at temperature T and external field H. Alternatively, the correlation length sets the scale of the typical largest fluctuations away from the two states with fully aligned spins, when $T < T_c$, and from the states with randomly orientated spins, when $T > T_c$.

In zero external field, the correlation length diverges in the vicinity of the critical temperature as a power law with critical exponent ν in terms of $|T - T_c|$, the distance of T from T_c,

$$\xi(T,0) \propto |T - T_c|^{-\nu} \quad \text{for } T \to T_c. \tag{2.43}$$

The correlation length is mathematically defined through the spin-spin correlation function

$$\begin{aligned} g(\mathbf{r}_i, \mathbf{r}_j) &= \langle (s_i - \langle s_i \rangle)(s_j - \langle s_j \rangle) \rangle \\ &= \langle s_i s_j \rangle - \langle s_i \rangle \langle s_j \rangle, \end{aligned} \tag{2.44}$$

which describes the correlations in the fluctuations of the spins s_i and s_j, at positions $\mathbf{r}_i$ and $\mathbf{r}_j$, around their average values $\langle s_i \rangle$ and $\langle s_j \rangle$. If the average of the product $\langle s_i s_j \rangle$ is equal to the product of the averages $\langle s_i \rangle \langle s_j \rangle$, the spin-spin correlation function is zero. In zero external field, this is the case for $T = \infty$, where spins point up and down randomly, and at $T = 0$, where all spins are aligned. Also, for a given temperature, the correlation function decreases with increasing external field as more and more spins align with the external field.

The spin-spin correlation function in the Ising model is related to the susceptibility. According to the fluctuation-dissipation theorem Equation (2.20)

$$Nk_BT\chi = \langle M^2 \rangle - \langle M \rangle^2$$

$$= \left\langle \sum_{k=1}^{N} s_k \sum_{j=1}^{N} s_j \right\rangle - \left\langle \sum_{k=1}^{N} s_k \right\rangle \left\langle \sum_{j=1}^{N} s_j \right\rangle$$

$$= \sum_{k=1}^{N} \sum_{j=1}^{N} (\langle s_k s_j \rangle - \langle s_k \rangle \langle s_j \rangle)$$

$$= \sum_{k=1}^{N} \sum_{j=1}^{N} g(\mathbf{r}_k, \mathbf{r}_j)$$

$$= N \sum_{j=1}^{N} g(\mathbf{r}_i, \mathbf{r}_j), \qquad (2.45)$$

where, in the last step, we exploit translational invariance $g(\mathbf{r}_k, \mathbf{r}_j) = g(\mathbf{r}_k + \mathbf{r}, \mathbf{r}_j + \mathbf{r})$, and choose an arbitrary site $\mathbf{r}_i$ as a reference position. Equation (2.45) is the analogue of the result in percolation describing how the site-site correlation function is related to the average cluster size.

Replacing the sum over all sites with an integral over the whole system, we have a so-called 'sum rule'

$$\int_V g(\mathbf{r}_i, \mathbf{r}_j) \, d\mathbf{r}_j = k_B T \chi. \qquad (2.46)$$

Note that Equation (2.46) relates the spin-spin correlation function to the free energy per spin via the susceptibility. Since the right-hand side of Equation (2.46) diverges in zero external field for $T \to T_c$, see Equation (2.41), the spin-spin correlation function cannot decay exponentially fast with distance $r = |\mathbf{r}_i - \mathbf{r}_j|$ at the critical point $(T_c, 0)$. Indeed, we find that at the critical point the correlation function decays like a power law

$$g(\mathbf{r}_i, \mathbf{r}_j) \propto r^{-(d-2+\eta)} \quad \text{for } (T, H) = (T_c, 0), \qquad (2.47)$$

where d is the dimension, and η is a critical exponent.

2.3.4 Critical temperature and external field

In percolation there is only a single control parameter, the occupation probability p, which must be tuned to p_c in order to observe critical behaviour. However, in the Ising model there are two control parameters, the temperature T and the external field H, which must be tuned to the critical point

$(T_c, 0)$ in order to observe critical behaviour. It is for this reason the critical exponents are defined in the vicinity of $T \to T_c$ and $H \to 0$.

The microscopic explanation for the phase transition at $(T_c, 0)$ can be understood qualitatively in terms of clusters of correlated spins. Figure 2.8 displays six microstates of the two-dimensional Ising model on a square lattice of size $L = 150$ for six different temperatures in zero external field, $H = 0$. At relatively high temperatures $T \gg T_c$, the spins are randomly orientated with no correlations. As the temperature is lowered, the spin-spin interactions are less suppressed so that larger and larger clusters of correlated spins form. However, as long as the correlation length is finite, there can be no net magnetisation in an infinite system since clusters of correlated spins are finite in size and are equally likely to point up or down throughout the system. In an infinite system, finite clusters of correlated spins are 'local'. Spontaneous magnetisation can therefore only arise through a non-local effect, namely the emergence of a macroscopic cluster of correlated spins associated with the divergence of the correlation length as the temperature approaches the critical temperature from above, $T \to T_c^+$.

At relatively low temperatures $T \ll T_c$, almost all the spins are aligned. In the low temperature phase, the correlation length sets the upper scale of fluctuations away from the fully aligned state. As the temperature is increased, the thermal noise becomes stronger and the correlation length increases. The net magnetisation persists as long as the correlation length remains finite but vanishes when the correlation length diverges as the temperature approaches the critical temperature from below, $T \to T_c^-$.

At $T = T_c$, a macroscopic cluster of correlated spins appears for the first time. This cluster is fractal and contains clusters of all sizes of opposite spins, which themselves contain clusters of all sizes of opposite spins, and so on, like droplets within droplets within droplets....

We caution the reader against identifying domains of aligned spins with clusters of correlated spins by giving two illustrative examples. First, in the panel corresponding to $T/T_c = 0.75$ in Figure 2.8, almost all spins are aligned. However, the spins are weakly correlated since there are only small fluctuations away from the average magnetisation, see Equation (2.44). Second, consider the Ising model on a triangular lattice at $T = \infty$. At this temperature, spins are independent and point up and down with equal probability. The spins are uncorrelated, but there nevertheless exists an infinite domain of aligned spins since the critical occupation probability for a triangular lattice is $p_c = 1/2$, see Table 1.1 on page 80.

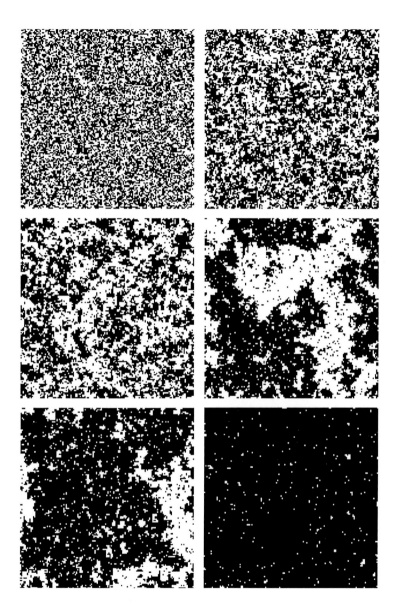

Fig. 2.8 Six microstates of the two-dimensional Ising model in zero external field on a square lattice of size $L = 150$ for relative temperatures $T/T_c = 4, 1.5, 1.2, 1, 0.99, 0.75$, from left to right and top to bottom, respectively. Sites containing up-spins are white while sites containing down-spins are black.

2.3.5 Symmetry breaking

The probabilities of finding the system in the microstates $\{s_i\}$ and $\{-s_i\}$ are

$$p_{\{s_i\}} = \frac{\exp(-\beta E_{\{s_i\}})}{Z}, \tag{2.48a}$$

$$p_{\{-s_i\}} = \frac{\exp(-\beta E_{\{-s_i\}})}{Z}, \tag{2.48b}$$

respectively, so that the ratio

$$\frac{p_{\{s_i\}}}{p_{\{-s_i\}}} = \exp\left[-\beta\left(E_{\{s_i\}} - E_{\{-s_i\}}\right)\right]. \tag{2.49}$$

In zero external field, the energy of a spin configuration is invariant if all the spins are reversed:

$$E_{\{s_i\}} - E_{\{-s_i\}} = 0, \quad \text{for } H = 0. \tag{2.50}$$

Consequently, the ratio $p_{\{s_i\}}/p_{\{-s_i\}}$ is unity and both configurations are equally probable. At the same time, however, the magnetisation changes sign if all the spins are reversed:

$$M_{\{s_i\}} = -M_{\{-s_i\}}. \tag{2.51}$$

Therefore, the average total magnetisation

$$\langle M \rangle = \sum_{\{s_i\}} p_{\{s_i\}} M_{\{s_i\}} = 0, \tag{2.52}$$

since the contribution from each spin configuration is cancelled out by its spin-reversed configuration. Have we now just proved that the average total magnetisation in the Ising model is always zero in zero external field, thereby destroying the possibility of a phase transition?

To answer this question negatively, we first consider the effect of introducing a small non-zero external field. Explicitly, the energy difference

$$E_{\{s_i\}} - E_{\{-s_i\}} = -2H \sum_{i=1}^{N} s_i = -2H M_{\{s_i\}}, \tag{2.53}$$

so that the ratio

$$\frac{p_{\{s_i\}}}{p_{\{-s_i\}}} = \exp(2\beta H M_{\{s_i\}}). \tag{2.54}$$

Without loss of generality, assume that $M_{\{s_i\}} > 0$. Taking the external field to zero before taking the thermodynamic limit, we find that

$$\lim_{N \to \infty} \lim_{H \to 0^{\pm}} \frac{p_{\{s_i\}}}{p_{\{-s_i\}}} = 1. \qquad (2.55)$$

On the other hand, taking the thermodynamic limit before taking the external field to zero, we find that

$$\lim_{H \to 0^{\pm}} \lim_{N \to \infty} \frac{p_{\{s_i\}}}{p_{\{-s_i\}}} = \begin{cases} \infty & \text{for } H \to 0^+ \\ 0 & \text{for } H \to 0^-. \end{cases} \qquad (2.56)$$

Thus with $H \to 0^+$, configurations with $M_{\{s_i\}} < 0$ have zero probability. Likewise, with $H \to 0^-$, configurations with $M_{\{s_i\}} > 0$ have zero probability. We describe configurations with zero probability as being inaccessible. The existence of a non-zero external field, however small, therefore breaks the symmetry among the spin configurations.

In summary, we have shown that the thermodynamic limit and the limit of vanishing external field are not interchangeable, that is,

$$\lim_{N \to \infty} \lim_{H \to 0^{\pm}} \langle M \rangle = 0, \qquad (2.57\text{a})$$

$$\lim_{H \to 0^{\pm}} \lim_{N \to \infty} \langle M \rangle \neq 0. \qquad (2.57\text{b})$$

Looking back at the definition of the average magnetisation per spin in Equation (2.39), we can now elaborate on the significance of the $H \to 0^{\pm}$ limit.

Consider a particular system of finite size at a temperature below T_c. The temporal[5] average of the magnetisation measured over a sufficiently long time would yield zero since the system spends an equal time in the $+m$ and $-m$ phases. However, if the system size is infinite, the system spends all of its time in only one of the phases, determined by initial conditions. In the thermodynamic limit, the ergodicity of the system is said to be 'spontaneously broken' for $T < T_c$, and it is this that gives rise to a non-zero magnetisation.

Therefore, the average magnetisation is zero for all finite systems. However, if the system size is infinite, the spontaneous symmetry breaking for $T < T_c$ guarantees a non-zero magnetisation.

In the thermodynamic limit, the application of a small external field confines the system to one of the two phases. The use of a residual ex-

---

[5] Equilibrium statistical mechanics contains no information on the dynamics. We are tacitly assuming that an equation of motion has been provided for the Ising model.

ternal field is thus a convenient device for specifying the initial conditions. Therefore, taking the thermodynamic limit before taking the limit of zero external field is the appropriate order for correctly evaluating thermodynamic averages, creating the possibility of a phase transition.

2.4 Ising Model in $d = 1$

We now turn our attention to the one-dimensional Ising model. The Ising model was originally formulated by Lenz in 1920 [Lenz, 1920] and Ising began working on the model for ferromagnetism under his supervision in 1922. For an account of the early history of the 'Lenz-Ising' model, see [Brush, 1967].

We will follow in the footsteps of Ising himself, who calculated exactly the partition function for the one-dimensional Ising model [Ising, 1925]. Once we have calculated the partition function explicitly, we can use the relations in Table 2.2 to calculate any thermodynamic quantity of interest. Hence we can show that in zero external field, there is no phase transition to an ordered ferromagnetic state at any finite temperature.

In $d = 1$ the total energy of the Ising model for N spins in a uniform external field H is

$$\begin{aligned} E_{\{s_i\}} &= -J \sum_{\langle ij \rangle} s_i s_j - H \sum_{i=1}^{N} s_i \\ &= -J \sum_{i=1}^{N} s_i s_{i+1} - H \sum_{i=1}^{N} s_i, \end{aligned} \quad (2.58)$$

where the sum over all distinct nearest-neighbour pairs reduces to a sum over all spins. In addition, we apply periodic boundary conditions such that $s_{N+1} = s_1$, see Figure 2.9.

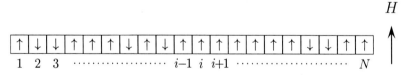

Fig. 2.9 A microstate of the one-dimensional Ising model on a lattice of size $L = N$ placed in a uniform external field, H. In $d = 1$, the nearest-neighbour spins of spin s_i are s_{i-1} and s_{i+1}. Periodic boundary conditions are imposed such that spins s_1 and s_N are nearest neighbours.

2.4.1 Partition function

There are various ways of calculating the partition function for the Ising model in $d = 1$. We employ the transfer matrix method.

$$\begin{aligned}
Z &= \sum_{\{s_i\}} \exp\left(-\beta E_{\{s_i\}}\right) \\
&= \sum_{\{s_i\}} \exp\left(\beta J \sum_{i=1}^{N} s_i s_{i+1} + \beta H \sum_{i=1}^{N} s_i\right) \\
&= \sum_{\{s_i\}} \exp\left(\beta J \sum_{i=1}^{N} s_i s_{i+1} + \beta \frac{H}{2} \sum_{i=1}^{N} (s_i + s_{i+1})\right) \quad (2.59) \\
&= \sum_{\{s_i\}} \exp\left(\beta J s_1 s_2 + \beta \frac{H}{2}(s_1 + s_2)\right) \cdots \exp\left(\beta J s_N s_1 + \beta \frac{H}{2}(s_N + s_1)\right).
\end{aligned}$$

We have recast the contribution from the external field so as to make the exponent symmetric in its indices. There are four possible configurations of the two spins s_i and s_{i+1}, and it is convenient to arrange these in a real and symmetric 2×2 transfer matrix, **T**, with entries

$$T_{s_i s_{i+1}} = \exp\left(\beta J s_i s_{i+1} + \beta \frac{H}{2}(s_i + s_{i+1})\right), \quad (2.60)$$

so that

$$\mathbf{T} = \begin{pmatrix} T_{+1+1} & T_{+1-1} \\ T_{-1+1} & T_{-1-1} \end{pmatrix} = \begin{pmatrix} \exp(\beta J + \beta H) & \exp(-\beta J) \\ \exp(-\beta J) & \exp(\beta J - \beta H) \end{pmatrix}. \quad (2.61)$$

After introducing the transfer matrix notation, the partition function becomes

$$\begin{aligned}
Z &= \sum_{\{s_i\}} T_{s_1 s_2} T_{s_2 s_3} \cdots T_{s_{N-1} s_N} T_{s_N s_1} \\
&= \sum_{s_1=\pm 1} \sum_{s_2=\pm 1} \cdots \sum_{s_N=\pm 1} T_{s_1 s_2} T_{s_2 s_3} \cdots T_{s_{N-1} s_N} T_{s_N s_1} \quad (2.62) \\
&= \sum_{s_1=\pm 1} \cdots \sum_{s_{N-1}=\pm 1} \left(\sum_{s_2=\pm 1} T_{s_1 s_2} T_{s_2 s_3}\right) \cdots \left(\sum_{s_N=\pm 1} T_{s_{N-1} s_N} T_{s_N s_1}\right).
\end{aligned}$$

In the last step, we have paired terms and summed over their common index, since each spin only occurs in two consecutive terms.

Recall that for an $n \times n$ matrix $\mathbf{A}$, the ijth entry in the matrix $\mathbf{A}^2$ is

$$A_{ij}^2 = \sum_{k=1}^{n} A_{ik} A_{kj}. \tag{2.63}$$

We can therefore rewrite the sum over paired terms as entries from their product matrix $\mathbf{T}^2$,

$$\begin{aligned} Z &= \sum_{s_1=\pm 1} \sum_{s_3=\pm 1} \cdots \sum_{s_{N-1}=\pm 1} T_{s_1 s_3}^2 T_{s_3 s_5}^2 \cdots T_{s_{N-3} s_{N-1}}^2 T_{s_{N-1} s_1}^2 \\ &= \sum_{s_1=\pm 1} \sum_{s_5=\pm 1} \cdots \sum_{s_{N-3}=\pm 1} T_{s_1 s_5}^4 T_{s_5 s_9}^4 \cdots T_{s_{N-7} s_{N-3}}^4 T_{s_{N-3} s_1}^4 \\ &= \sum_{s_1=\pm 1} T_{s_1 s_1}^N \\ &= \mathrm{Tr}\,(\mathbf{T}^N), \end{aligned} \tag{2.64}$$

where we have continued the process of pairing terms until we are left with a single sum. The final expression in Equation (2.64) is the trace of $\mathbf{T}^N$, that is, the sum over the diagonal elements of the matrix $\mathbf{T}^N$.

A calculation of $\mathbf{T}^N$ would typically be laborious. However, for the trace we only need the diagonal elements rather than the whole matrix itself. We borrow a result from linear algebra which states that for the real and symmetric 2×2 matrix $\mathbf{T}$ there exists a 2×2 unitary matrix $\mathbf{U}$, such that

$$\mathbf{U}^{-1} \mathbf{T} \mathbf{U} = \begin{pmatrix} \lambda_+ & 0 \\ 0 & \lambda_- \end{pmatrix}, \tag{2.65}$$

where $\lambda_+ > \lambda_-$ are the eigenvalues of the transfer matrix $\mathbf{T}$. To find these eigenvalues, we set the determinant to zero,

$$\det(\mathbf{T} - \lambda \mathbf{I}) = 0, \tag{2.66}$$

where $\mathbf{I}$ is the 2×2 identity matrix, so that the characteristic equation reads

$$\lambda^2 - \exp(\beta J)\left[\exp(\beta H) + \exp(-\beta H)\right]\lambda + \exp(2\beta J) - \exp(-2\beta J) = 0 \tag{2.67}$$

with solutions

$$\lambda_\pm = \exp(\beta J) \left(\cosh \beta H \pm \sqrt{\sinh^2 \beta H + \exp(-4\beta J)} \right) \tag{2.68}$$

corresponding to the two eigenvalues of $\mathbf{T}$.

Using the identity $\mathbf{UU}^{-1} = \mathbf{I}$ and the commutative property of the trace operation, $\text{Tr}(\mathbf{AB}) = \text{Tr}(\mathbf{BA})$, the partition function is therefore

$$\begin{aligned}Z &= \text{Tr}\left(\mathbf{T}^N\right) \\ &= \text{Tr}\left(\mathbf{TUU}^{-1}\mathbf{TUU}^{-1}\cdots\mathbf{TUU}^{-1}\right) \\ &= \text{Tr}(\underbrace{(\mathbf{U}^{-1}\mathbf{TU})(\mathbf{U}^{-1}\mathbf{TU})\cdots(\mathbf{U}^{-1}\mathbf{TU})}_{N\,\text{factors}}) \\ &= \text{Tr}\begin{pmatrix}\lambda_+^N & 0 \\ 0 & \lambda_-^N\end{pmatrix} \\ &= \lambda_+^N + \lambda_-^N. \end{aligned} \qquad (2.69)$$

Note that for $J = 0$, the eigenvalues are $\lambda_+ = 2\cosh\beta H$ and $\lambda_- = 0$ so that we recover the partition function for a system of non-interacting spins, see Equation (2.25).

For zero external field $H = 0$, the eigenvalues are $\lambda_+ = 2\cosh\beta J$ and $\lambda_- = 2\sinh\beta J$ and the partition function

$$Z(T,0) = (2\cosh\beta J)^N(1+\tanh^N\beta J) \to (2\cosh\beta J)^N \text{ for } N \to \infty \quad (2.70)$$

where the large N behaviour follows from $\lambda_-/\lambda_+ = \tanh\beta J < 1$.

2.4.2 Free energy

The total free energy is evaluated in limit $N \to \infty$, using that $\lambda_-/\lambda_+ < 1$:

$$\begin{aligned} F &= -k_B T \ln Z \\ &= -k_B T \ln\left(\lambda_+^N\left[1+\left(\frac{\lambda_-}{\lambda_+}\right)^N\right]\right) \\ &= -k_B T \ln \lambda_+^N \quad \text{for } N \to \infty \\ &= -Nk_B T\left[\beta J + \ln\left(\cosh\beta H + \sqrt{\sinh^2\beta H + \exp(-4\beta J)}\right)\right]. \end{aligned} \quad (2.71)$$

Figure 2.10(a) displays the free energy per spin versus the temperature and the external field. The free energy per spin is analytic everywhere except at $(T, H) = (0, 0)$ where a cusp develops.

The singular point is clearly visible in Figure 2.10(b) where, for fixed temperatures, the free energy per spin is plotted versus the external field. For $T > 0$ the graphs are analytic everywhere, but for $T = 0$ a cusp exists at $H = 0$.

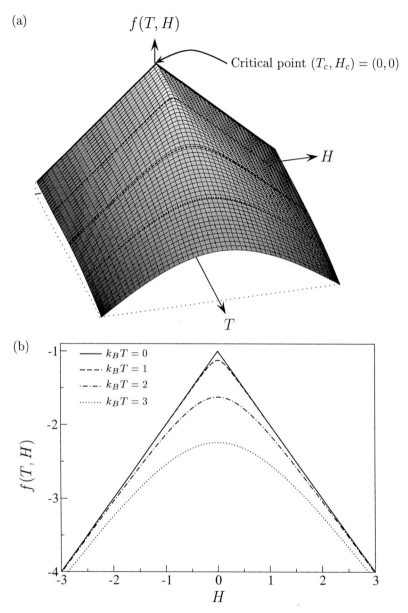

Fig. 2.10 Exact solution of the free energy per spin for the one-dimensional Ising model. (a) The free energy per spin, $f(T, H)$, versus the temperature, T, and the external field, H. (b) The free energy per spin, $f(T, H)$, versus the external field, H, for four different temperatures $k_B T = 3, 2, 1, 0$. The free energy is analytic everywhere except at the critical point, $(T_c, H_c) = (0, 0)$.

Note that for $J = 0$ we recover the free energy for a system of non-interacting spins, see Equation (2.26).

In zero external field $H = 0$, the free energy per spin for large N

$$f(T,0) = -k_B T \ln(2 \cosh \beta J) \to \begin{cases} -k_B T \ln 2 & \text{for } T \to \infty \\ -J & \text{for } T \to 0. \end{cases} \quad (2.72)$$

For high temperatures, all the spins are randomly orientated. The energy is zero and the free energy is determined by the maximisation of the entropy. For low temperatures, all the spins are aligned. The entropy is zero and the free energy is determined by the minimisation of the energy, see Equation (2.38).

2.4.3 Magnetisation and susceptibility

The average magnetisation per spin is

$$\begin{aligned} m(T, H) &= -\left(\frac{\partial f}{\partial H}\right)_T \\ &= k_B T \, \frac{\beta \sinh \beta H + \frac{2\beta \sinh \beta H \cosh \beta H}{2\sqrt{\sinh^2 \beta H + \exp(-4\beta J)}}}{\cosh \beta H + \sqrt{\sinh^2 \beta H + \exp(-4\beta J)}} \\ &= \frac{\sinh \beta H}{\sqrt{\sinh^2 \beta H + \exp(-4\beta J)}}. \end{aligned} \quad (2.73)$$

Figure 2.11(a) displays the magnetisation per spin versus the temperature and external field. The surface is minus the slope of the surface of the free energy per spin as a function of the external field for a given temperature.

Note that for $J = 0$ we recover the magnetisation for a system of non-interacting spins, see Equation (2.28). However, for $J \neq 0$ the average magnetisation per spin in vanishing external field

$$m_0(T) = \lim_{H \to 0^{\pm}} m(T, H) = \begin{cases} 0 & \text{for } T > 0 \\ \pm 1 & \text{for } T = 0. \end{cases} \quad (2.74)$$

In Figure 2.11(b), $m_0(T)$ lies at the intersection between the vertical line $H = 0$ and the graphs of m. There is no spontaneous magnetisation in zero external field for any finite temperature in the one-dimensional Ising model. However, at $T = 0$ the spins align spontaneously.

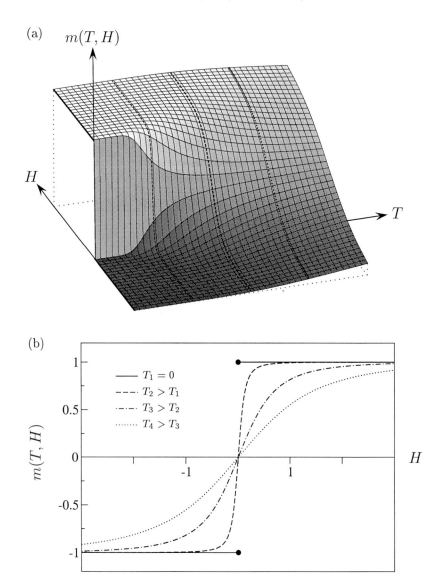

Fig. 2.11 Exact solution of the average magnetisation per spin for the one-dimensional Ising model. (a) The average magnetisation per spin, $m(T, H)$, versus the temperature, T, and the external field, H. (b) The average magnetisation per spin, $m(T, H)$, versus the external field, H, for temperatures $T_4 > T_3 > T_2 > T_1 = 0$. At high temperatures, a strong external field must be applied to align the spins (dotted line). When the temperature is lowered, the spins align more easily (dashed lines). At zero temperature, all spins align with the external field (solid line). When $H = 0$, there is no spontaneous magnetisation for any finite temperature but at $T = 0$, the spins align spontaneously.

The susceptibility per spin is the sensitivity of the average magnetisation per spin to changes in the external field at a fixed temperature. Hence, graphically, the susceptibility per spin is the slope of the surface of the magnetisation per spin in Figure 2.11(a) as a function of the external field for a given temperature:

$$\chi(T,H) = \left(\frac{\partial m}{\partial H}\right)_T$$

$$= \beta \frac{\cosh\beta H \sqrt{\sinh^2\beta H + \exp(-4\beta J)} - \sinh\beta H \frac{\sinh\beta H \cosh\beta H}{\sqrt{\sinh^2\beta H + \exp(-4\beta J)}}}{\sinh^2\beta H + \exp(-4\beta J)}$$

$$= \beta \frac{\cosh\beta H \exp(-4\beta J)}{[\sinh^2\beta H + \exp(-4\beta J)]^{3/2}}. \tag{2.75}$$

Note that for $J = 0$, we recover the susceptibility for a system of non-interacting spins, see Equation (2.29).

Figure 2.12(a) displays the susceptibility per spin versus the external field for four different temperatures. In zero external field, the susceptibility per spin increases with decreasing temperature and diverges for $T \to 0$ as can been seen by inspecting the slope of the graphs at $H = 0$ in Figure 2.11.

The graph of $\chi(T,0)$ as a function of k_BT/J is displayed in Figure 2.12(b). For $J \neq 0$, the susceptibility in zero external field

$$\chi(T,0) = \beta \exp(2\beta J) \to \begin{cases} \beta & \text{for } T \to \infty \\ \beta \exp(2\beta J) & \text{for } T \to 0, \end{cases} \tag{2.76}$$

that is, for $T \to \infty$ we recover Equation (2.30), while for $T \to 0$ the susceptibility diverges exponentially fast.

The variance of the total magnetisation per spin is, according to the fluctuation-dissipation theorem in Equation (2.20) given by

$$k_BT\chi(T,H) = \frac{\cosh\beta H \exp(-4\beta J)}{[\sinh^2\beta H + \exp(-4\beta J)]^{3/2}}. \tag{2.77}$$

In zero external field, the variance increases with decreasing temperature. Explicitly,

$$k_BT\chi(T,0) = \exp(2\beta J). \tag{2.78}$$

For $J \neq 0$, the fluctuations diverge for temperatures approaching zero, unlike the fluctuations in a system of non-interacting spins, see Equation (2.31).

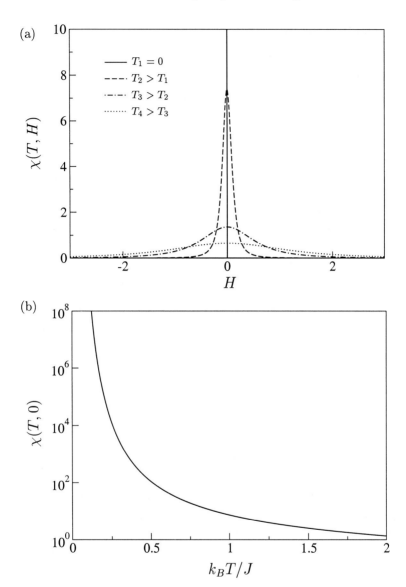

Fig. 2.12 Exact solution of the susceptibility per spin for the one-dimensional Ising model. (a) The susceptibility per spin, $\chi(T,H)$, versus the external field, H, for temperatures $T_4 > T_3 > T_2 > T_1 = 0$. In zero external field, the susceptibility per spin increases with decreasing temperature. (b) The susceptibility per spin in zero external field, $\chi(T,0)$, versus the ratio, $k_B T/J$. When $k_B T/J \to 0$, the coupling constant is much greater than the thermal energy, and the susceptibility per spin diverges exponentially fast. When $k_B T/J \to \infty$, the thermal energy is much greater than the coupling constant, and the susceptibility per spin approaches zero.

2.4.4 Energy and specific heat

For simplicity we restrict ourselves to zero external field where $\ln Z(T,0) = N \ln(2\cosh\beta J)$, see Equation (2.70). The average energy per spin

$$\varepsilon(T,0) = -\frac{1}{N}\frac{\partial \ln Z(T,0)}{\partial \beta}$$
$$= -J\tanh\beta J$$
$$\to \begin{cases} 0 & \text{for } T \to \infty \\ -J & \text{for } T \to 0; \end{cases} \qquad (2.79)$$

see Figure 2.13(a). For $T \to \infty$, all spins become randomly orientated and the average total energy approaches zero. For $T \to 0$, all spins align and the average total energy approaches $-J$.

The specific heat in zero external field is the derivative of the average energy per spin with respect to temperature. Using the chain rule for differentiating

$$\frac{\partial}{\partial T} = \frac{\partial}{\partial \beta}\frac{\partial \beta}{\partial T} = -\frac{1}{k_B T^2}\frac{\partial}{\partial \beta}, \qquad (2.80)$$

we find that the specific heat

$$c(T,0) = \frac{\partial \varepsilon(T,0)}{\partial T}$$
$$= \frac{\partial \varepsilon}{\partial \beta}\frac{\partial \beta}{\partial T}$$
$$= \frac{J^2}{k_B T^2}\operatorname{sech}^2 \beta J. \qquad (2.81)$$

Therefore, unlike the susceptibility, the specific heat does not diverge for $T \to 0$, but instead has a maximum around $\beta J \approx 1$, see Figure 2.13(b). The specific heat tends to zero for $k_B T \gg J$, where the thermal energy is much greater than the coupling constant, and for $k_B T \ll J$, where the coupling constant is much greater than the thermal energy.

The variance of the total energy per spin in zero external field is, according to the fluctuation-dissipation theorem in Equation (2.22) given by

$$k_B T^2 c(T,0) = J^2 \operatorname{sech}^2 \beta J, \qquad (2.82)$$

which tends to its maximum value J^2 when $T \to \infty$. The variance of the total energy per spin in zero external field remains finite for all temperatures.

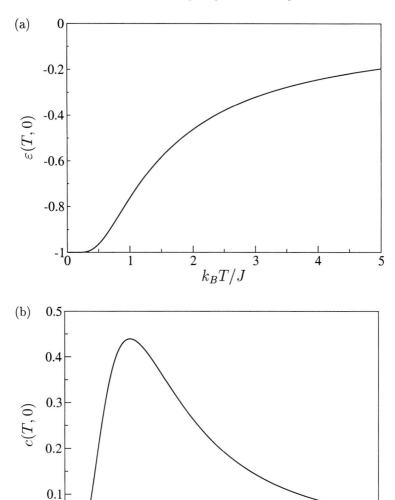

Fig. 2.13 Exact solution of the average energy per spin and the specific heat for the one-dimensional Ising model. (a) The average energy per spin in zero external field, $\varepsilon(T,0)$, versus the ratio, k_BT/J. (b) The specific heat in zero external field, $c(T,0)$, versus the ratio, k_BT/J. The specific heat has a maximum around $k_BT = J$. The specific heat tends to zero for $k_BT/J \to \infty$, where the thermal energy is much greater than the coupling constant, and for $k_BT/J \to 0$, where the coupling constant is much greater than the thermal energy.

2.4.5 Correlation function

Recall that the spin-spin correlation function

$$g(\mathbf{r}_i, \mathbf{r}_j) = \langle (s_i - \langle s_i \rangle)(s_j - \langle s_j \rangle) \rangle$$
$$= \langle s_i s_j \rangle - \langle s_i \rangle \langle s_j \rangle \tag{2.83}$$

describes the correlations in the fluctuations of s_i and s_j around their average values $\langle s_i \rangle$ and $\langle s_j \rangle$.

The Ising model is translationally invariant and the average value of a spin does not depend on its position implying $\langle s_i \rangle = \langle s_j \rangle = m$. In addition, the Ising model is rotationally invariant so that the correlation function only depends on the distance between spins, $r = |\mathbf{r}_i - \mathbf{r}_j|$, measured in lattice spacings. Hence, the spin-spin correlation function is

$$g(\mathbf{r}_i, \mathbf{r}_j) = \langle s_i s_{i+r} \rangle - m^2. \tag{2.84}$$

For simplicity, we calculate the spin-spin correlation function in zero external field $H = 0$ and we replace m by $m_0(T)$:

$$g(\mathbf{r}_i, \mathbf{r}_j) = \langle s_i s_{i+r} \rangle - m_0^2(T) \quad \text{for } H = 0. \tag{2.85}$$

At $T = 0$, all spins align and hence the correlation function $g(\mathbf{r}_i, \mathbf{r}_j) = 0$; there are no fluctuations away from the average magnetisation.

In the $d = 1$ Ising model, $m_0(T) = 0$ for any finite temperature $T > 0$, see Equation (2.74), so that the spin-spin correlation function reduces to

$$g(\mathbf{r}_i, \mathbf{r}_j) = \langle s_i s_{i+r} \rangle \qquad \text{for } d = 1, T > 0, H = 0$$
$$= \frac{1}{Z} \sum_{\{s_i\}} \exp(-\beta E_{\{s_i\}}) s_i s_{i+r} \quad \text{for } d = 1, T > 0, H = 0. \tag{2.86}$$

It is advantageous to assume, for the moment, that the nearest-neighbour interactions depend on position. In zero external field, the total energy takes the form

$$E_{\{s_i\}} = -\sum_{i=1}^{N} J_i s_i s_{i+1}. \tag{2.87}$$

In the limit of large N, the partition function for this system is

$$Z = \prod_{i=1}^{N} 2 \cosh \beta J_i. \tag{2.88}$$

This result is a generalisation of Equation (2.70) in the limit of large N.

Since $s_k s_k = 1$ and

$$\frac{\partial}{\partial J_k} \exp\left(\beta \sum_{i=1}^{N} J_i s_i s_{i+1}\right) = \beta s_k s_{k+1} \exp\left(\beta \sum_{i=1}^{N} J_i s_i s_{i+1}\right) \quad (2.89)$$

we find

$$\begin{aligned}
g(\mathbf{r}_i, \mathbf{r}_j) &= \frac{1}{Z} \sum_{\{s_i\}} \exp\left(\beta \sum_{i=1}^{N} J_i s_i s_{i+1}\right) s_i s_{i+r} \\
&= \frac{1}{Z} \sum_{\{s_i\}} \exp\left(\beta \sum_{i=1}^{N} J_i s_i s_{i+1}\right) s_i s_{i+1} s_{i+1} \cdots s_{i+r-1} s_{i+r-1} s_{i+r} \\
&= \frac{1}{Z} \sum_{\{s_i\}} \frac{1}{\beta^r} \frac{\partial^r}{\partial J_i \partial J_{i+1} \cdots \partial J_{i+r-1}} \exp\left(\beta \sum_{i=1}^{N} J_i s_i s_{i+1}\right) \\
&= \frac{1}{Z \beta^r} \frac{\partial^r}{\partial J_i \partial J_{i+1} \cdots \partial J_{i+r-1}} \sum_{\{s_i\}} \exp\left(\beta \sum_{i=1}^{N} J_i s_i s_{i+1}\right) \\
&= \frac{1}{Z \beta^r} \frac{\partial^r}{\partial J_i \partial J_{i+1} \cdots \partial J_{i+r-1}} \prod_{i=1}^{N} 2 \cosh \beta J_i \\
&= \frac{(2 \cosh \beta J)^{N-r} (2\beta \sinh \beta J)^r}{(2 \cosh \beta J)^N \beta^r} \quad \text{after restoring } J_i = J \\
&= \tanh^r \beta J \\
&= \exp\left[r \ln(\tanh \beta J)\right]. \quad (2.90)
\end{aligned}$$

Therefore, in the one-dimensional Ising model the spin-spin correlation function only depends on the distance between spins $r = |\mathbf{r}_i - \mathbf{r}_j|$, and decays exponentially with the correlation length in zero external field

$$g(\mathbf{r}_i, \mathbf{r}_j) = \exp(-r/\xi) \quad \text{for } d=1, T>0, H=0, \quad (2.91)$$

where the correlation length

$$\xi(T, 0) = -\frac{1}{\ln(\tanh \beta J)} = \frac{1}{\ln(\lambda_+/\lambda_-)} \quad \text{for } d=1, \quad (2.92)$$

since the eigenvalues in zero external field are $\lambda_\pm = \exp(\beta J) \pm \exp(-\beta J)$. In Figure 2.14 we display the correlation length in zero external field $\xi(T, 0)$ versus the temperature for the one-dimensional Ising model.

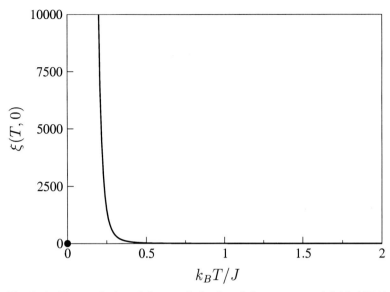

Fig. 2.14 Exact solution of the correlation length in zero external field, $\xi(T,0)$, versus the ratio, k_BT/J. When $k_BT/J \to 0$, the coupling constant is much greater than the thermal energy and the correlation length diverges. However, at $T=0$, all the spins are aligned and the correlation length $\xi(0,0) = 0$. When $k_BT/J \to \infty$, the thermal energy is much greater than the coupling constant and the correlation length tends to zero.

For $T \to \infty$, the ratio $\lambda_+/\lambda_- \to \infty$ and hence the correlation length tends to zero. For $J \neq 0$ and $T \to 0^+$, we find

$$\begin{aligned}
\xi(T,0) &= \frac{1}{\ln[\lambda_+/\lambda_-]} \\
&= \frac{1}{\ln\left[(1+\exp(-2\beta J))/(1-\exp(-2\beta J))\right]} \\
&\approx \frac{1}{\ln\left[(1+\exp(-2\beta J))(1+\exp(-2\beta J))\right]} \qquad \text{for } T \to 0^+ \\
&\approx \frac{1}{\ln[1+2\exp(-2\beta J)]} \qquad \text{for } T \to 0^+ \\
&\to \frac{1}{2}\exp(2\beta J) \qquad \text{for } T \to 0^+. \qquad (2.93)
\end{aligned}$$

In summary, for $J \neq 0$ the limits of the correlation length are

$$\xi(T,0) \to \begin{cases} 0 & \text{for } T \to \infty \\ \frac{1}{2}\exp(2\beta J) & \text{for } T \to 0^+ \\ 0 & \text{for } T = 0. \end{cases} \qquad (2.94)$$

At high temperatures, the spins are randomly orientated and thus uncorrelated. The correlation length is zero and there are no fluctuations away from states with randomly orientated spins. As the temperature is decreased, spins align to form clusters of larger and larger size limited only by the correlation length. For temperatures approaching zero, the correlation length diverges. Clusters of all sizes form and there are fluctuations of all scales away from states with randomly orientated spins. However, at $T = 0$ all spins are aligned and the correlation length is zero.

In contrast, a system of non-interacting spins $J = 0$ is characterised by a temperature independent $\xi = 0$, since $g(\mathbf{r}_i, \mathbf{r}_j) = \langle s_i s_j \rangle = \langle s_i \rangle \langle s_j \rangle = 0$, which is consistent with taking the limit $J \to 0$ in Equation (2.90).

We demonstrate explicitly for $d = 1$ that, in zero external field, the spin-spin correlation function is related to the susceptibility by the 'sum rule' in Equation (2.46). Summing the correlation function over all positions $\mathbf{r}_j$,

$$\sum_{\mathbf{r}_j} g(\mathbf{r}_i, \mathbf{r}_j) = \sum_{\mathbf{r}_j} (\tanh \beta J)^{|\mathbf{r}_i - \mathbf{r}_j|}$$

$$= \cdots + \tanh^2 \beta J + \tanh \beta J + 1 + \tanh \beta J + \tanh^2 \beta J + \cdots$$

$$= \frac{1 + \tanh \beta J}{1 - \tanh \beta J}$$

$$= \exp(2\beta J)$$

$$= k_B T \chi(T, 0), \tag{2.95}$$

where, in the last step, we use Equation (2.76) for the susceptibility per spin in zero external field.

2.4.6 *Critical temperature*

We conclude our analysis of the one-dimensional Ising model by remarking on the role played by the temperature $T = 0$ in zero external field. Approaching this value, the susceptibility and the correlation length diverge. This is intimately related to the onset of spontaneous magnetisation at $T = 0$. We can therefore identify the critical point $(T_c, H_c) = (0, 0)$, at which the phase transition takes place.

Note the correspondence between the control parameters in percolation and the Ising model. As $T \geq 0$ and $p \leq 1$, the critical point is accessible from one direction only in the one-dimensional systems. In percolation, the correlation length diverges as the occupation probability p approaches 1 from below and an infinite cluster forms at $p = 1$. Likewise, in the Ising model, the correlation length diverges as the temperature T approaches 0 from above and an infinite cluster of aligned spins forms at $T = 0$.

In both cases, the correlation length describes the typical size of the largest correlated cluster. The correlation length sets the upper scale of the fluctuations away from the empty configuration in percolation, and the upper scale of the fluctuations away from the randomly orientated configurations in the Ising model.

In one-dimensional percolation, it is obvious why $p_c = 1$ for there to be an infinite cluster since every site must be occupied. In the one-dimensional Ising model, there is a thermodynamic reason why there can only be an infinite cluster with non-zero magnetisation at $T_c = 0$.

To investigate whether the single domain of aligned spins is stable against thermal fluctuations for $T > 0$, we calculate the difference between the free energy for a single domain of aligned spins, $F_{\text{1-dom}}$, and the free energy for two domains of oppositely aligned spins, $F_{\text{2-dom}}$, see Figure 2.15. We will show that for any finite temperature $T > 0$, $F_{\text{2-dom}} < F_{\text{1-dom}}$ when $N \to \infty$.

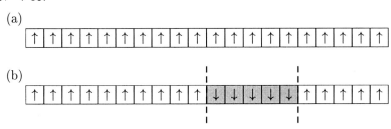

Fig. 2.15 The one-dimensional Ising model with periodic boundary conditions in zero external field. (a) One of the two microstates with a single spin domain. The free energy of the one-dimensional Ising model with a single domain $F_{\text{1-dom}} = -NJ - k_BT \ln 2$. (b) One of the $2N(N-1)$ equally probable microstates with two spin domains. The two domain walls indicated by the dashed lines cost $4J$ in energy so that the total energy $E = -NJ + 4J$. The free energy of the one-dimensional Ising model with two domains $F_{\text{2-dom}} = -NJ + 4J - k_BT \ln 2N(N-1)$.

There are two microstates with a single spin domain, each with energy $-NJ$. The entropic contribution to the free energy is $-k_BT \ln 2$. Hence, the associated free energy for a single spin domain $F_{\text{1-dom}} = -NJ - k_BT \ln 2$, see Figure 2.15(a).

To make two domains of oppositely aligned spins, two domain walls must be inserted with an energy cost of $4J$, see Figure 2.15(b). There are $2N(N-1)$ equally probable microstates with two domain walls. The free energy is then $F_{\text{2-dom}} = -NJ + 4J - k_BT \ln 2N(N-1)$.

The difference between the free energies

$$\begin{aligned}F_{\text{2-dom}} - F_{\text{1-dom}} &= -NJ + 4J - k_BT\ln 2N(N-1) + NJ + k_BT\ln 2\\ &= 4J - k_BT\ln N(N-1)\\ &\approx 4J - 2k_BT\ln N \quad \text{for large } N.\end{aligned} \quad (2.96)$$

A single domain of aligned spins is unstable against thermal fluctuations if $F_{\text{2-dom}} < F_{\text{1-dom}}$, or, equivalently, if

$$\frac{2J}{k_BT} < \ln N. \quad (2.97)$$

Thus, for any finite temperature, it is energetically favourable to insert at least two domain walls when $N \to \infty$. However, the ensemble average over all these microstates gives zero magnetisation. Thus the critical temperature $T_c = 0$ for the one-dimensional Ising model.

Note that by setting $F_{\text{2-dom}} = F_{\text{1-dom}}$, Equation (2.96) estimates a maximum domain size $N = \exp(2J/k_BT)$ at temperature T, which is consistent with the correlation length in the small T limit in Equation (2.94).

In conclusion, there is no phase transition at any finite temperature in the one-dimensional Ising model.

2.5 Mean-Field Theory of the Ising Model

It is relatively easy to solve the Ising model analytically when ignoring fluctuations. Such an approach is generically referred to as mean field, and is often the first port of call when a non-trivial problem is encountered. Mean-field theory yields considerable insight into the behaviour of the Ising model by providing a qualitatively correct description of the phase transition in $d > 1$. Indeed, the theory correctly predicts a phase transition at a finite temperature T_c in zero external field.

In order to explicitly expose where the fluctuations enter the Ising model, we first rewrite the interaction energy as follows:

$$\begin{aligned}E_{\text{int}} &= -J\sum_{\langle ij\rangle} s_i s_j\\ &= -J\sum_{\langle ij\rangle}(s_i - \langle s_i\rangle + \langle s_i\rangle)(s_j - \langle s_j\rangle + \langle s_j\rangle) \quad (2.98)\\ &= -J\sum_{\langle ij\rangle}\left[(s_i - \langle s_i\rangle)\langle s_j\rangle + (s_j - \langle s_j\rangle)\langle s_i\rangle + \langle s_i\rangle\langle s_j\rangle + (s_i - \langle s_i\rangle)(s_j - \langle s_j\rangle)\right].\end{aligned}$$

When fluctuations around the average magnetisation per spin are small, we can neglect the second-order term[6] $(s_i - \langle s_i \rangle)(s_j - \langle s_j \rangle)$. This in turn simplifies the problem by discounting spin-spin interactions.

Since the Ising model is translationally invariant, we introduce the notation $m = \langle s_i \rangle = \langle s_j \rangle$ for the average magnetisation per spin and the total energy in the mean-field Ising model is

$$\begin{aligned} E_{\{s_i\}} &\approx -J \sum_{\langle ij \rangle} \left[(s_i + s_j)m - m^2\right] - H \sum_{i=1}^{N} s_i \\ &= -2Jm \sum_{\langle ij \rangle} s_i + J \sum_{\langle ij \rangle} m^2 - H \sum_{i=1}^{N} s_i \\ &= -2Jm \frac{z}{2} \sum_{i=1}^{N} s_i + J \frac{Nz}{2} m^2 - H \sum_{i=1}^{N} s_i \\ &= -(Jzm + H) \sum_{i=1}^{N} s_i + \frac{NJz}{2} m^2. \end{aligned} \quad (2.99)$$

Therefore, the mean-field Ising model with coordination number z becomes a system of non-interacting spins immersed in an effective field of strength $(Jzm + H)$ plus a constant term $NJzm^2/2$. The effective field is made up of an 'internal' field, Jzm, resulting from the z nearest neighbours each contributing a field of strength Jm, and the external field H.

Note that if the interaction strength $J = 0$, we recover the energy of a system of non-interacting spins discussed in Section 2.2.

2.5.1 *Partition function and free energy*

As usual, the starting point is to calculate the partition function, which is easily done since the spins are non-interacting:

$$\begin{aligned} Z &= \sum_{\{s_i\}} \exp\left((\beta Jzm + \beta H) \sum_{i=1}^{N} s_i - \beta NJzm^2/2\right) \\ &= \exp\left(-\beta NJzm^2/2\right) \sum_{\{s_i\}} \prod_{i=1}^{N} \exp\left[(\beta Jzm + \beta H) s_i\right] \\ &= \exp\left(-\beta NJzm^2/2\right) \left[2\cosh(\beta Jzm + \beta H)\right]^N. \end{aligned} \quad (2.100)$$

---

[6]In Section 2.14, we address when it may be valid to neglect the second-order term.

The free energy per spin is

$$f = -\frac{1}{N}k_B T \ln\left[\exp\left(-\beta N Jzm^2/2\right)\left[2\cosh\left(\beta Jzm + \beta H\right)\right]^N\right]$$
$$= \frac{Jzm^2}{2} - k_B T \ln\left[2\cosh\left(\beta Jzm + \beta H\right)\right]. \quad (2.101)$$

It might appear that the free energy is a function of the temperature T, the external field H, and the average magnetisation per spin m. However, the order parameter is not a free variable. In equilibrium, it is constrained to take values which minimise the free energy. In this way, Equation (2.101) implicitly determines the order parameter m.

2.5.2 Magnetisation and susceptibility

To calculate the average magnetisation per spin m we keep in mind that it is a function of the temperature T, and the external field H. Thus we find

$$m = -\left(\frac{\partial f}{\partial H}\right)_T$$
$$= -Jzm\left(\frac{\partial m}{\partial H}\right)_T + k_B T \frac{2\sinh(\beta Jzm + \beta H)}{2\cosh(\beta Jzm + \beta H)}\left(\beta Jz\left(\frac{\partial m}{\partial H}\right)_T + \beta\right)$$
$$= Jz\left(\frac{\partial m}{\partial H}\right)_T[\tanh(\beta Jzm + \beta H) - m] + \tanh(\beta Jzm + \beta H). \quad (2.102)$$

To solve Equation (2.102), we rewrite it as

$$\left(Jz\left(\frac{\partial m}{\partial H}\right)_T + 1\right)[\tanh(\beta Jzm + \beta H) - m] = 0. \quad (2.103)$$

The first bracket is always positive, since the susceptibility per spin, $(\partial m/\partial H)_T$, is positive, see Equation (2.20). Therefore the magnetisation per spin is a solution to

$$m(T, H) = \tanh(\beta Jzm + \beta H), \quad (2.104)$$

with the requirement that it also minimises the free energy.

Equation (2.104) implicitly defines the average magnetisation per spin as a function of temperature and external field. Taking the limit of the external field to zero, the average magnetisation per spin

$$m_0(T) = \tanh\left[\beta Jzm_0(T)\right] = \tanh\left[\frac{T_c}{T}m_0(T)\right], \quad (2.105)$$

where we have introduced the temperature

$$T_c = \frac{Jz}{k_B}. \tag{2.106}$$

A graphical way of solving Equation (2.105) is to plot the graphs of $m_0(T)$ and $\tanh[(T_c/T)m_0(T)]$ versus $m_0(T)$ for various values of temperature T. The solutions lie at the intersections of the two graphs, see Figure 2.16.

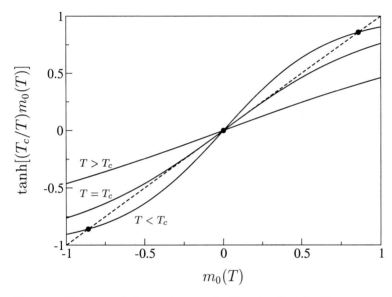

Fig. 2.16 Exact graphical procedure to solve Equation (2.105) with respect to the magnetisation per spin in zero external field for the mean-field Ising model. The graphs of $\tanh[(T_c/T)m_0(T)]$ versus the average magnetisation per spin in zero external field, $m_0(T)$, for temperatures, $T > T_c$, $T = T_c$ and $T < T_c$. The intersections between these graphs and $y = m_0(T)$ (dashed line) are the solutions to Equation (2.105) with respect to $m_0(T)$. For $T \geq T_c$, there is only one solution $m_0(T) = 0$. For $T < T_c$, there are three solutions but only the two non-zero solutions are stable.

Solutions that minimise the free energy are stable. For all temperatures, $m_0(T) = 0$ is a solution. This solution is stable and unique for $T \geq T_c$, although only marginally so at $T = T_c$, and is unstable for $T < T_c$, where two stable non-zero solutions appear for the first time. Mean-field theory therefore predicts a phase transition at $T = T_c$ from a disordered phase with zero average magnetisation above T_c to an ordered phase with non-zero average magnetisation below T_c, see Figure 2.17(a). The critical temperature for the mean-field theory of the Ising model is therefore $T_c = Jz/k_B$.

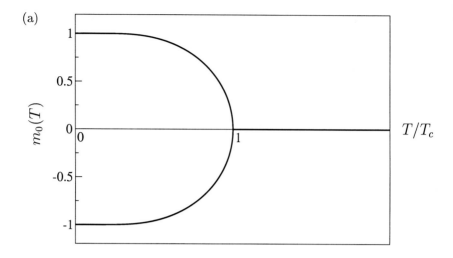

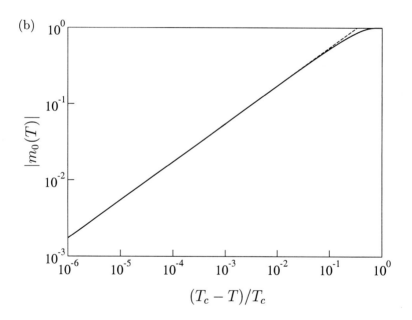

Fig. 2.17 Exact solution of the magnetisation per spin in zero external field for the mean-field Ising model. (a) The average magnetisation per spin in zero external field, $m_0(T)$, versus the relative temperature T/T_c. For $T \geq T_c, m_0(T) = 0$ but then picks up abruptly for $T < T_c$. (b) The absolute average magnetisation per spin in zero external field, $|m_0(T)|$, versus $(T_c - T)/T_c$ for $T < T_c$ (solid line). For $T \to T_c^-$, the order parameter $m_0(T) \propto \pm(T_c - T)^\beta$ with $\beta = 1/2$. The dashed straight line has slope 1/2.

Solving Equation (2.104) with respect to $m(T, H)$ under the constraint that it minimises the free energy in Equation (2.101), we determine the free energy per spin as a function of temperature and external field, see Figure 2.18. The free energy is analytic everywhere, except along the line $(T, 0)$ with $0 \leq T \leq T_c$, terminating at the critical point $(T_c, 0)$ where a cusp exists. Note that along the line $(T, 0)$ with $0 \leq T < T_c$, the left and right first derivatives of the free energy with respect to the external field are non-zero with opposite signs. This line of so-called first-order transitions ends at the critical point $(T_c, 0)$ where the first derivatives are zero.

Figure (2.19)(a) displays the magnetisation per spin as a function of temperature and external field for the mean-field Ising model. The effect of the free energy per spin losing analyticity at the critical point is clearly visible, since, graphically, the magnetisation per spin is minus the slope of the free energy per spin as a function of external field for a given temperature. Figure 2.19(b) is a cut along the plane $H = 0$. For $T < T_c$, a discontinuous first-order phase transition occurs when switching the direction of the external field through $H = 0$. For $T = T_c$, the continuous second-order phase transition occurs, where the first derivative of the magnetisation per spin with respect to the external field diverges.

To investigate the continuous but abrupt pick-up of the order parameter from zero average magnetisation, we expand the right-hand side of Equation (2.105) around $m_0(T) = 0$,

$$m_0(T) = \tanh\left[\frac{T_c}{T}m_0(T)\right] = \frac{T_c}{T}m_0(T) - \frac{1}{3}\left(\frac{T_c}{T}m_0(T)\right)^3 + \cdots. \qquad (2.107)$$

Keeping the first two non-zero terms and rearranging,

$$m_0(T)\left[1 - \frac{T_c}{T} + \frac{1}{3}\left(\frac{T_c}{T}\right)^3 m_0^2(T)\right] = 0. \qquad (2.108)$$

For $T < T_c$, the two non-trivial solutions are

$$m_0(T) = \pm\sqrt{3\left(\frac{T}{T_c}\right)^3}\sqrt{\frac{T_c - T}{T}} \quad \text{for } T \to T_c^-. \qquad (2.109)$$

In summary, the order parameter

$$m_0(T) = \begin{cases} 0 & \text{for } T \geq T_c \\ \pm\sqrt{3/T_c}(T_c - T)^\beta & \text{for } T \to T_c^-, \end{cases} \qquad (2.110)$$

where $\beta = 1/2$ for the mean-field theory of the Ising model.

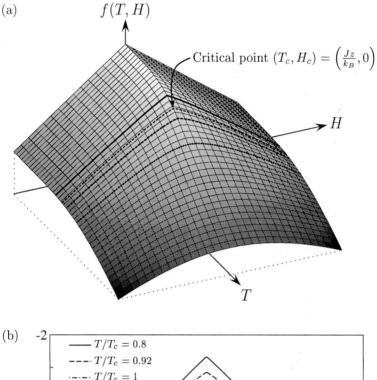

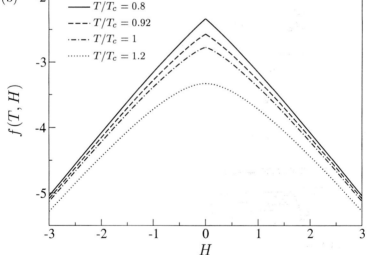

Fig. 2.18 Exact solution of the free energy per spin for the mean-field Ising model. (a) The free energy per spin, $f(T,H)$, versus the temperature, T, and the external field, H. The free energy is analytic everywhere except along the line $(T,0)$ for $0 \leq T \leq T_c$, terminating at the critical point $(T_c, H_c) = (Jz/k_B, 0)$. (b) The free energy per spin, $f(T,H)$, versus the external field, H, for temperatures $T/T_c = 1.2, 1, 0.92, 0.8$.

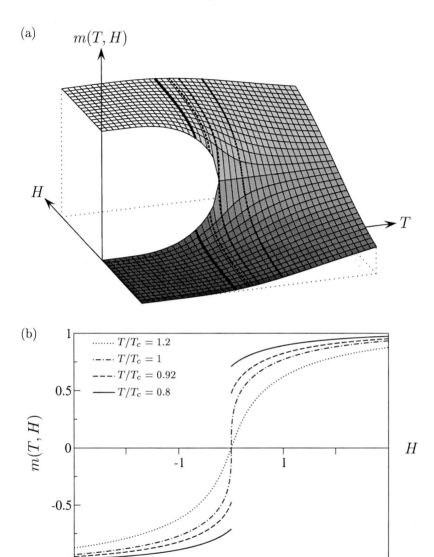

Fig. 2.19 Exact solution of the magnetisation per spin for the mean-field Ising model. (a) The magnetisation per spin, $m(T, H)$, versus the temperature, T, and the external field, H. The magnetisation per spin is a continuous function of the external field for $T \geq T_c$ but with an infinite slope at the critical point $(T_c, H_c) = (Jz/k_B, 0)$. The magnetisation per spin is a discontinuous function at $H = 0$ for $T < T_c$, revealing the line of first-order transitions associated with the free energy being non-analytic. (b) The magnetisation per spin, $m(T, H)$, versus the external field, H, for temperatures $T/T_c = 1.2, 1, 0.92, 0.8$.

To determine an implicit equation for the susceptibility per spin, we take the derivative of Equation (2.104) with respect to the external field keeping the temperature constant,

$$\left(\frac{\partial m}{\partial H}\right)_T = \text{sech}^2\left(\frac{T_c}{T}m + \beta H\right)\left[\frac{T_c}{T}\left(\frac{\partial m}{\partial H}\right)_T + \beta\right]. \quad (2.111)$$

Rearranging terms, the susceptibility per spin in zero external field becomes

$$\chi(T,0) = \lim_{H\to 0}\left(\frac{\partial m}{\partial H}\right)_T$$
$$= \frac{\beta\,\text{sech}^2[(T_c/T)m_0(T)]}{1 - (T_c/T)\text{sech}^2[(T_c/T)m_0(T)]}$$
$$= \frac{1}{k_B}\frac{1}{T\cosh^2[(T_c/T)m_0(T)] - T_c}. \quad (2.112)$$

The susceptibility per spin in zero external field diverges when $T \to T_c$, see Figure 2.20(a).

To investigate how the susceptibility diverges as T approaches T_c in zero external field, we consider the two limits $T \to T_c^\pm$ separately, see Equation (2.110). When $T \to T_c^-$, the order parameter $m_0(T)$ approaches zero. Using the Taylor expansion $\cosh^2 x \approx 1 + x^2$, see Appendix A, we find

$$\chi(T,0) = \frac{1}{k_B}\frac{1}{T\cosh^2[(T_c/T)m_0(T)] - T_c}$$
$$= \frac{1}{k_B}\frac{1}{T[1 + (T_c/T)^2 m_0^2(T)] - T_c}$$
$$= \begin{cases} \frac{1}{k_B}(T - T_c)^{-\gamma^+} & \text{for } T > T_c \\ \frac{1}{2k_B}(T_c - T)^{-\gamma^-} & \text{for } T \to T_c^-. \end{cases} \quad (2.113)$$

Hence, the susceptibility per spin in zero external field diverges as $T \to T_c^\pm$ as a power law with exponents $\gamma^\pm$ in terms of the distance of T from T_c:

$$\chi(T,0) \propto |T - T_c|^{-\gamma^\pm} \quad \text{for } T \to T_c^\pm, \quad (2.114)$$

see Figure 2.20(b). For the mean-field Ising model, $\gamma^\pm = 1$. We have explicitly demonstrated the general result that the critical exponents take the same value below and above T_c. In addition, the ratio of the amplitudes, $\Gamma^\pm$, which appear as prefactors in the power-law divergence, is universal,

$$\frac{\Gamma^+}{\Gamma^-} = 2. \quad (2.115)$$

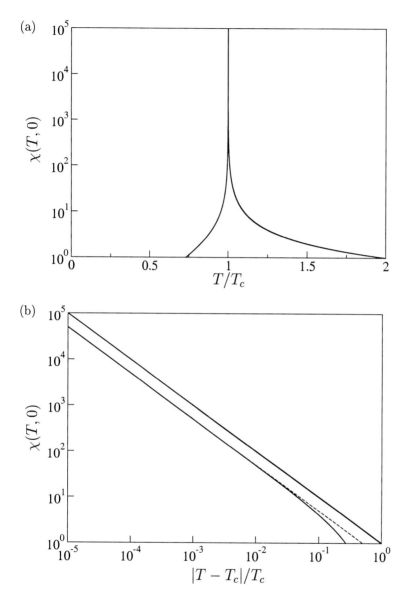

Fig. 2.20 Exact solution of the susceptibility per spin for the mean-field Ising model. (a) The susceptibility per spin in zero external field, $\chi(T,0)$, versus the relative temperature T/T_c. The susceptibility per spin diverges as T approaches the critical temperature T_c. (b) The susceptibility per spin in zero external field, $\chi(T,0)$, versus the reduced temperature, $|T-T_c|/T_c$, for $T<T_c$ (lower solid line) and for $T>T_c$ (upper solid line). For $T \to T_c^\pm$, the susceptibility per spin diverges as a power law with exponent -1 in terms of $|T-T_c|$, the distance of T from T_c. The dashed straight lines have slope -1.

Figure 2.21(a) displays the magnetisation per spin $m(T, H)$ versus the relative temperature T/T_c for various external fields. In the presence of an external field, the magnetisation becomes non-zero at $T = T_c$.

In Figure 2.21(b), we plot the absolute average magnetisation per spin $|m(T_c, H)|$ at $T = T_c$ as a function of an applied positive external field. Just as the critical exponent β describes the pick-up of the magnetisation in the vicinity of T_c^- in zero external field, the critical exponent δ describes the magnetisation for small external fields at T_c.

We expand the right-hand side of Equation (2.104) at $T = T_c$ where $\beta_c Jz = 1$ in a small external field, keeping terms to third order

$$m(T_c, H) = \tanh(m + \beta_c H)$$
$$= m + \beta_c H - \frac{1}{3}(m + \beta_c H)^3 + \cdots$$
$$= m + \beta_c H - \frac{1}{3}m^3 + \mathcal{O}(m^2 H, mH^2, H^3). \quad (2.116)$$

After rearranging, we find that the mean-field Ising model predicts

$$m(T_c, H) = \text{sign}(H)\,(3\beta_c |H|)^{1/3} \propto \text{sign}(H)\,|H|^{1/\delta}, \quad (2.117)$$

with $\delta = 3$. Note that $m^2 H \propto m^5, mH^2 \propto m^7, H^3 \propto m^9$ so that Equation (2.116) is indeed an expansion to the third order in m. The dashed straight line in Figure 2.21(b) has slope $1/\delta = 1/3$.

2.5.3 Energy and specific heat

The average energy per spin in zero external field

$$\varepsilon(T, 0) = -\frac{1}{N}\frac{\partial \ln Z(T, 0)}{\partial \beta}$$
$$= \frac{Jz}{2}m_0^2(T) - \frac{2\sinh[\beta Jzm_0(T)]}{2\cosh[\beta Jzm_0(T)]}Jzm_0(T)$$
$$= -\frac{Jz}{2}m_0^2(T)$$
$$= \begin{cases} 0 & \text{for } T \geq T_c \\ -\frac{3}{2}k_B(T_c - T) & \text{for } T \to T_c^-, \end{cases} \quad (2.118)$$

which is consistent with taking the ensemble average of Equation (2.99) directly. Figure 2.22(a) displays the average energy per spin in zero external field. A feature of the mean-field description is that the energy per spin in zero external field is zero for all $T \geq T_c$.

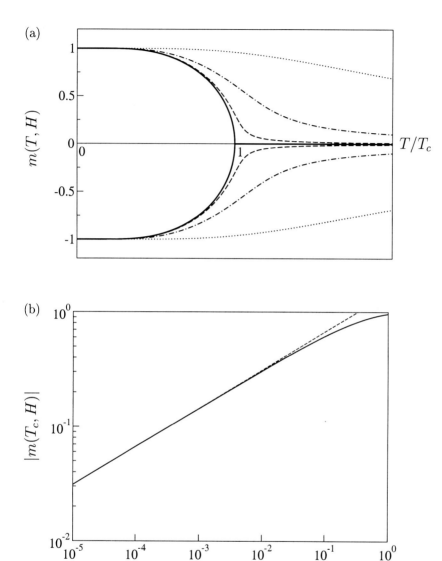

Fig. 2.21 Exact solution of the magnetisation per spin for the mean-field Ising model. (a) The average magnetisation per spin, $m(T, H)$, versus the relative temperature, T/T_c, for various external fields. (b) The absolute average magnetisation per spin at $T = T_c$, $|m(T_c, H)|$, versus the external field, H. For $H \to 0^\pm$, the average magnetisation per spin increases (decreases) as a power law with exponent $1/\delta = 1/3$ in terms of $|H|$. The dashed straight line has slope $1/3$.

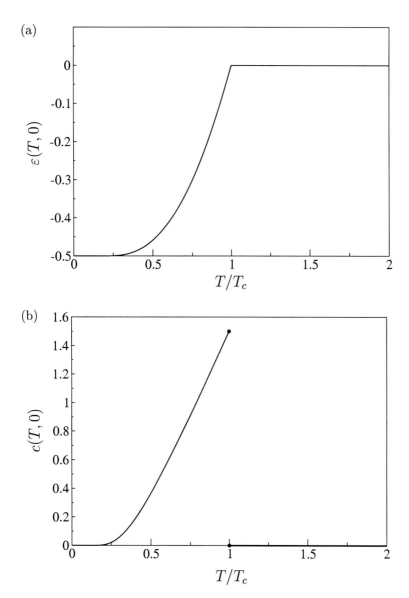

Fig. 2.22 Exact solution of the average energy per spin and the specific heat for the mean-field Ising model. (a) The average energy per spin in zero external field, $\varepsilon(T,0)$, versus the relative temperature, T/T_c. Note the average energy per spin is zero for $T \geq T_c$ and non-differentiable for $T = T_c$. (b) The specific heat in zero external field, $c(T,0)$, versus the relative temperature, T/T_c.

The specific heat in zero external field

$$c(T,0) = \frac{\partial \varepsilon(T,0)}{\partial T}$$

$$= k_B \left(\frac{T_c}{T}\right)^2 \frac{m_0^2(T)}{\cosh^2[(T_c/T)m_0(T)] - T_c/T}$$

$$= \begin{cases} 0 & \text{for } T \geq T_c \\ \frac{3}{2}k_B & \text{for } T \to T_c^-. \end{cases} \quad (2.119)$$

Figure 2.22(b) displays the specific heat in zero external field. The specific heat does not diverge in the mean-field theory of the Ising model but jumps discontinuously from zero for $T \geq T_c$ to non-zero for $T \to T_c^-$.

The critical exponents $\alpha^\pm$ are defined through

$$c(T,0) \propto |T - T_c|^{-\alpha^\pm} \quad \text{for } T \to T_c^\pm. \quad (2.120)$$

Therefore, we conclude that for the mean-field theory of the Ising model $\alpha^- = 0$. The critical exponent α^+ associated with the limit $T \to T_c^+$ is not defined in the simple mean-field treatment presented here. However, more elaborate mean-field approaches [Pathria, 1996] yield

$$\alpha^\pm = 0. \quad (2.121)$$

We summarise the success of mean-field theory for the Ising model. Most importantly, it predicts a second-order phase transition at a finite critical temperature and zero external field, which is correct for $d > 1$. There is also a line of first-order transitions $(T, 0)$ for $0 \leq T < T_c$ terminating at the critical point. The increase of the critical temperature $T_c = Jz/k_B$ with the coupling constant and the coordination number is qualitatively correct. Therefore, just as for percolation, the critical point depends on lattice details and is not a universal quantity. However, the critical exponents $\alpha = 0, \beta = 1/2, \gamma = 1$ and $\delta = 3$ do not depend on the coupling constant and the coordination number and are indeed universal. Finally, mean-field theory predicts universal amplitude ratios, as in Equation (2.115).

2.6 Landau Theory of the Ising Model

The above derivations of the mean-field critical exponents take advantage of Taylor expansions around $T = T_c$ where the order parameter $m_0(T)$ vanishes in zero external field. An implicit equation for the order parameter was derived by taking the partial derivative of the free energy with respect to the external field, see Equation (2.104). In equilibrium, a system will minimise its free energy for a given temperature and external field, and therefore the order parameter can be determined by minimising the free energy with respect to the order parameter. Thus we solve the condition

$$\left(\frac{\partial f}{\partial m}\right)_{T,H} = 0, \tag{2.122}$$

only keeping solutions that represent minima of the free energy.

We exploit this route to re-derive Equation (2.104). Starting from the free energy in Equation (2.101), we find

$$\left(\frac{\partial f}{\partial m}\right)_{T,H} = Jz\left[m - \tanh(\beta Jzm + \beta H)\right]. \tag{2.123}$$

Setting the partial derivative equal to zero, we recover Equation (2.104). To find out which solutions to Equation (2.104) represent minima of the free energy, we could apply the same arguments as in Section 2.5.2.

2.6.1 Free energy

An alternative and more fundamental approach to characterising the minima of the free energy is to perform a Taylor expansion of the free energy itself in powers of the order parameter. Of course, such an expansion is only valid when the order parameter is small, that is, in the vicinity of the critical temperature and zero external field. To see how this works in practice, after introducing $f_0 = -k_B T \ln 2$ as the entropic (high temperature) part of the free energy per spin, Equation (2.101) becomes

$$f = f_0 + \frac{Jzm^2}{2} - k_B T \ln\left[\cosh\left(\beta Jzm + \beta H\right)\right]. \tag{2.124}$$

We now Taylor expand to fourth order the logarithmic term on the right-hand side of Equation (2.124) around $(T_c, 0)$ first using $\cosh x = 1 + x^2/2! + x^4/4! + \cdots$ and then $\ln(1+x) = x - x^2/2 + \cdots$, see Appendix A, implying

$$\begin{aligned}
\ln(\cosh x) &= \ln\left(1 + \frac{1}{2!}x^2 + \frac{1}{4!}x^4 + \cdots\right) \\
&= \frac{1}{2!}x^2 + \frac{1}{4!}x^4 - \frac{1}{2}\left(\frac{1}{2!}x^2 + \frac{1}{4!}x^4\right)^2 + \cdots \\
&= \frac{1}{2}x^2 - \frac{1}{12}x^4 + \mathcal{O}(x^6).
\end{aligned} \tag{2.125}$$

Substituting the expansion with $x = (T_c/T)m + \beta H$ into Equation (2.124) we find

$$f = f_0 + \frac{Jzm^2}{2} - k_B T \left[\frac{1}{2}\left(\frac{T_c}{T}m + \beta H\right)^2 - \frac{1}{12}\left(\frac{T_c}{T}m + \beta H\right)^4 \right]. \quad (2.126)$$

Collecting terms in increasing powers of m up to fourth order, the free energy becomes

$$f = f_0 - \frac{T_c}{T}Hm + \frac{1}{2}k_B\frac{T_c}{T}(T - T_c)m^2 + \frac{1}{12}k_B T\left(\frac{T_c}{T}\right)^4 m^4. \quad (2.127)$$

Since $H \propto m^3$, see Equation (2.117), we have dropped the terms proportional to H^2, Hm^3, H^2m^2, H^3m, and H^4, which are all of higher order than m^4. Therefore, in the vicinity of $(T_c, 0)$ where $T_c/T \to 1$, the free energy to fourth order in the magnetisation reduces to

$$f = f_0 - Hm + \frac{1}{2}k_B(T - T_c)m^2 + \frac{1}{12}k_B T m^4$$
$$= f_0 - Hm + a_2(T - T_c)m^2 + a_4 m^4 \quad \text{for } (T, H) \to (T_c, 0), \quad (2.128)$$

where

$$f_0 = -k_B T \ln 2, \quad (2.129a)$$

$$a_2 = \frac{1}{2}k_B, \quad (2.129b)$$

$$a_4 = \frac{1}{12}k_B T. \quad (2.129c)$$

Note that the sign of the coefficient of m depends on the direction of the external field. The coefficient of m^2, namely $a_2(T - T_c)$, changes sign at $T = T_c$, while the coefficient of m^4 is positive. The average magnetisation per spin $m(T, H)$ is now determined by minimising the free energy in Equation (2.128). However, as a result of the Taylor expansion of the free energy, we lose information about the detailed behaviour of the system for the full temperature range. To see this, we need only compare Equations (2.124) and (2.128) for the free energy per spin in the mean-field and Landau theories of the Ising model, respectively. The difference is displayed graphically in Figure 2.23. Nevertheless, Equation (2.128) preserves all the information required to extract the critical exponents that determine the behaviour of the mean-field Ising model in the vicinity of the critical temperature and small external fields.

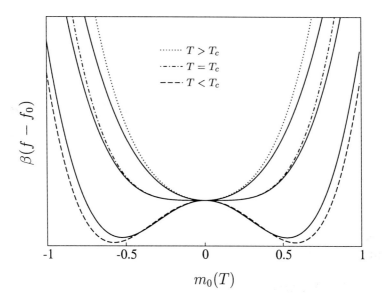

Fig. 2.23 The reduced free energy per spin, $\beta(f - f_0)$, versus the average magnetisation per spin in zero external field, $m_0(T)$, for the mean-field Ising model Equation (2.124) (solid lines) and the Landau theory of the Ising model Equation (2.128) for temperatures $T > T_c$ (dotted line), $T = T_c$ (dotted-dashed line), and $T < T_c$ (dashed line). The free energy per spin in the Landau theory of the Ising model is a Taylor expansion to fourth order in the order parameter of the free energy per spin in the mean-field Ising model.

2.6.2 *Magnetisation and susceptibility*

The starting point for all calculations is the Taylor expanded free energy per spin

$$f = f_0 - Hm + a_2(T - T_c)m^2 + a_4 m^4, \qquad (2.130)$$

which is valid in the vicinity of the critical temperature and small external fields. Figure 2.24(a) displays the free energy per spin $f - f_0$ in zero external field versus m_0.

To find the average magnetisation per spin $m(T, H)$, we look for solutions of Equation (2.122),

$$-H + 2a_2(T - T_c)m + 4a_4 m^3 = 0, \qquad (2.131)$$

that minimise the free energy.

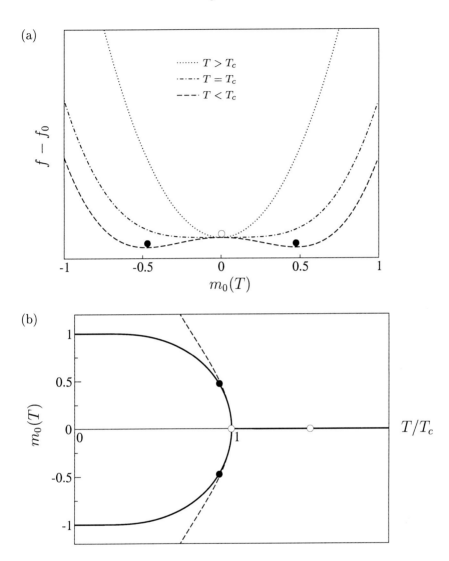

Fig. 2.24 The free energy per spin and the magnetisation per spin in zero external field for the Landau theory of the Ising model. (a) The free energy per spin, $f - f_0$, versus the average magnetisation per spin in zero external field, m_0. The solid circles show the position of the minima of the free energy of the corresponding graph. For $T \geq T_c$, the minimal free energy implies $m_0 = 0$ but at $T = T_c$, the trivial solution $m_0 = 0$ is marginally stable. However, for $T < T_c$, the minimal free energy implies $m_0 \neq 0$. (b) The average magnetisation per spin in zero external field, $m_0(T)$, versus the relative temperature T/T_c for the Landau theory prediction in Equation (2.133) (dashed lines) and the mean-field Ising model (solid lines). The Landau theory of the Ising model is a valid approximation to the mean-field Ising model for small order parameter.

In zero external field, Equation (2.131) reduces to

$$2m_0(T)[a_2(T - T_c) + 2a_4 m_0^2(T)] = 0. \qquad (2.132)$$

For all temperatures $m_0(T) = 0$ is a solution. When $T > T_c$, the square bracket is always positive so that the only solution is $m_0(T) = 0$. When $T = T_c$, the first term in the square bracket vanishes so that the only solution is $m_0(T) = 0$. However, when $T < T_c$, two additional solutions appear, $m_0(T) = \pm\sqrt{a_2(T_c - T)/2a_4}$. These two solutions minimise the free energy, see Figure 2.24(a).

In summary, the magnetisation per spin in zero external field,

$$m_0(T) = \begin{cases} 0 & \text{for } T \geq T_c \\ \pm\sqrt{a_2(T_c - T)/2a_4} & \text{for } T \to T_c^- \end{cases}$$

$$= \begin{cases} 0 & \text{for } T \geq T_c \\ \pm\sqrt{3/T_c}\,(T_c - T)^\beta & \text{for } T \to T_c^-, \end{cases} \qquad (2.133)$$

shown with dashed lines in Figure 2.24(b). Thus we recover the mean-field results of Equation (2.110), predicting a phase transition where the order parameter picks up abruptly as a power law with critical exponent $\beta = 1/2$ in terms of the distance of T below T_c.

To determine the susceptibility per spin, we take the partial derivative of Equation (2.131) with respect to H,

$$-1 + 2a_2(T - T_c)\left(\frac{\partial m}{\partial H}\right)_T + 12a_4 m^2 \left(\frac{\partial m}{\partial H}\right)_T = 0. \qquad (2.134)$$

To investigate how the susceptibility per spin diverges as T approaches T_c in zero external field, we solve Equation (2.134) for $(\partial m/\partial H)_T$ and substitute m with $m_0(T)$. Considering the two limits $T \to T_c^\pm$ separately, see Equation (2.133), we find

$$\chi(T, 0) = 1/(2a_2(T - T_c) + 12a_4 m_0^2)$$

$$= \begin{cases} 1/2a_2(T - T_c) & \text{for } T \to T_c^+ \\ 1/\left[2a_2(T - T_c) + 12a_4(a_2(T_c - T)/2a_4)\right] & \text{for } T \to T_c^- \end{cases}$$

$$= \begin{cases} \frac{1}{k_B}(T - T_c)^{-\gamma^+} & \text{for } T \to T_c^+ \\ \frac{1}{2k_B}(T_c - T)^{-\gamma^-} & \text{for } T \to T_c^-, \end{cases} \qquad (2.135)$$

(using $2a_2 = k_B$, see Equation (2.129b)) with the critical exponents $\gamma^\pm = 1$, in agreement with the mean-field results in Equation (2.113).

To extract the critical exponent δ that describes the magnetisation for small external fields at T_c, we evaluate Equation (2.131) at T_c with the coefficients given in Equation (2.129) and find

$$-H + \frac{1}{3}k_B T_c m^3 = 0 \quad \text{for } T = T_c. \tag{2.136}$$

After rearranging, we recover the mean-field result of Equation (2.117)

$$m(T_c, H) = \text{sign}(H)\left(3\beta_c |H|\right)^{1/3} \propto \text{sign}(H)|H|^{1/\delta}, \tag{2.137}$$

with the critical exponent $\delta = 3$.

2.6.3 Specific heat

The specific heat is related to the second partial derivative of the free energy with respect to temperature. Therefore, we first substitute Equation (2.133) into Equation (2.130) to obtain the free energy per spin as a function of temperature,

$$\begin{aligned}
f &= \begin{cases} f_0 & \text{for } T > T_c \\ f_0 + a_2(T - T_c)(a_2(T_c - T)/2a_4) + a_4 a_2^2 (T - T_c)^2/4a_4^2 & \text{for } T \to T_c^- \end{cases} \\
&= \begin{cases} f_0 & \text{for } T > T_c \\ f_0 - a_2^2(T - T_c)^2/4a_4 & \text{for } T \to T_c^- \end{cases} \\
&= \begin{cases} f_0 & \text{for } T > T_c \\ f_0 - \frac{3}{4T_c}k_B(T - T_c)^2 & \text{for } T \to T_c^-, \end{cases}
\end{aligned} \tag{2.138}$$

such that the specific heat in zero external field is

$$c(T, 0) = -T\left(\frac{\partial^2 f}{\partial T^2}\right)\bigg|_{H=0} = \begin{cases} 0 & \text{for } T > T_c \\ \frac{3}{2}k_B & \text{for } T \to T_c^-, \end{cases} \tag{2.139}$$

consistent with the results in Equation (2.119). Thus we recover the mean-field exponent $\alpha^- = 0$.

2.7 Landau Theory of Continuous Phase Transitions

In the mean-field theory of the Ising model we had an explicit form for the free energy valid for all temperatures and external fields, see Equation (2.101). However, for the purpose of extracting the critical exponents characterising the phase transition, it was sufficient to work with the Taylor expansion of the free energy in terms of the order parameter (the magnetisation per spin) close to the critical point $(T_c, 0)$ where the order parameter is small, see Equation (2.130).

Figure 2.25 is a sketch of the free energy per spin $f - f_0$ versus the average magnetisation for $H < 0$ (top row), $H = 0$ (middle row), and $H > 0$ (bottom row) for temperatures $T < T_c$ (first column), $T = T_c$ (second column), and $T > T_c$ (third column).

In zero external field, there is a continuous second-order phase transition at $T = T_c$; the average magnetisation picks up abruptly from zero magnetisation for $T \geq T_c$ to non-zero magnetisation for $T < T_c$.

For temperatures below the critical temperature, there is a discontinuous first-order phase transition at $H = 0$; the average magnetisation is a discontinuous function at $H = 0$ when changing the sign of the external field.

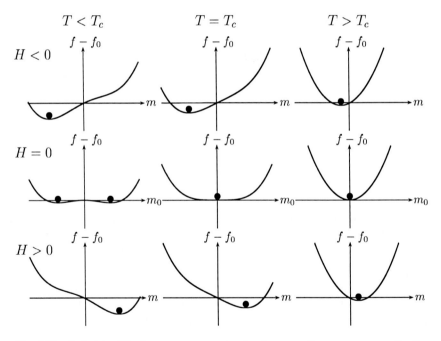

Fig. 2.25 A sketch of the free energy per spin, $f - f_0$, versus the average magnetisation per spin, m. The balls indicate the position of the minimum of the free energy. The three panels in the top row are for negative external field and temperatures $T < T_c, T = T_c, T > T_c$, respectively. The three panels in the middle row are for zero external field and temperatures $T < T_c, T = T_c, T > T_c$, respectively. The three panels in the bottom row are for positive external field and temperatures $T < T_c, T = T_c, T > T_c$, respectively.

The Taylor expansion of the mean-field free energy is a specific example of a general phenomenological approach to continuous phase transitions devised by Landau in 1937 [Landau, 1937]. Landau was awarded the Nobel prize in physics 1962 for 'his pioneering theories for condensed matter, especially liquid helium' explaining the fluid-superfluid phase transition in ^{4}He. He argued that if the free energy is analytic near the critical point, then it can be expanded in terms of the order parameter which is small in the vicinity of the phase transition

$$f(T, H; \phi) = \sum_{k=0}^{\infty} \alpha_k(T, H) \phi^k \quad \text{for } T \to T_c, H \to 0, \quad (2.140)$$

where ϕ denotes a general order parameter and $\alpha_k(T, H)$ are coefficients that depend on the control parameters. For example, in the Ising model these control parameters are the temperature and the external field. We remind the reader that the order parameter is defined implicitly by minimising the free energy and is thus not an independent variable in the same way that the temperature and the external field are.

Symmetry arguments can be used to constrain the coefficients $\alpha_k(T, H)$. For example, for the Ising model in zero external field, the free energy is an even function of the order parameter, since in the absence of any external field the spins are equally likely to be pointing up or down, on average. This up-down symmetry is mathematically expressed as

$$f(T, 0; \phi) = f(T, 0; -\phi), \quad (2.141)$$

from which it immediately follows that no odd powers of ϕ may appear in Equation (2.140), so that $\alpha_k(T, 0) = 0$ for odd k in zero external field.

In the vicinity of the phase transition, the order parameter is small. Thus we expect the higher-order terms in the expansion of the free energy to be negligible. Furthermore, our experience with a mean-field Ising model leads us to expect three extrema for the free energy when $T < T_c$.

Therefore, in zero external field, the simplest possible form of the free energy that can describe a continuous phase transition is a fourth-order polynomial in even powers of the order parameter

$$f(T, 0; \phi) = \alpha_0(T, 0) + \alpha_2(T, 0)\phi_0^2 + \alpha_4(T, 0)\phi_0^4. \quad (2.142)$$

When $T > T_c$, the order parameter is zero and $\alpha_0(T, 0)$ is the only term remaining in the expansion of the free energy. Thus we identify $\alpha_0(T, 0)$ as the entropic part of the free energy since the average energy is zero.

Furthermore, the free energy has a unique extremum at $\phi_0 = 0$ which is a stable minimum. It follows that the coefficients $\alpha_2(T, 0)$ and $\alpha_4(T, 0)$ are positive for $T > T_c$.

When $T < T_c$, the order parameter is non-zero and all three terms are present in the expansion of the free energy. The free energy has three extrema, one unstable maximum at zero order parameter, and two stable minima at non-zero order parameters, symmetric around zero. It follows that the coefficient $\alpha_2(T, 0)$ is negative and $\alpha_4(T, 0)$ is positive for $T < T_c$.

At $T = T_c$, the free energy has a unique extremum at $\phi_0 = 0$ which is a marginally stable minimum. Since $\alpha_2(T, 0)$ is positive for $T > T_c$ and negative for $T < T_c$, it follows that $\alpha_2(T, 0)$ must be zero at $T = T_c$. However, $\alpha_4(T, 0)$ remains positive to ensure that the extremum is a minimum.

Assuming that the coefficients $\alpha_k(T, H)$ are analytic around $(T_c, 0)$, they can themselves be expanded in powers of $(T - T_c)$ and H. For temperatures close to the critical temperature and small external fields, it is sufficient to keep only the leading-order term.

The leading-order terms for $\alpha_0(T, H)$ and $\alpha_4(T, H)$ are the zeroth-order terms $\tilde{\alpha}_0$, for which the sign is irrelevant, and $\tilde{\alpha}_4$, which is positive.

The leading-order term for $\alpha_2(T, H)$ is the first-order term,

$$\alpha_2(T, H) = \tilde{\alpha}_2(T - T_c) \quad \text{for } T \to T_c, H \to 0, \tag{2.143}$$

where $\tilde{\alpha}_2$ is positive.

If we restore a small external field, the coefficients $\alpha_k(T, H)$ with odd k become non-zero. The leading-order term for $\alpha_1(T, H)$ is the first-order term

$$\alpha_1(T, H) = \tilde{\alpha}_1 H \quad \text{for } T \to T_c, H \to 0. \tag{2.144}$$

In summary, in the Landau theory of continuous (second-order) phase transitions with order parameter ϕ, the simplest form of the free energy is

$$f(T, H; \phi) = \tilde{\alpha}_0 + \tilde{\alpha}_1 H\phi + \tilde{\alpha}_2(T - T_c)\phi^2 + \tilde{\alpha}_4 \phi^4, \tag{2.145}$$

where $\tilde{\alpha}_0$ is the entropic part of the free energy, $\tilde{\alpha}_1 = -1$, $\tilde{\alpha}_2 > 0$, and the coefficient of ϕ^2 changes sign at T_c, and $\tilde{\alpha}_4 > 0$. The coefficients in Equation (2.128) are consistent with the general considerations of the phenomenological Landau theory of continuous phase transitions.

Following the steps in Sections 2.6.2 and 2.6.3 one could once again derive the mean-field exponents for the Ising model, $\alpha = 0, \beta = 1/2, \gamma = 1$, and $\delta = 3$.

2.8 Ising Model in $d = 2$

The exact results obtained in the one-dimensional and mean-field Ising models were derived from an explicit calculation of the partition function, and hence the free energy, as a function of temperature and external field. In two dimensions, the partition function in zero external field was first obtained by Onsager in 1944 [Onsager, 1944]. This analytical solution implies that the pick-up of the order parameter at the critical temperature is described by the critical exponent $\beta = 1/8$, in disagreement with the mean-field result $\beta = 1/2$. The derivation of the partition function is beyond the scope of this book and we only state the result. We instead present numerical results for the two-dimensional Ising model that provide evidence for a phase transition, and we further support its existence at a non-zero critical temperature with a simple physical argument.

In practice, the measurement of a thermodynamic quantity yields a temporal average. The definition of the Ising model in Equation (2.5) does not, however, contain any information about its temporal evolution. Nevertheless, we can obtain good estimates for the average value of an observable defined in Equation (2.9) by applying simulation techniques that are designed to sample microstates $\{s_i\}$ according to the Boltzmann distribution given in Equation (2.8). The following numerical results come from simulations of the Ising model in zero external field on a square lattice with periodic boundary conditions using the so-called Metropolis algorithm [Metropolis et al., 1953] far away from the critical point, and the Wolff algorithm [Wolff, 1989] close to the critical point. The reader is invited to consult Appendix H for details of how the Metropolis algorithm works.

2.8.1 *Partition function*

For completeness, we state the partition function in zero external field for the two-dimensional Ising model derived by Onsager [Onsager, 1944]

$$Z(T, 0) = [2\cosh(2K)\exp(\mathcal{I})]^N, \tag{2.146}$$

where $K = J/(k_B T)$ is the reduced (dimensionless) coupling constant and

$$\mathcal{I}(\kappa) = \frac{1}{2\pi}\int_0^\pi \ln\frac{1 + \sqrt{1 - \kappa^2 \sin^2\varphi}}{2}\, d\varphi, \tag{2.147a}$$

$$\kappa = \frac{2\sinh 2K}{\cosh^2 2K}. \tag{2.147b}$$

2.8.2 Magnetisation and susceptibility

For an infinite number of spins in zero external field, the ergodicity of the system is spontaneously broken for $T < T_c$, and the system is confined to either the $+m$ or $-m$ phase, see Section 2.3.5. However, all simulations are necessarily performed with a finite number of spins. A finite system remains ergodic for $T < T_c$. Since a finite system in zero external field is equally likely to be in the $+m$ or $-m$ phase, the measured average magnetisation would always be zero, even for $T < T_c$.

One of the simplest ways of emulating a system confined to a particular phase is to measure the absolute value of the magnetisation. However, the measured average absolute magnetisation per spin in zero external field $\langle |M(T,0)| \rangle /N$ would always be non-zero, even for $T > T_c$. But, in the limit of infinite system size, $\langle |M(T,0)| \rangle /N$ approaches the exact solution for the average magnetisation per spin for $H \to 0^+$ [Yang, 1952],

$$m_0(T) = \begin{cases} 0 & \text{for } T \geq T_c \\ \left(1 - [\sinh(2J/k_B T)]^{-4}\right)^{1/8} & \text{for } T < T_c \end{cases}$$

$$= \begin{cases} 0 & \text{for } T \geq T_c \\ \left[4\sqrt{2}\ln(1+\sqrt{2})\right]^{1/8} \left(\frac{T_c - T}{T_c}\right)^{1/8} & \text{for } T \to T_c^-, \end{cases} \quad (2.148)$$

with a critical temperature, in units of the interaction strength J, for a square lattice [Kramers and Wannier, 1941]

$$\frac{k_B T_c}{J} = \frac{2}{\ln(1+\sqrt{2})} = 2.269185\ldots. \quad (2.149)$$

Figure 2.26(a) shows the numerically measured average absolute magnetisation per spin for a finite system of size $L = 100$ versus the relative temperature in zero external field. For finite system sizes, $\langle |M| \rangle /N > 0$ for $T \geq T_c$. This is a finite-size effect. However, for increasing system sizes, the pick-up in $\langle |M| \rangle /N$ approaches T_c, and for $L \to \infty$, $\langle |M| \rangle /N = 0$ for $T \geq T_c$. For $T < T_c$, the numerical results are consistent with the analytical expression when $L \gg \xi$.

In Figure 2.26(b) we present the numerical data for the average absolute magnetisation per spin versus the reduced temperature for $T < T_c$ in a double-logarithmic plot to focus on the region $T \to T_c^-$. For comparison, alongside we plot the analytical solution which appears as a straight line with slope $\beta = 1/8$ in the limit of $T \to T_c^-$, see Equation (2.148).

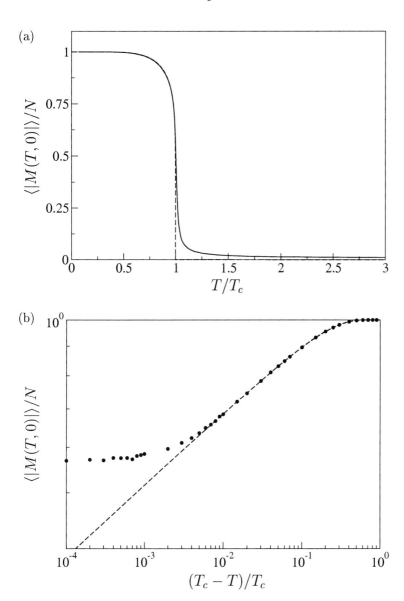

Fig. 2.26 Exact solution in the thermodynamic limit (dashed lines) and numerical results for the two-dimensional Ising model on a square lattice of size $L = 100$. (a) The average absolute magnetisation per spin in zero external field, $\langle |M(T,0)|\rangle/N$, versus the relative temperature T/T_c. For $L \to \infty$, $\langle |M(T,0)|\rangle/N = 0$ for $T \geq T_c$ but then picks up abruptly as $\langle |M(T,0)|\rangle/N \propto (T_c - T)^\beta$ for $T \to T_c^-$. (b) The average absolute magnetisation per spin in zero external field, $\langle |M(T,0)|\rangle/N$, versus $(T_c - T)/T_c$ for $T < T_c$. For $T \to T_c^-$, the dashed line has slope $1/8$.

For temperatures $(T_c - T)/T \gtrsim 5 \times 10^{-3}$, the numerical results are consistent with the analytical expression. However, for $(T_c - T)/T \lesssim 2 \times 10^{-3}$, the measured magnetisation deviates from the analytical expression and saturates at a non-zero value when $L \ll \xi$. This is a finite-size effect.

The susceptibility per spin is the slope of the graph for the average magnetisation per spin as a function of the external field. The fluctuation-dissipation theorem relates the susceptibility to the variance of the total magnetisation, see Equation (2.20). To obtain numerical results for the susceptibility per spin in zero external field, we measure the variance of the total magnetisation in zero external field and use the relation

$$k_B T \chi = \frac{1}{N}\left(\langle M^2 \rangle - \langle M \rangle^2\right). \quad (2.150)$$

Figure 2.27(a) shows the numerically measured susceptibility per spin for a finite system of size $L = 100$ versus the relative temperature. Since the average magnetisation per spin hardly changes with external field in the limits $T \to \infty$ and $T \to 0$, the corresponding susceptibility per spin is zero. However, as the critical temperature is approached, the magnetisation per spin becomes more and more sensitive to changes in the external field. At the critical temperature, there is a sharp peak in the susceptibility per spin. In an infinite system, the susceptibility per spin diverges at $T = T_c$, but in a finite system the divergence is capped.

Figure 2.27(b) shows the numerically measured susceptibility per spin versus the absolute value of the reduced temperature in a double-logarithmic plot with solid circles referring to $T \to T_c^+$ and open circles to $T \to T_c^-$. The divergence in the vicinity of the critical point is consistent with a power law in terms of the distance of T from T_c. No analytical solution is yet available for the susceptibility per spin because the partition function is not known for the two-dimensional Ising model in non-zero external field. However, in the vicinity of the critical point

$$\chi(T,0) = \Gamma^\pm \left|\frac{T - T_c}{T_c}\right|^{-\gamma} \quad \text{for } T \to T_c^\pm, \quad (2.151)$$

where the critical exponent $\gamma = 7/4$ and the amplitudes $\Gamma^+ = 0.4241971\ldots$ and $\Gamma^- = 0.01125380\ldots$ for $T \to T_c^\pm$, respectively [Barouch et al., 1973]. The graphs for these analytical results are plotted alongside the numerical data as dashed lines in Figure 2.27(b). Both the exponent $\gamma = 7/4$ and the amplitude ratio $\Gamma^+/\Gamma^- = 37.69365\ldots$ for the two-dimensional Ising model are in disagreement with the mean-field results $\gamma = 1$ and $\Gamma^+/\Gamma^- = 2$.

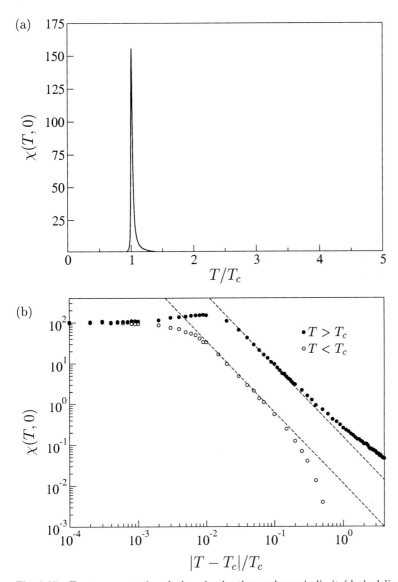

Fig. 2.27 Exact asymptotic solutions in the thermodynamic limit (dashed lines) and numerical results for the two-dimensional Ising model on a square lattice of size $L = 100$. (a) The susceptibility per spin in zero external field, $\chi(T,0)$, versus the relative temperature, T/T_c. For $T \to T_c$, the susceptibility per spin diverges but for finite lattices, the divergence is capped. (b) The susceptibility per spin in zero external field, $\chi(T,0)$, versus the reduced temperature, $|T - T_c|/T_c$, for $T < T_c$ (open circles) and for $T > T_c$ (solid circles). When $L \to \infty$, the susceptibility per spin diverges as a power law $\chi(T,0) \propto |T - T_c|^{-\gamma}$ for $T \to T_c$ with $\gamma = 7/4$. The dashed lines have slope $-7/4$.

2.8.3 Energy and specific heat

Figure 2.28(a) shows the numerically measured average energy per spin for a finite system of size $L = 100$ versus the relative temperature. For $T \to \infty$, the spins are non-interacting and the average energy per spin is zero. For $T \to 0$, the spins are fully aligned and the average energy per spin is $-2J$. On close inspection, the reader with keen eyesight will notice a kink in the graph at $T = T_c$.

The specific heat is the slope of the graph for the average energy per spin as a function of temperature. However, to obtain numerical results for the specific heat in zero external field, we measure the variance of the total energy and use the relation

$$k_B T^2 c = \frac{1}{N}\left(\langle E^2\rangle - \langle E\rangle^2\right). \tag{2.152}$$

Figure 2.28(b) shows the numerically measured specific heat for a finite system of size $L = 100$ versus the relative temperature. Since the average energy per spin hardly changes with temperature in the limits $T \to \infty$ and $T \to 0$, the corresponding specific heat is zero. The kink in the graph for the average energy per spin at $T = T_c$ reveals itself as a sharp peak in the specific heat. In an infinite system, the specific heat diverges at $T = T_c$, but in a finite system the divergence is capped.

For comparison, alongside we plot the analytical solutions for the energy per spin and the specific heat in an infinite system:

$$\varepsilon(T,0) = -J\coth(2K)\left[1 + \frac{2}{\pi}\kappa' F(\kappa)\right], \tag{2.153}$$

$$\frac{c(T,0)}{k_B} = \frac{4}{\pi}K^2\coth^2(2K)\left[F(\kappa) - E(\kappa) - \mathrm{sech}^2(2K)\left(\frac{\pi}{2} + \kappa' F(\kappa)\right)\right] \tag{2.154}$$

where $K = J/(k_B T)$ is the reduced coupling constant, κ given by Equation (2.147b), $\kappa' = 2\tanh^2 2K - 1$ and $F(\kappa)$ and $E(\kappa)$ are the complete elliptic integrals of the first and second kind, respectively:

$$F(\kappa) = \int_0^{\pi/2} \frac{1}{\sqrt{1 - \kappa^2 \sin^2\varphi}}\, d\varphi, \tag{2.155a}$$

$$E(\kappa) = \int_0^{\pi/2} \sqrt{1 - \kappa^2 \sin^2\varphi}\, d\varphi. \tag{2.155b}$$

The numerical results agree well with the analytical solutions.

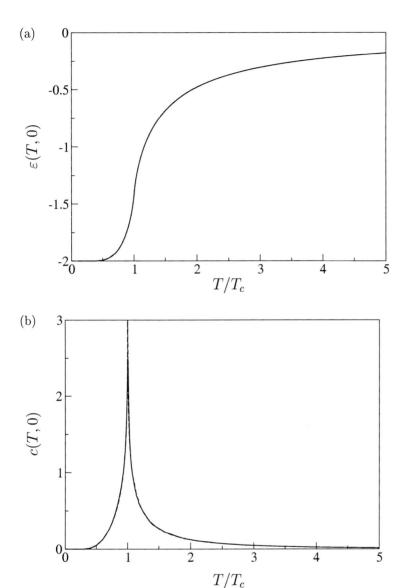

Fig. 2.28 Exact solution in the thermodynamic limit (dashed lines) and numerical results for the energy and specific heat in zero external field versus the relative temperature for the two-dimensional Ising model in zero external field on a square lattice of size $L = 100$. (a) The average energy per spin in zero external field, $\varepsilon(T, 0)$, versus the relative temperature, T/T_c. (b) The specific heat in zero external field, $c(T, 0)$, versus the relative temperature, T/T_c. The specific heat diverges as T approaches the critical temperature T_c. For finite systems, the divergence is capped.

Note that at $T = T_c$ the variable $\kappa = 1$, and the elliptic integral

$$F(1) = \int_0^{\pi/2} \frac{1}{\sqrt{1-\sin^2\varphi}}\, d\varphi = \left[\ln\left(\frac{1}{\cos\varphi} + \tan\varphi\right)\right]_0^{\pi/2} \quad (2.156)$$

diverges logarithmically in the upper limit $\varphi = \pi/2$. Because κ' changes sign at $T = T_c$, there is a kink in the graph for the average energy per spin. Furthermore, the divergence for the specific heat for $T \to T_c$ is logarithmic

$$\frac{c(T,0)}{k_B} = -\frac{8}{\pi} K_c^2 \ln\left|\frac{T-T_c}{T_c}\right| + \text{constant} \quad \text{for } T \to T_c^{\pm}. \quad (2.157)$$

The logarithmic divergence for the specific heat is associated with a critical exponent $\alpha = 0$ since the logarithm function satisfies the limit

$$-\ln\left|\frac{T-T_c}{T_c}\right| = \lim_{\alpha \to 0} \frac{1}{\alpha}\left(\left|\frac{T-T_c}{T_c}\right|^{-\alpha} - 1\right). \quad (2.158)$$

2.8.4 Critical temperature

Based on the analysis of the one-dimensional model, Ising incorrectly concluded in his PhD thesis of 1924 that there was no phase transition in any dimension for $T > 0$ [Ising, 1925]. However, Peierls argued in 1936 that the two-dimensional Ising model would show a spontaneous magnetisation for sufficiently low temperatures [Peierls, 1936]. Indeed, in 1941 Kramers and Wannier calculated exactly the critical temperature of the Ising model on a square lattice, see Equation (2.149) [Kramers and Wannier, 1941]. We can gain some insight into why a phase transition occurs at $T > 0$ for $d > 1$ by using a simple heuristic argument based on Peierls' idea.

Consider the Ising model on a two-dimensional square lattice with periodic boundary conditions. To investigate whether a single domain of aligned spins is stable against thermal fluctuations for $T > 0$, we calculate the difference between the free energy for a single spin domain of aligned spins, $F_{\text{1-dom}}$, and the free energy for two spin domains of oppositely aligned spins, $F_{\text{2-dom}}$, see Figure 2.29.

Assume that the interface between the two domains separates n spins. The difference in energy between the single and two domain configurations is $2Jn$, while the difference in entropy is $k_B \ln \Omega(n)$, where $\Omega(n)$ is the total number of interfaces separating n spins. To estimate an upper bound for $\Omega(n)$, consider an interface as a self-avoiding random walk of length n on the square lattice. There are at most three different choices for each step

(a) 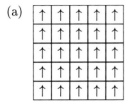 (b)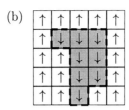

Fig. 2.29 The two-dimensional Ising model on a square lattice with periodic boundary conditions in zero external field. (a) One of the two microstates with a single spin domain. The free energy $F_{\text{1-dom}} = -NJz/2 - k_B T \ln 2$. (b) One of the $2\Omega(n)$ microstates with two spin domains, where $\Omega(n)$ is the total number of different interfaces separating n spins. The interface between two spin domains separating n spins costs $2Jn$ in energy. The free energy $F_{\text{2-dom}} = -NJz/2 + 2Jn - k_B T \ln 2\Omega(n)$.

that do not retrace the interface, and therefore a total of 3^n interfaces. Since many of these are neither self-avoiding nor closed, this number is a gross overestimate. Finally, there are N different positions for the centre of mass of a given interface, implying $\Omega(n) < N3^n$. For a compact object in two dimensions, the perimeter is proportional to the square root of the area. Therefore, for the two domains to be of comparable size, the interface must be of length $n \propto \sqrt{N}$. The discussion above leads to

$$F_{\text{2-dom}} - F_{\text{1-dom}} = 2Jn - k_B T \ln \Omega(n) > 2J\sqrt{N} - k_B T \ln N 3^{\sqrt{N}}. \quad (2.159)$$

The system is stable against the formation of two domains of comparable size if $F_{\text{2-dom}} > F_{\text{1-dom}}$, that is,

$$k_B T < \frac{2J}{\ln 3 + \ln N/\sqrt{N}}. \quad (2.160)$$

In the thermodynamic limit $N \to \infty$, we thus obtain a lower bound for the critical temperature as

$$\frac{k_B T_c}{J} > \frac{2}{\ln 3} \approx 1.820. \quad (2.161)$$

In the one-dimensional Ising model, the energy cost of the interface is independent of the size of the domain and is therefore outweighed by the increase in entropy for all temperatures $T > 0$ in the thermodynamic limit. However, in the two-dimensional Ising model, the energy cost of the interface is proportional to its length so that it will outweigh the increase in entropy for sufficiently low temperatures. It is exactly this feature that enables the two-dimensional Ising model to display spontaneous magnetisation for $T < T_c$. For a rigorous proof, see [Griffiths, 1972].

2.9 Widom Scaling Ansatz

The failure of mean-field theory in low dimensions motivates its replacement with a more general framework. Just as for the cluster number density in percolation, we now search for a general scaling ansatz for the free energy per spin that compactly summarises the behaviour of the Ising model in the vicinity of the critical point. With a scaling ansatz for the free energy per spin, we would be able to derive a scaling ansatz for all thermodynamic quantities and establish scaling relations among the critical exponents.

It is convenient to introduce the dimensionless reduced temperature, t, and the dimensionless reduced external field, h:

$$t = \frac{T - T_c}{T_c}, \tag{2.162a}$$

$$h = \frac{H}{k_B T} = \beta H, \tag{2.162b}$$

such that the limits

$$t \to 0^{\pm} \quad \Leftrightarrow \quad T \to T_c^{\pm}, \tag{2.163a}$$

$$h \to 0^{\pm} \quad \Leftrightarrow \quad H \to 0^{\pm}. \tag{2.163b}$$

In reduced variables, the critical point $(T, H) = (T_c, 0)$ is $(t, h) = (0, 0)$.

The free energy $f(t, h)$ describes a two-dimensional surface which is analytic everywhere except along a line $(t, 0)$ for $-1 \leq t \leq 0$, terminating at the critical point $(0, 0)$ where a cusp exists, see e.g. Figure 2.18. The free energy can be decomposed into regular (analytic) and singular (non-analytic) parts, $f(t, h) = f_r(t, h) + f_s(t, h)$, and it is the latter that is responsible for the cusp. Since we are interested in the critical behaviour of the Ising model we are only concerned with the singular part of the free energy which contains information about critical exponents, scaling functions, associated amplitudes, and so on. Specifically, it is the singular part of the free energy $f_s(t, h)$ for which we construct the scaling ansatz.

We are more familiar with the behaviour of the magnetisation in the vicinity of the critical point; therefore we first discuss how to encapsulate its limiting behaviour in a compact scaling ansatz. Since the magnetisation is a partial derivative of the free energy with respect to the external field, we are then in a position to propose a scaling ansatz for the free energy.

Recall that the magnetisation is an odd function of the external field

$$m(t, h) = -m(t, -h). \tag{2.164a}$$

In addition, the magnetisation per spin in the limit of zero external field

$$\lim_{h \to 0^\pm} m(t,h) \propto \begin{cases} 0 & \text{for } t \geq 0 \\ \pm |t|^\beta & \text{for } t \to 0^-, \end{cases} \quad (2.164b)$$

and for small external fields at $t = 0$

$$m(0,h) \propto \text{sign}(h)|h|^{1/\delta} \quad \text{for } h \to 0^\pm. \quad (2.164c)$$

The symmetry and the limiting behaviours can be compactly summarised in the following Widom scaling ansatz for the magnetisation [Widom, 1965]:

$$m(t,h) = |t|^\beta \mathcal{M}_\pm \left(h/|t|^\Delta\right) \quad \text{for } t \to 0^\pm, h \to 0, \quad (2.165)$$

where Δ is known as the gap exponent and $\mathcal{M}_+$ and $\mathcal{M}_-$ are the scaling functions for the magnetisation per spin in the two regimes $t > 0$ and $t < 0$, respectively. Note that while the magnetisation per spin on the left-hand side of Equation (2.165) is a function of the reduced temperature t and the reduced external field h, the scaling functions on the right-hand side is only a function of the ratio $h/|t|^\Delta$.

To recover the known behaviour of the magnetisation per spin in Equations (2.164), the scaling functions $\mathcal{M}_\pm$ must satisfy certain constraints and the gap exponent Δ must be related to known exponents.

By symmetry, the scaling functions are odd functions

$$\mathcal{M}_\pm(x) = -\mathcal{M}_\pm(-x), \quad (2.166a)$$

and in the limit of zero external field

$$\lim_{x \to 0^\pm} \mathcal{M}_+(x) = 0,$$

$$\lim_{x \to 0^\pm} \mathcal{M}_-(x) = \pm \text{ non-zero constant}. \quad (2.166b)$$

Finally, when $t \to 0$ in small external field, the argument of the scaling functions $x = h/|t|^\Delta \to \pm\infty$. Therefore, we require

$$\mathcal{M}_\pm(x) \propto \text{sign}(x)|x|^{1/\delta} \quad \text{for } x \to \pm\infty,$$

$$\Delta = \beta\delta, \quad (2.166c)$$

to ensure that

$$m(t,h) \propto |t|^\beta \text{sign}(h) \left(|h|/|t|^\Delta\right)^{1/\delta} \quad \text{for } h \to 0^\pm, h/|t|^\Delta \to \pm\infty$$

$$\propto \text{sign}(h)|h|^{1/\delta} \quad \text{for } h \to 0^\pm, h/|t|^\Delta \to \pm\infty. \quad (2.167)$$

Figure 2.30 is a sketch of the scaling functions for the magnetisation per spin $\mathcal{M}_\pm$, summarising their constraints graphically.

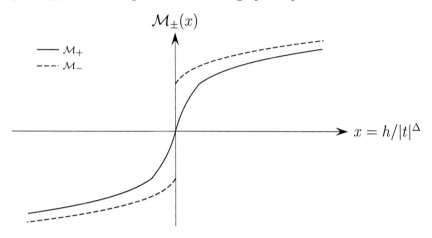

Fig. 2.30 A sketch of the scaling functions, $\mathcal{M}_\pm(x)$, for the magnetisation per spin in the regimes $t > 0$ and $t < 0$, respectively, as a function of the rescaled variable, $x = h/|t|^\Delta$. The scaling functions are odd functions, $\mathcal{M}_\pm(x) = -\mathcal{M}_\pm(-x)$. In the limit of small arguments, $\lim_{x \to 0^\pm} \mathcal{M}_+(x) = 0$ (solid line) and $\lim_{x \to 0^\pm} \mathcal{M}_-(x) = \pm$ non-zero constant (dashed lines). In the limits of large arguments, $x \to \pm\infty$, the scaling functions $\mathcal{M}_\pm(x) \propto \text{sign}(x)|x|^{1/\delta}$. When $(t, h) \to (0, 0)$, data for the magnetisation per spin, $m(t, h)$, will collapse onto the graphs for the scaling functions when plotting the transformed magnetisation per spin, $m(t, h)|t|^{-\beta}$, versus the rescaled variable, $x = h/|t|^\Delta$.

2.9.1 Scaling ansatz for the free energy

We now propose a scaling ansatz for the singular part of the free energy per spin which must be consistent with the magnetisation per spin in Equation (2.165), and the singular behaviour of the susceptibility per spin and the specific heat. Widom argued that

$$f_s(t, h) = |t|^{2-\alpha} \mathcal{F}_\pm \left(h/|t|^\Delta \right) \quad \text{for } t \to 0^\pm, h \to 0, \tag{2.168}$$

where $\mathcal{F}_+$ and $\mathcal{F}_-$ are the scaling functions for the free energy per spin in the two regimes $t > 0$ and $t < 0$, respectively [Widom, 1965]. Note that while the free energy on the left-hand side of Equation (2.168) is a function of the reduced temperature and the reduced external field, the scaling function on the right-hand side is only a function of the ratio $h/|t|^\Delta$. In Section 2.15 we will justify the Widom scaling ansatz by exploiting scale invariance at the critical point within the real-space renormalisation group theory.

2.9.2 Scaling ansatz for the specific heat

Since we have the operator identity

$$\frac{\partial}{\partial T} = \frac{1}{T_c}\frac{\partial}{\partial t}, \qquad (2.169)$$

we find for the specific heat,

$$\begin{aligned} c(t,h) &= -T\frac{1}{T_c^2}\left(\frac{\partial^2 f_s}{\partial t^2}\right)_h \\ &= -\frac{T}{T_c}\frac{(2-\alpha)(1-\alpha)}{T_c}|t|^{-\alpha}\mathcal{F}_\pm\left(h/|t|^\Delta\right) + \mathcal{O}(h,h^2) \\ &= |t|^{-\alpha}\mathcal{C}_\pm\left(h/|t|^\Delta\right) \quad \text{for } t \to 0^\pm, h \to 0, \end{aligned} \qquad (2.170)$$

where the scaling functions for the specific heat is given by

$$\mathcal{C}_\pm(x) = -\frac{(2-\alpha)(1-\alpha)}{T_c}\mathcal{F}_\pm(x). \qquad (2.171)$$

In the limit of zero external field

$$c(t,0) = |t|^{-\alpha}\mathcal{C}_\pm(0) \quad \text{for } t \to 0^\pm. \qquad (2.172)$$

This explains why the exponent $2-\alpha$ appears in the scaling ansatz for the free energy in Equation (2.168).

2.9.3 Scaling ansatz for the magnetisation

To confirm that Equation (2.168) does contain the correct scaling ansatz for the magnetisation, we differentiate with respect to the external field H at fixed temperature. Using the operator identity

$$\frac{\partial}{\partial H} = \frac{1}{k_B T}\frac{\partial}{\partial h}, \qquad (2.173)$$

we find for the magnetisation per spin,

$$\begin{aligned} m(t,h) &= -\frac{1}{k_B T}\left(\frac{\partial f_s}{\partial h}\right)_t \\ &= -\frac{1}{k_B T}|t|^{2-\alpha-\Delta}\mathcal{F}'_\pm\left(h/|t|^\Delta\right) \\ &= |t|^{2-\alpha-\Delta}\mathcal{M}_\pm\left(h/|t|^\Delta\right) \quad \text{for } t \to 0^\pm, h \to 0, \end{aligned} \qquad (2.174)$$

where the scaling functions for the magnetisation per spin is given by

$$\mathcal{M}_\pm(x) = -\frac{1}{k_B T}\mathcal{F}'_\pm(x). \tag{2.175}$$

Taking the limit of zero external field,

$$\lim_{h\to 0^\pm} m(t,h) = |t|^{2-\alpha-\Delta} \lim_{x\to 0^\pm}\mathcal{M}_\pm(x) \propto \begin{cases} 0 & \text{for } t\to 0^+ \\ \pm|t|^{2-\alpha-\Delta} & \text{for } t\to 0^- \end{cases} \tag{2.176}$$

and we identify the scaling relation

$$2 - \alpha - \Delta = \beta. \tag{2.177}$$

2.9.4 Scaling ansatz for the susceptibility

Using the operator identity in Equation (2.173), the susceptibility per spin

$$\begin{aligned}\chi(t,h) &= \frac{1}{k_B T}\left(\frac{\partial m}{\partial h}\right)_t \\ &= -\frac{1}{k_B^2 T^2}|t|^{2-\alpha-2\Delta}\mathcal{F}''_\pm\left(h/|t|^\Delta\right) \\ &= |t|^{2-\alpha-2\Delta}\mathcal{X}_\pm\left(h/|t|^\Delta\right) \quad \text{for } t\to 0^\pm, h\to 0,\end{aligned} \tag{2.178}$$

where the scaling functions for the susceptibility per spin is given by

$$\mathcal{X}_\pm(x) = -\frac{1}{k_B^2 T^2}\mathcal{F}''_\pm(x). \tag{2.179}$$

Taking the limit of zero external field,

$$\chi(t,0) = |t|^{2-\alpha-2\Delta}\mathcal{X}_\pm(0), \tag{2.180}$$

and we conclude that

$$\begin{aligned}\mathcal{X}_\pm(0) &= \text{non-zero constants}, \\ 2 - \alpha - 2\Delta &= -\gamma.\end{aligned} \tag{2.181}$$

2.9.5 Scaling ansatz for the spin-spin correlation function

It is also possible to write down a scaling ansatz for the spin-spin correlation function, which does not immediately follow from the free-energy per spin,

$$g(r,t,h) \propto r^{-(d-2+\eta)}\mathcal{G}_\pm(r/\xi, h/|t|^\Delta) \quad \text{for } t\to 0^\pm, h\to 0, \tag{2.182}$$

where $r = |\mathbf{r}_i - \mathbf{r}_j|$ is the distance between the spins, η is a critical exponent, ξ is the correlation length and $\mathcal{G}_+$ and $\mathcal{G}_-$ are the scaling functions in the two regimes $t > 0$ and $t < 0$, respectively [Kadanoff, 1966]. In zero external field, the spin-spin correlation function

$$g(r,t,0) \propto r^{-(d-2+\eta)} \mathcal{G}_\pm(r/\xi, 0) \quad \text{for } t \to 0^\pm. \tag{2.183}$$

The dimensionality appears explicitly in the scaling ansatz for the spin-spin correlation function. The one-dimensional result for the spin-spin correlation function Equation (2.91) is recovered with $\eta = 1$ and $\mathcal{G}_-(r/\xi, 0) = \exp(-r/\xi)$, where the correlation length ξ is given by Equation (2.92). The mean-field result is recovered by the scaling functions

$$\mathcal{G}_\pm(r/\xi, 0) \propto \begin{cases} \text{constant} & \text{for } r \ll \xi \\ (r/\xi)^{\eta+(d-3)/2} \exp(-r/\xi) & \text{for } r \gg \xi. \end{cases} \tag{2.184}$$

In Section 2.15 we will justify the scaling ansatz for the spin-spin correlation function by exploiting scale invariance at the critical point within the real-space renormalisation group theory.

2.10 Scaling Relations

We note that the critical exponents are not independent. The divergences of the specific heat and the susceptibility in zero external field $h = 0$ as $t \to 0$ are described by α and γ, respectively. The pick-up of the order parameter $m(t,0)$ at $t = 0$ is described by β, while δ describes how the order parameter $m(0,h)$ vanishes when the external field $h \to 0$. Finally, η is related to the power-law decay of the spin-spin correlation function at the critical point, while ν describes the divergence of the correlation length in zero external field as $t \to 0$. The critical exponents α, β, γ, and δ are a feature of the non-analyticity of the free energy at the critical point and are characterised by the geometry of the free energy surface in the vicinity of the cusp; they are therefore related through scaling relations.

Combining Equations (2.166c), (2.177) and (2.181) and eliminating Δ yields two scaling relations, namely

$$\beta\delta = \beta + \gamma, \tag{2.185}$$

$$\alpha + 2\beta + \gamma = 2. \tag{2.186}$$

Using a simple scaling argument, a third scaling relation follows from considering the singular part of free energy per spin in zero external field

$$f_s(t,0) \propto |t|^{2-\alpha} \quad \text{for } t \to 0. \tag{2.187}$$

The free energy per spin is the density of the free energy and therefore scales with inverse volume[7]

$$f_s(t,0) \propto L^{-d}, \qquad (2.188)$$

so that the free energy density within a length scale ξ

$$f_s(t,0) \propto \xi^{-d} \propto |t|^{\nu d} \quad \text{for } t \to 0. \qquad (2.189)$$

Therefore, we conclude that

$$2 - \alpha = \nu d. \qquad (2.190)$$

The scaling relation in Equation (2.190) is a so-called hyperscaling relation since it involves the dimensionality d. Note that by substituting the mean-field values of $\alpha = 0$ and $\nu = 1/2$ into the hyperscaling relation, we find $d = 4$. Therefore, the scaling relation in Equation (2.190) is only valid for $d \leq 4$. We will return to the significance of $d = 4$ shortly.

A final scaling relation can be derived using the scaling ansatz for the spin-spin correlation function combined with the 'sum rule'

$$k_B T \chi(t,h) = \int_V g(\mathbf{r}_i, \mathbf{r}_j) d\mathbf{r}_j. \qquad (2.191)$$

In zero external field, the left-hand side

$$k_B T_c \chi(t,0) \propto |t|^{-\gamma} \quad \text{for } t \to 0, \qquad (2.192)$$

while the right-hand side

$$\begin{aligned}
\int_V g(\mathbf{r}_i, \mathbf{r}_j) d\mathbf{r}_j &\propto \int_0^\infty r^{-(d-2+\eta)} \mathcal{G}_\pm(r/\xi, 0) r^{d-1} \, dr \\
&= \int_0^\infty r^{1-\eta} \mathcal{G}_\pm(r/\xi, 0) \, dr && \text{for } t \to 0^\pm \\
&= \int_0^\infty (u\xi)^{1-\eta} \mathcal{G}_\pm(\tilde{r}, 0) \, \xi \, du && \text{with } r = u\xi \\
&= \xi^{2-\eta} \int_0^\infty u^{1-\eta} \mathcal{G}_\pm(u,0) \, du \\
&= |t|^{-\nu(2-\eta)} \int_0^\infty u^{1-\eta} \mathcal{G}_\pm(u,0) \, du && \text{for } t \to 0. \quad (2.193)
\end{aligned}$$

---

[7]For a thorough discussion, see e.g. [Goldenfeld, 1992].

Ising Model

The integral is just a number that depends on which side $t = 0$ is approached from. After comparing with Equation (2.192) we conclude that

$$\gamma = \nu(2 - \eta). \qquad (2.194)$$

In summary, only two of the six critical exponents are independent since they obey the four scaling relations

$$\beta\delta = \beta + \gamma, \qquad (2.195a)$$
$$\alpha + 2\beta + \gamma = 2, \qquad (2.195b)$$
$$\gamma = \nu(2 - \eta), \qquad (2.195c)$$
$$2 - \alpha = \nu d \quad \text{for } d \leq 4. \qquad (2.195d)$$

Note that the first three relations are valid in all dimensions. The hyperscaling relation in Equation (2.195d) involving the dimensionality is only valid for $d \leq 4$.

2.11 Widom Scaling Form and Critical Exponents in $d = 1$

The free energy per spin for the one-dimensional Ising model has a cusp at the critical point $(T_c, H_c) = (0, 0)$, see Figure 2.10, indicative of a phase transition. This is reflected in the magnetisation per spin, which is the first partial derivative of the free energy per spin with respect to the external field, since it takes the value -1 if approached from $H \to 0^-$, and $+1$ if approached from $H \to 0^+$, see Figure 2.11. Thus, clearly the free energy per spin is not analytic at the critical point. Also, the susceptibility per spin, which is the second partial derivative of the free energy per spin with respect to the external field, diverges at the critical point, see Equation (2.76). Graphically, it is the slope of the graph of the magnetisation per spin as a function of external field for a given temperature.

We would therefore like to characterise the behaviour of these thermodynamic quantities with critical exponents, in the same way as we did for the mean-field and two-dimensional Ising model.

Furthermore, the Widom scaling ansatz for the singular part of the free energy per spin in Equation (2.168) introduced in Section 2.9 should also encompass the behaviour of the exact solution for the one-dimensional Ising model in the neighbourhood of the critical point.

However, we must point out some important differences. First, since the critical temperature is $T_c = 0$, it is impossible to explore the behaviour of the magnetisation per spin for $T \to T_c^-$, as required by Equation (2.39) for the purposes of defining the associated critical exponent β for the order parameter. Meanwhile, Equation (2.76) and Equation (2.94) show that the susceptibility per spin and the correlation length diverge exponentially rather than as a power law in the vicinity of the critical point. Finally, the specific heat does not diverge at all, see Equation (2.81) and Figure 2.13.

In spite of these reservations, let us nevertheless attempt to write the free energy per spin in the form of a general scaling ansatz Equation (2.168). To this end, we first note that the part of the exact free energy per spin in Equation (2.71) responsible for the cusp is the second term containing the logarithm. Let us therefore ignore the regular part and concentrate on the reduced singular part of the free energy βf_s, by studying the behaviour of

$$\beta f_s(T, H) = -\ln\left(\cosh\beta H + \sqrt{\sinh^2 \beta H + \exp(-4\beta J)}\right) \qquad (2.196)$$

near the critical point $(T_c, H_c) = (0, 0)$. The reduced temperature in Equation (2.162a) is not well defined, but we can define a reduced 'temperature' $t = \exp(-k\beta J)$, where $k > 0$ is a positive constant such that when $T \to T_c^+$, $t \to 0^+$. With the reduced external field the above equation becomes

$$\beta f_s(t, h) = -\ln\left(\cosh h + \sqrt{\sinh^2 h + t^{4/k}}\right). \qquad (2.197)$$

To home in on the cusp of the free energy per spin, we choose small values of h and use the Taylor expansions for the hyperbolic cosines and sines, see Appendix A, and find that

$$\beta f_s(t, h) = -\ln\left(1 + \sqrt{h^2 + t^{4/k}}\right) \quad \text{for } h \to 0$$
$$\to -\sqrt{h^2 + t^{4/k}} \quad \text{for } h \to 0, t \to 0^+, \qquad (2.198)$$

where in the last line we have used the Taylor expansion $\ln(1 + x) \to x$ for $x \to 0$, see Appendix A. Finally, we can therefore rewrite the reduced singular part of the free energy per spin in the following scaling form

$$\beta f_s(t, h) = -\sqrt{h^2 + t^{4/k}} \quad \text{for } h \to 0, t \to 0^+$$
$$= -t^{2/k}\sqrt{\left(h/t^{2/k}\right)^2 + 1} \quad \text{for } h \to 0, t \to 0^+$$
$$= t^{2/k}\mathcal{F}_{1d}\left(h/t^{2/k}\right) \quad \text{for } h \to 0, t \to 0^+, \qquad (2.199)$$

where we have recognised that the term underneath the square root is only a function of the ratio $h/t^{2/k}$. Therefore, a comparison with Equation (2.168) yields $2-\alpha = 2/k$ and the gap exponent $\Delta = 2/k$. From the scaling relation in Equation (2.177) it follows that $\beta = 0$ and Equation (2.181) implies $\gamma = 2/k$. Invoking the hyperscaling relation in Equation (2.195d) we find $\nu = 2/k$ and hence Equation (2.195c) implies $\eta = 1$, while Equation (2.195a) can only be consistent if $\delta = \infty$. However, it is instructive to extract the exponents directly from the scaling form in Equation (2.199) and the exact solutions previously derived in Section 2.4.

Taking a partial derivative of the free energy with respect to the external field, using the identity in Equation (2.173), yields the magnetisation

$$m(t,h) = t^0 \mathcal{M}_{1d}(h/t^{2/k}) \quad \text{for } h \to 0, t \to 0^+, \tag{2.200}$$

where the scaling function $\mathcal{M}_{1d} = (\partial \mathcal{F}_{1d}/\partial h)_t$. This suggests that the critical exponent for the magnetisation, in terms of the reduced 'temperature' t, takes the value $\beta = 0$. This result can be checked directly against the exact expression for the magnetisation in Equation (2.73) which, when rewritten in terms of the reduced 'temperature' and the reduced external field, reads

$$\begin{aligned} m(t,h) &= \frac{\sinh h}{\sqrt{\sinh^2 h + t^{4/k}}} \\ &= \frac{h}{\sqrt{h^2 + t^{4/k}}} \quad \text{for } h \to 0 \\ &= \frac{h/t^{2/k}}{\sqrt{\left(h/t^{2/k}\right)^2 + 1}} \quad \text{for } h \to 0 \\ &= t^0 \mathcal{M}_{1d}\left(h/t^{2/k}\right) \quad \text{for } h \to 0, \end{aligned} \tag{2.201}$$

as required, and indeed we confirm that the scaling function $\mathcal{M}_{1d}$ is the partial derivative of the scaling function $\mathcal{F}_{1d}$ in Equation (2.199).

To derive the critical exponent δ, characterising how the magnetisation at $t = 0$ vanishes as the reduced external field tends to zero, we find

$$m(0,h) = \frac{\sinh h}{\sqrt{\sinh^2 h}} = \text{sign}(h) = \text{sign}(h)\,|h|^0 = \text{sign}(h)\,|h|^{1/\infty} \tag{2.202}$$

which is consistent with the critical exponent $\delta = \infty$.

Taking a second partial derivative of the singular part of the free energy per spin with respect to the reduced field yields the divergence of the susceptibility per spin

$$\chi(t,h) = \beta t^{-2/k} \chi_{1d}\left(h/t^{2/k}\right) \quad \text{for } h \to 0, t \to 0^+, \tag{2.203}$$

where the scaling function $\chi_{1d} = \left(\partial^2 \mathcal{F}_{1d}/\partial h^2\right)_t$. The critical exponent for the susceptibility per spin, in terms of the reduced 'temperature' t, is therefore $\gamma = 2/k$. Again, this is necessarily consistent with the behaviour of the susceptibility per spin in Equation (2.76):

$$\chi(t,0) = \beta \exp(2\beta J) = \beta t^{-2/k}. \tag{2.204}$$

From the expression for the exponential divergence of correlation length given in Equation (2.94) we may immediately read off the critical exponent:

$$\xi(t,0) \to \frac{1}{2}\exp(2\beta J) = \frac{1}{2} t^{-2/k} \quad \text{for } t \to 0^+; \tag{2.205}$$

therefore, in terms of the reduced 'temperature' t, the exponent $\nu = 2/k$.

Comparing the correlation function in Equation (2.91) with that in Equation (2.47), we conclude that $d - 2 + \eta = 0$, since the correlation function decays purely exponentially in one dimension. Therefore the critical exponent $\eta = 1$.

In summary, in the one-dimensional Ising model we have the critical exponents $\alpha = 2 - 2/k$, $\beta = 0$, $\gamma = 2/k$, $\delta = \infty$, $\eta = 1$ and $\nu = 2/k$. The exponents satisfy the scaling relations in Equation (2.195).

However, we stress once again that these exponents are somewhat artificial, in that they are not associated with a power-law behaviour in $T - T_c$, but with a power-law behaviour in the reduced 'temperature' $t = \exp(-k\beta J)$, that also approaches zero as $T \to T_c$. Furthermore, the three critical exponents, α, γ, and ν, depend on the choice of the positive but otherwise arbitrary constant k.

2.12 Non-Universal Critical Temperatures

The critical temperature of the critical point $(T_c, 0)$ where the Ising model undergoes a phase transition depends on the underlying lattice details.

In Table 2.3 we list the critical temperature for various lattice types and dimensions. The current best estimates for the critical temperature, which are not known exactly, are given in decimal form. The second column lists the coordination number, z (the number of nearest neighbours) for a given lattice. In the mean-field approximation, the critical temperature is proportional to the coordination number. In general, increasing the coordination number within a given dimension increases the critical temperature. Hence, the critical temperature depends on the lattice details; it is non-universal.

Table 2.3 The critical temperature in zero external field, $k_B T_c/J$, in units of J for various lattice types and dimensions in the Ising model. The second column lists the coordination number, z, for a given lattice.

Lattice	z	$k_B T_c/J$
$d = 1$ line	2	0
$d = 2$ hexagonal	3	$2/\ln(2+\sqrt{3})$[a]
square	4	$2/\ln(1+\sqrt{2})$[b] ≈ 2.269185
triangular	6	$4/\ln 3$[a]
$d = 3$ diamond	4	2.70[c]
simple cubic	6	4.51152[d]
body-centred cubic	8	6.40[e]
face-centred cubic	12	9.79[e]
Mean-field	z	z

[a][Baxter, 1982].
[b][Kramers and Wannier, 1941].
[c][Gaunt and Sykes, 1973].
[d][Arisue et al., 2004].
[e][Sykes et al., 1972].

2.13 Universal Critical Exponents

The critical exponents are universal. They only depend on the dimensionality of the Ising model and are entirely insensitive to the underlying lattice details. Table 2.4 lists the critical exponents for various dimensions. For a review of numerical simulations in $d = 3$, see [Binder and Luijten, 2001].

Table 2.4 The values of the critical exponents for the Ising model in dimensions $d = 1, 2, 3$, and $d \geq 4$, and in the mean-field theory of the Ising model. The critical exponents in $d = 3$ are not known exactly but the current best numerical results are listed with the uncertainty on the last digit(s) given by the figure(s) in the brackets. Two of the critical exponents have been measured numerically and the remaining critical exponents are evaluated from the scaling relations in Equation (2.195).

Exponent: Quantity	$d = 1$[a]	$d = 2$	$d = 3$	$d \geq 4$	Mean-field
α : $c(t,0) \propto \|t\|^{-\alpha}$	$2 - 2/k$	$0\,(\log)$	$0.111(2)$	0	$0\,(\text{dis})$
β : $m(t,0) \propto (-t)^\beta$	0	$1/8$	$0.3262(13)$[b]	$1/2$	$1/2$
γ : $\chi(t,0) \propto \|t\|^{-\gamma}$	$2/k$	$7/4$	$1.237(3)$	1	1
δ : $m(0,h) \propto \text{sign}(h)\|h\|^{1/\delta}$	∞	15	$4.792(18)$	3	3
ν : $\xi(t,0) \propto \|t\|^{-\nu}$	$2/k$	1	$0.6297(8)$[b]	$1/2$	$1/2$
η : $g(r,t,0) \propto r^{-(d-2+\eta)}\mathcal{G}_\pm(r/\xi,0)$	1	$1/4$	$0.036(5)$	0	0

[a]Using the reduced 'temperature' $t = \exp(-kJ/k_B T)$, where $k > 0$ is a constant.
[b][Binder and Luijten, 2001].

Exact values are known in one and two dimensions, in four dimensions and above, and in mean-field theory. The critical exponents for $d \geq 4$ remain unchanged and thus the upper critical dimension for the Ising model $d_u = 4$. Note that the critical exponents for $d > 4$ and the mean-field theory are identical.

2.14 Ginzburg Criterion

The agreement between mean-field exponents and $d > 4$ exponents is a result of the unimportance of fluctuations in higher dimensions. We can explore this idea quantitatively in the context of mean-field theory to show that for $d > 4$, fluctuations can be neglected.

The energy of the Ising model can be rewritten as

$$E_{\{s_i\}} = \underbrace{-(Jzm + H)\sum_{i=1}^{N} s_i + \frac{NJzm^2}{2}}_{\text{mean-field energy}} + \Delta E_{\{s_i\}}, \qquad (2.206)$$

where the spin-spin interaction ignored in Equation (2.99) is contained in

$$\Delta E_{\{s_i\}} = -\frac{Jz}{2}\sum_{i=1}^{N}(s_i - \langle s_i \rangle)(s_j - \langle s_j \rangle). \qquad (2.207)$$

The left-hand side is the sum over the spin-spin correlation function which is related to the susceptibility per spin via the 'sum rule', see Equation (2.46),

$$|\langle \Delta E_{\{s_i\}} \rangle| = \frac{Jz}{2}\sum_{i=1}^{N} g(\mathbf{r}_i, \mathbf{r}_j) = \frac{Jz}{2}k_B T \chi. \qquad (2.208)$$

Mean-field theory suppresses fluctuations by setting $\Delta E_{\{s_i\}} = 0$. However, it is only justified to neglect this term if, on average, it is much smaller in magnitude than the mean-field energy. Hence, for the mean-field theory to be valid, this requires, in zero external field, that

$$\frac{Jz}{2}k_B T \chi \ll \left|\left\langle -Jzm\sum_{i=1}^{N} s_i + \frac{NJzm^2}{2}\right\rangle\right|$$

$$= \left|-NJzm\langle s_i\rangle + \frac{NJzm^2}{2}\right|$$

$$= \frac{Jz}{2}Nm^2. \qquad (2.209)$$

Therefore, fluctuations can be safely neglected if

$$Nk_BT\chi \ll (Nm)^2 = \langle M \rangle^2, \tag{2.210}$$

which is known as the Ginzburg criterion. By the fluctuation-dissipation theorem, the quantity $Nk_BT\chi$ is the variance of the total magnetisation $\langle M^2 \rangle - \langle M \rangle^2$. Therefore the Ginzburg criterion simply requires that the relative fluctuations of the magnetisation are negligible, that is,

$$\frac{\sqrt{\langle M^2 \rangle - \langle M \rangle^2}}{\langle M \rangle} \ll 1. \tag{2.211}$$

The correlation length sets the upper linear scale of the fluctuations and diverges as $\xi(t,0) \propto |t|^{-\nu}$ for $t \to 0$. Consider the fluctuations within the volume ξ^d as $t \to 0^-$, restricting the validity of the following argument to $t < 0$. The susceptibility per unit volume diverges as t approaches 0 so that

$$Nk_BT\chi \propto k_BT_c(-t)^{-\gamma}\xi^d \quad \text{for } t \to 0^-. \tag{2.212}$$

The square of the average total magnetisation within volume ξ^d

$$\langle M \rangle^2 \propto (-t)^{2\beta}\xi^{2d} \quad \text{for } t \to 0^-. \tag{2.213}$$

Therefore, the Ginzburg criterion in Equation (2.210) reads

$$k_BT_c(-t)^{-\gamma}\xi^d \ll (-t)^{2\beta}\xi^{2d} \quad \text{for } t \to 0^- \tag{2.214}$$

or equivalently

$$k_BT_c(-t)^{-\gamma-2\beta}\xi^{-d} \propto (-t)^{-\gamma-2\beta+\nu d} \ll 1 \quad \text{for } t \to 0^-. \tag{2.215}$$

Hence, in the limit $t \to 0^-$, the exponent of $(-t)$ must be positive so that

$$d > \frac{\gamma + 2\beta}{\nu}. \tag{2.216}$$

Substituting the mean-field exponents $\beta = 1/2, \gamma = 1$, and $\nu = 1/2$ implies that $d > 4$. Thus we can interpret $d_u = 4$ as the upper critical dimension for the Ising model. Above the upper critical dimension, the mean-field theory for the Ising model gives correct predictions for the critical exponents. Of course, in the real world experiments are performed in three or less dimensions. As a result, the Ginzburg criterion is guaranteed to fail sufficiently close to the critical point in all practical situations.[8]

---

[8] Nevertheless, in type I superconductors, e.g., mean-field theory is a good approximation even close to the critical point. Ginzburg was awarded the Nobel prize in physics 2003 'for pioneering contributions to the theory of superconductors and superfluids.'

2.15 Real-Space Renormalisation

Just as in percolation, we are interested in the behaviour of the Ising model in the vicinity of the critical point $(T_c, 0)$ where the system undergoes a phase transition. For simplicity, we consider the Ising model with $H = 0$. Apart from the lattice spacing, there is only one relevant length scale, namely the correlation length. As the critical point is approached the correlation length diverges. The Ising model becomes scale invariant at $(T_c, 0)$, where the correlation length is infinite. The Ising model is also scale invariant at $T = 0$ and $T = \infty$, where the correlation length is zero. The scale invariance of the Ising model at these points is related to the self-similarity of the associated microstates.

At $T = 0$, there are no fluctuations away from the two microstates with fully aligned spins and $\xi = 0$. The fluctuations increase with temperature and diverge as the temperature approaches T_c from below. At $T = \infty$, there are no fluctuations away from the microstates with randomly orientated spins and $\xi = 0$. The fluctuations increase with decreasing temperature and diverge as the temperature approaches T_c from above. At $T = T_c$, the system is delicately poised in a non-trivial self-similar state between two trivially self-similar states corresponding to $T = 0$ and $T = \infty$.

Our task is to relate the self-similarity at $T = 0, T_c, \infty$ in zero external field to fixed points of a renormalisation transformation. In doing so, we justify the Widom scaling ansatz, the scaling ansatz of the correlation function and the universality of critical exponents.

2.15.1 Kadanoff's block spin transformation

Kadanoff argued that since spins are correlated over scales up to the correlation length, it may be plausible to regard spins within regions up to ξ in size as behaving like a single block spin [Kadanoff, 1966]. In this spirit, Kadanoff outlined a real-space renormalisation procedure over scales $b \lesssim \xi$:

(1) Divide the lattice into blocks, I, of linear size b (in terms of the lattice constant) with each block containing b^d spins, see Figure 2.31(a).
(2) Replace each block I of spins with a single block spin, s_I, according to some coarse graining rule which is some function of the spins within block I, see Figure 2.31(b).
(3) Rescale all lengths by the dimensionless scale factor b to restore the original lattice spacing, see Figure 2.31(c).

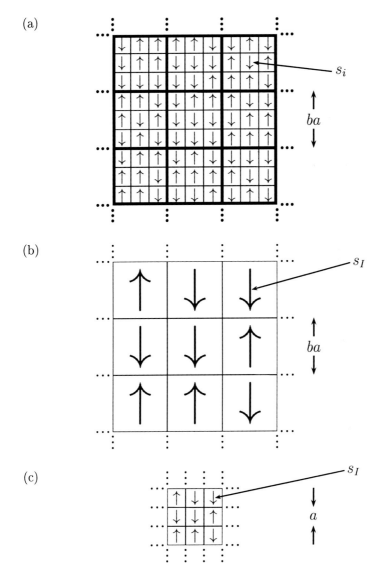

Fig. 2.31 Real-space renormalisation group transformation of the two-dimensional Ising model on a square lattice where spins s_i are located on sites of size a. (a) The lattice is divided into blocks of size ba, each containing b^2 spins. (b) Each block is coarse grained according to some rule and all of its spins are replaced by a single block spin, s_I, located in a block of size ba. (c) All length scales are reduced by the factor b. The block size is thereby reduced to a, the same size as the sites in the original lattice: block spins become spins. We therefore obtain a rescaled version of the original lattice where, because of coarsening, sites have spins s_I.

The effect of the coarse graining is to average out fluctuations on scales less than size b which are not important for the large scale behaviour of the system. There is no unique way to perform the coarse graining. One possibility would be to employ a majority rule, where the block spin takes the value of the majority of spins within the block I, that is, $s_I = \text{sign}\left(\sum_{i \in I} s_i\right) = \pm 1$. Strictly speaking, this rule is only defined for a block I with an odd number of spins. In the event of a tie for a block with an even number of spins, one might simply choose the value of the block spin at random. Another possibility would be to employ a decimation rule, where the block spin takes the value of an arbitrarily assigned spin within the block, for example, $s_I = $ (top right spin s_i).

In Figure 2.32, we put the real-space renormalisation group transformation to work with the majority rule on blocks of size 3×3 in zero external field. The renormalisation transformation is applied from top to bottom on three systems with reduced temperatures $t < 0, t = 0, t > 0$. For $t < 0$ in the left-hand column, the renormalisation transformation induces a flow towards the trivial fixed point associated with $T = 0$. For $t > 0$ in the right-hand column, the renormalisation transformation induces a flow towards the trivial fixed point associated with $T = \infty$. For $t = 0$ in the middle column, there is no flow associated with the renormalisation transformation since the system is at the critical point. Systems at $T = 0, T_c, \infty$ are invariant with respect to the renormalisation transformation.

The real-space renormalisation reduces all lengths, including the correlation length, by a factor b. If the system is not at the critical point, the correlation length is finite and becomes shorter with each application of the renormalisation transformation. The reduction in the correlation length is associated with a flow away from the critical point. In terms of the reduced variables (t, h), which gives the distance from the critical point, the flow can be described as $(t, h) \mapsto (t', h')$. If the system is at the critical point, the correlation length is infinite and is therefore unaffected by the renormalisation transformation. From the flow in the vicinity of the critical point, we note that if $t = 0$, $t' = 0$, while if $t > 0$, $t' > t$, and finally if $t < 0$, $t' < t$. Since the flow is directed away from the critical point, we deduce that $t' \propto t$, to first order. Similarly, by symmetry one can argue that $h' \propto h$, to first order.

Therefore, in the vicinity of the critical point, to first order, $t' = \lambda_t(b) t$ where $\lambda_t(b)$ is a proportionality constant that depends on the block size b. If $b > 1$, then $\lambda_t(b) > 1$ while $\lambda_t(1) = 1$. Renormalising twice with blocks of sizes b_1 and b_2 should be equivalent to renormalising once with a block

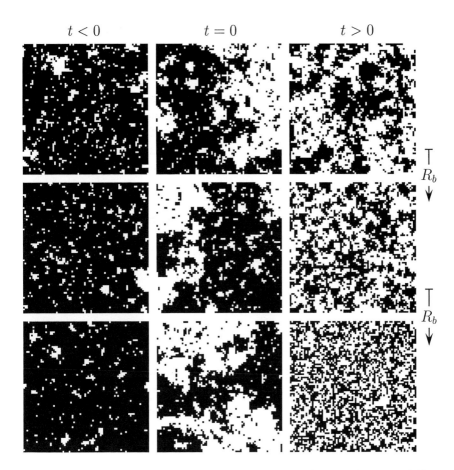

Fig. 2.32 Real-space renormalisation group transformation of the two-dimensional Ising model on a square lattice. The panels are windows of size $\ell = 80$ inside larger lattices. The three panels in the top row correspond to lattices in zero external field with reduced temperatures $t < 0$, $t = 0$, $t > 0$ from left to right. In each of the three columns, the renormalisation transformation, R_b, is carried out twice from top to bottom, revealing large scale behaviour. Coarsening is achieved by employing the majority rule with $b = 3$.

of size $b_1 b_2$. Therefore, $t'' = \lambda_t(b_2) t' = \lambda_t(b_2) \lambda_t(b_1) t = \lambda_t(b_1 b_2) t$, implying that the proportionality constant satisfies the functional equation

$$\lambda_t(b_2) \lambda_t(b_1) = \lambda_t(b_1 b_2), \tag{2.217a}$$

$$\lambda_t(1) = 1, \tag{2.217b}$$

and similarly for the proportionality constant of the reduced external field.

The unique functional solution to this equation is a power law, see Appendix C, so that

$$t' = b^{y_t} t \quad \text{for } t \to 0^{\pm}, \text{ with } y_t > 0, \quad (2.218a)$$
$$h' = b^{y_h} h \quad \text{for } h \to 0^{\pm}, \text{ with } y_h > 0. \quad (2.218b)$$

The exponents y_t and y_h are positive since the flow is directed away from the critical point. The exponent y_t describing the flow of the temperature away from the critical temperature is in fact related to the critical exponent ν. Upon renormalisation, the correlation length is reduced by a factor b,

$$\xi' = \frac{\xi}{b}. \quad (2.219)$$

The correlation length $\xi(t,0) = \text{constant}\, |t|^{-\nu}$ as $t \to 0^{\pm}$, so that

$$\text{constant}\, |t'|^{-\nu} = \frac{\text{constant}\, |t|^{-\nu}}{b} \quad \text{for } t \to 0^{\pm}, h = 0. \quad (2.220)$$

Substituting $t' = b^{y_t} t$ and rearranging, we find $b^{1-\nu y_t} = 1$. Since $b > 1$ is arbitrary, the exponent must be zero and we conclude that

$$y_t = \frac{1}{\nu}. \quad (2.221)$$

2.15.2 Kadanoff's block spin and the free energy

With respect to the partition function, the coarse graining amounts to summing over all those configurations $\{s_i\}$ in the original lattice which are consistent with a particular block spin configuration $\{s_I\}$ in the renormalised lattice. The calculation of the partition function is then completed by summing over all possible block spin configurations,

$$Z(t,h,N) = \sum_{\{s_i\}} \exp\left(-\beta E_{\{s_i\}}\right)$$
$$= \sum_{\{s_I\}} \sum_{\substack{\text{configurations } \{s_i\} \\ \text{consistent with } \{s_I\}}} \exp\left(-\beta E_{\{s_i\}}\right). \quad (2.222)$$

The sum associated with the coarse graining includes only those configurations $\{s_i\}$ that lead to a particular block spin configuration $\{s_I\}$. For example, in the majority rule on a block of 3×3 spins, a block spin that points upwards is consistent with any combination that contains a majority of 5 or more up spins. We demonstrate explicitly that the two-stage sum includes all of the 2^N microstates. Denote by $n_{\pm}$ the number of microstates within a particular block that is consistent with $S_I = \pm 1$, respectively.

The number of microstates $n_\pm$ depends on the coarse graining rule, but they always add up to the total number of microstates within a block, $n_+ + n_- = 2^{b^d}$. A particular block spin configuration $\{s_I\}$ consisting of j up block spins and therefore $N' - j$ down block spins is associated with $n_+^j n_-^{N'-j}$ microstates in the original lattice. For each $j = 0, \ldots, N'$, there is a total of $\binom{N'}{j}$ block spin combinations, so that a sum over all possible block spin configurations yields

$$\sum_{j=0}^{N'} \binom{N'}{j} n_+^j n_-^{N'-j} = (n_+ + n_-)^{N'} = \left(2^{b^d}\right)^{N/b^d} = 2^N. \qquad (2.223)$$

Defining the energy of the Ising model in the renormalised lattice, $E'_{\{s_I\}}$, through the equation

$$\exp\left(-\beta E'_{\{s_I\}}\right) = \sum_{\substack{\text{configurations } \{s_i\} \\ \text{consistent with } \{s_I\}}} \exp\left(-\beta E_{\{s_i\}}\right), \qquad (2.224)$$

and substituting into Equation (2.222) we have that the partition function remains invariant under the real-space renormalisation transformation

$$\begin{aligned} Z(t, h, N) &= \sum_{\{s_I\}} \exp\left(-\beta E'_{\{s_I\}}\right) \\ &= Z(t', h', N'). \end{aligned} \qquad (2.225)$$

We have assumed that the functional form of the renormalised partition function is the same as the original partition function. Accordingly, it is only a function of the number of block spins N', and the renormalised reduced control parameters of temperature t' and external field h'. We will see shortly that while the partition function does remain invariant under renormalisation, a slightly more general framework is required to validate the assumed invariance of its functional form.

Since the partition function remains invariant under renormalisation, so does the total free energy. However, the free energy per spin renormalises according to

$$\begin{aligned} f(t, h) &= -\frac{1}{N} k_B T \ln Z(t, h, N) \\ &= -b^{-d} \frac{1}{N'} k_B T \ln Z(t', h', N') \\ &= b^{-d} f(t', h'). \end{aligned} \qquad (2.226)$$

Substituting Equations (2.218) into Equation (2.226) gives

$$f(t,h) = b^{-d}f(b^{y_t}t, b^{y_h}h) \quad \text{for } t \to 0^{\pm}, h \to 0. \tag{2.227}$$

The singular part of the free energy per spin transforms as a generalised homogeneous function in the vicinity of the critical point. Although the free energy per spin consists of regular and singular parts, it is the latter that is responsible for the critical behaviour. For the purposes of obtaining the Widom scaling ansatz in Equation (2.168), we concentrate on the singular part of the free energy per spin which, from Equation (2.227), obeys

$$f_s(t,h) = b^{-d}f_s(b^{y_t}t, b^{y_h}h) \quad \text{for } t \to 0^{\pm}, h \to 0. \tag{2.228}$$

Equation (2.228) implies the Widom scaling ansatz. The right-hand side is a function of two variables but can be recast as a function of one variable by setting the block size $b = |t|^{-1/y_t} \propto \xi$ in Equation (2.228),

$$\begin{aligned} f_s(t,h) &= \left[|t|^{-1/y_t}\right]^{-d} f_s\left(\left[|t|^{-1/y_t}\right]^{y_t} t, \left[|t|^{-1/y_t}\right]^{y_h} h\right) \\ &= |t|^{\nu d} f_s\left(t/|t|, h/|t|^{y_h/y_t}\right) \\ &= |t|^{\nu d} f_s\left(\pm 1, h/|t|^{y_h/y_t}\right) \quad \text{for } t \to 0^{\pm}, h \to 0. \end{aligned} \tag{2.229}$$

By comparing with Equation (2.168) we make the identifications

$$2 - \alpha = \nu d, \tag{2.230a}$$

$$\Delta = y_h/y_t, \tag{2.230b}$$

$$\mathcal{F}_{\pm}(h/|t|^{\Delta}) = f_s\left(\pm 1, h/|t|^{y_h/y_t}\right), \tag{2.230c}$$

so that we recover the Widom scaling ansatz

$$f_s(t,h) = |t|^{2-\alpha} \mathcal{F}_{\pm}\left(h/|t|^{\Delta}\right) \quad \text{for } t \to 0^{\pm}, h \to 0. \tag{2.231}$$

The two branches of the scaling function for the free energy per spin for $t \to 0^{\pm}$ appear naturally as a result of the first argument in the free energy per spin on the right-hand side of Equation (2.230c).

In summary, the renormalised partition function takes the same form as the original partition function but with rescaled parameters. This implies that the free energy per spin is a generalised homogeneous function. Together with Equations (2.218), this provides an explanation for the Widom scaling ansatz in Equation (2.168) and the hyperscaling relation.

2.15.3 Kadanoff's block spin and the correlation function

Using the Kadanoff block spin picture, we can relate the spin-spin correlation function in the renormalised lattice to that of the original lattice, thereby deriving the scaling form of the spin-spin correlation function in Equation (2.182).

For simplicity, we restrict ourselves to the majority rule. Defining the magnetisation per spin within a block

$$m_I = \frac{1}{b^d} \sum_{i \in I} s_i, \qquad (2.232)$$

the block spin is given by

$$s_I = \text{sign}(m_I) = \frac{m_I}{|m_I|}. \qquad (2.233)$$

The distance between two block spins in the renormalised lattice is $r' = |\mathbf{r}_I - \mathbf{r}_J|$ measured in lattice spacing ba, while the distance between two spins in the original lattice is $r = |\mathbf{r}_i - \mathbf{r}_j|$ measured in lattice spacing a. The spin-spin correlation function in the renormalised lattice is therefore

$$\begin{aligned} g(r',t',h') &= \langle s_I s_J \rangle - \langle s_I \rangle \langle s_J \rangle \\ &= \frac{1}{|m_I||m_J|} [\langle m_I m_J \rangle - \langle m_I \rangle \langle m_J \rangle] \\ &= \frac{1}{|m_I||m_J|} \frac{1}{b^d b^d} \sum_{i \in I} \sum_{j \in J} [\langle s_i s_j \rangle - \langle s_i \rangle \langle s_j \rangle] \\ &= \frac{1}{|m_I||m_J|} g(r,t,h). \end{aligned} \qquad (2.234)$$

Since spins within a block behave like a single block spin, all the $b^d b^d$ spin-spin correlation functions in the double sum take the same value. Hence, in the last step, the factors of b^d cancel.

The absolute value of the average magnetisation per spin within a block

$$|m_I| \propto b^{-\beta/\nu} \quad \text{for } b < \xi. \qquad (2.235)$$

Since length scales are reduced by a factor b, $r' = r/b$, and we find after substituting Equations (2.218) and (2.235) into Equation (2.234) that

$$g(r,t,h) \propto b^{-2\beta/\nu} g(r/b, b^{y_t} t, b^{y_h} h) \quad \text{for } t \to 0^\pm, h \to 0. \qquad (2.236)$$

The spin-spin correlation function is a generalised homogeneous function in the vicinity of the critical point. Equation (2.236) implies the scaling form of the spin-spin correlation function. The right-hand side is a function of three variables but can be recast as a function of two variables by setting the block size $b = |t|^{-1/y_t} \propto \xi$ in Equation (2.236),

$$\begin{aligned}
g(r,t,h) &\propto |t|^{2\beta/y_t \nu} g\left(r/|t|^{-1/y_t}, t/|t|, h/|t|^{y_h/y_t}\right) \\
&= |t|^{2\beta} g\left(r/|t|^{-1/y_t}, \pm 1, h/|t|^{y_h/y_t}\right) \\
&= |t|^{2\beta} \left(r/|t|^{-1/y_t}\right)^{-2\beta y_t} \tilde{g}\left(r/|t|^{-1/y_t}, \pm 1, h/|t|^{y_h/y_t}\right) \\
&= r^{-2\beta y_t} \tilde{g}\left(r/|t|^{-1/y_t}, \pm 1, h/|t|^{y_h/y_t}\right) \\
&= r^{-2\beta y_t} \tilde{g}\left(r/\xi, \pm 1, h/|t|^{y_h/y_t}\right) \quad \text{for } t \to 0^{\pm}, h \to 0. \quad (2.237)
\end{aligned}$$

We have cancelled out the $|t|^{2\beta}$ factor to reveal explicitly the scaling with distance r. This is achieved by pulling out a suitable power of the first argument $r/|t|^{-1/y_t}$ from within the function g, such that $g = r^{-2\beta y_t} |t|^{-2\beta} \tilde{g}$. By comparing with Equation (2.182) we make the identifications

$$d - 2 + \eta = 2\beta y_t, \quad (2.238a)$$

$$\nu = 1/y_t, \quad (2.238b)$$

$$\Delta = y_h/y_t, \quad (2.238c)$$

$$\mathcal{G}_{\pm}\left(r/\xi, h/|t|^{\Delta}\right) = \tilde{g}\left(r/\xi, \pm 1, h/|t|^{\Delta}\right), \quad (2.238d)$$

so that we recover the scaling ansatz for the spin-spin correlation function

$$g(r,t,h) \propto r^{-(d-2+\eta)} \mathcal{G}_{\pm}\left(r/\xi, h/|t|^{\Delta}\right) \quad \text{for } t \to 0^{\pm}, h \to 0. \quad (2.239)$$

The two branches of the scaling function for the correlation function for $t \to 0^{\pm}$ appear naturally as a result of the second argument in the function on the right-hand side of Equation (2.238d).

In summary, the spin-spin correlation function is a generalised homogeneous function. Together with Equations (2.218), this implies the scaling ansatz for the spin-spin correlation function and the hyperscaling relation.

It is instructive to apply the renormalisation transformation to the $d = 1$ and $d = 2$ Ising model in zero external field. The $d = 1$ case is exact and is consistent with the above presentation. However, the $d = 2$ case will steer us towards a more general framework for the real-space renormalisation group transformation developed by Wilson [Wilson, 1971a; Wilson, 1971b].

2.15.4 Renormalisation in $d = 1$

Consider the $d = 1$ Ising model of N spins with periodic boundary conditions in zero external field [Nelson and Fisher, 1975]. The partition function

$$Z(K_1, N) = \sum_{\{s_i\}} \exp(-\beta E_{\{s_i\}})$$

$$= \sum_{\{s_i\}} \exp\left(K_1 \sum_{i=1}^{N} s_i s_{i+1}\right), \qquad (2.240)$$

where the reduced nearest-neighbour coupling constant,[9] K_1, is given by

$$K_1 = \beta J = \frac{J}{k_B T}. \qquad (2.241)$$

First, we divide the lattice into blocks of size $b = 2$ each containing two spins, see Figure 2.33(a). Second, we replace each block of spins with a single block spin s_I which takes the value of the odd spin, see Figure 2.33(b). This choice is arbitrary and constitutes a decimation coarsening rule. Third, all length scales are reduced by the factor $b = 2$ to restore the original lattice spacing. We are left with a renormalised system with $N' = N/2$ spins where $\{s_I\}$ are the odd spins in the original lattice, see Figure 2.33(c).

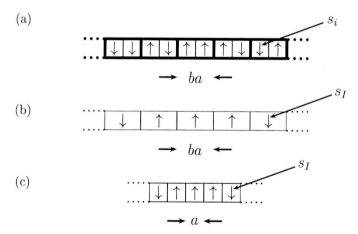

Fig. 2.33 Real-space renormalisation group transformation of the one-dimensional Ising model. (a) The lattice is divided into blocks, each containing $b = 2$ spins; hence, we are tacitly assuming that N is even. (b) Each block is coarse grained and replaced with a single block spin, s_I, which takes the value of the odd spin. (c) All length scales are reduced by the factor b to obtain a renormalised version of the original lattice.

---
[9]The 'reduced coupling constants' are often called simply the 'coupling constants'.

To determine the partition function for the renormalised system, we sum out (integrate over) even spins in the original lattice so that only the odd spins in each block survives. Since each spin has two nearest neighbours, each spin appears twice in the exponent. Collecting each even spin in a single term, we find

$$Z(K_1, N) = \sum_{\substack{\text{odd} \\ \text{spins}}} \sum_{\substack{\text{even} \\ \text{spins}}} \exp\left(K_1 \sum_{i=1}^{N} s_i s_{i+1}\right)$$

$$= \sum_{\substack{\text{odd} \\ \text{spins}}} \sum_{\substack{\text{even} \\ \text{spins}}} \exp\left(K_1[s_1 s_2 + s_2 s_3]\right) \cdots \exp\left(K_1[s_{N-1} s_N + s_N s_1]\right)$$

$$= \sum_{\substack{\text{odd} \\ \text{spins}}} 2\cosh\left(K_1[s_1 + s_3]\right) \cdots 2\cosh\left(K_1[s_{N-1} + s_1]\right), \quad (2.242)$$

where the coarse graining sum over each of the even spins is readily performed. For example, for the spin s_2 that couples to spins s_1 and s_3,

$$\sum_{s_2 = \pm 1} \exp\left(K_1 s_2 [s_1 + s_3]\right) = 2\cosh\left(K_1 [s_1 + s_3]\right). \quad (2.243)$$

The pair of spins (s_1, s_3) can be in one of $2^2 = 4$ microstates. However, the right-hand side of this equation takes only two different values because of symmetry, and may be written with two appropriately defined renormalised (reduced) coupling constants K_0' and K_1' in the form

$$2\cosh\left(K_1 [s_1 + s_3]\right) = \exp\left(K_0' + K_1' s_1 s_3\right). \quad (2.244)$$

The two simultaneous equations that determine the renormalised coupling constants are

$$2\cosh 2K_1 = \exp\left(K_0' + K_1'\right) \quad \text{for } s_1 = s_3, \quad (2.245\text{a})$$
$$2 = \exp\left(K_0' - K_1'\right) \quad \text{for } s_1 = -s_3. \quad (2.245\text{b})$$

Solving for K_0' and K_1' in terms of K_1, we find

$$K_0' = \ln\left(2\sqrt{\cosh 2K_1}\right), \quad (2.246\text{a})$$
$$K_1' = \frac{1}{2}\ln(\cosh 2K_1). \quad (2.246\text{b})$$

Therefore, expressing the partition function in terms of the renormalised coupling constants

$$Z(K_1, N) = \sum_{\substack{\text{odd} \\ \text{spins}}} \exp\left(K_0' + K_1' s_1 s_3\right) \cdots \exp\left(K_0' + K_1' s_{N-1} s_1\right)$$

$$= \exp(N' K_0') \sum_{\{s_I\}} \exp\left(K_1' \sum_{I=1}^{N'} s_I s_{I+1}\right)$$

$$= \exp(N' K_0') \, Z(K_1', N'). \qquad (2.247)$$

In the penultimate line, the odd spins $s_1, s_3, \ldots, s_{N-1}$ in the original lattice are relabelled as s_I in the renormalised lattice, with $I = 1, \ldots, N'$. Factorising out the constant $\exp(N' K_0')$, we recognise that the summation term takes the same form as the partition function for the original system but with a reduced number of spins $N' = N/b$ and a renormalised coupling constant $K_1' < K_1$ given by Equation (2.246b).

The entire expression in the last line of Equation (2.247) is the partition function for the renormalised system. The total free energy remains the same after renormalisation and, in units of $k_B T$, is given by

$$-\ln Z(K_1, N) = -N' K_0' - \ln Z(K_1', N'), \qquad (2.248)$$

where the term $N' K_0'$ appears explicitly as a free energy offset. However, when calculating probabilities and ensemble averages, the constant $\exp(N' K_0')$ cancels out and plays no role.

How does the renormalised system behave compared to the original system? The renormalisation of the coupling constant in Equation (2.246b) implies that

$$K_1' < K_1 \quad \text{for } K_1 > 0. \qquad (2.249)$$

This manifests itself in Figure 2.34(a) in that the graph of the renormalised coupling constant K_1' lies below the dashed line $K_1' = K_1$ for all $K_1 > 0$. Therefore, the coupling between nearest-neighbour spins in the renormalised lattice is weaker than between nearest-neighbour spins in the original lattice. The renormalisation procedure can be readily applied again and again with the effect that the coupling constant becomes weaker and weaker.

Successive applications of the renormalisation procedure induces a flow in the coupling constant towards the fixed point $K_1^* = 0$, see Figure 2.34(b).

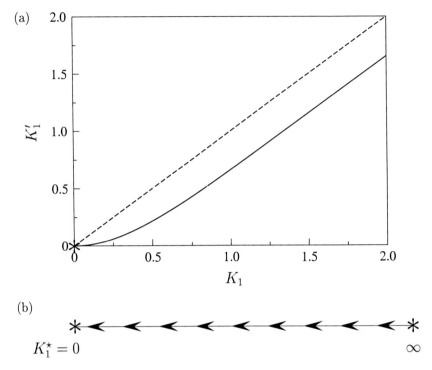

Fig. 2.34 (a) The renormalised coupling constant, K_1', in the one-dimensional Ising model in zero external field versus the coupling constant in the original lattice, K_1. The fixed points (∗) lie at the intersections between the graph for K_1' and the dashed line $K_1' = K_1$. The fixed point $K_1^\star = \infty$ implied by Equation (2.246b) is not visible. (b) The associated renormalisation group transformation flow in K_1-space. In the renormalised lattice, nearest-neighbour spins couple with strength $K_1' < K_1$. The fixed point $K_1^\star = \infty$ is unstable. For $0 < K_1 < \infty$, applying the renormalisation group transformation will induce a flow towards the stable fixed point $K_1^\star = 0$.

This fixed point, commonly known as the high-temperature (weak coupling) fixed point associated with $T = \infty$, corresponds to a system of non-interacting spins. A system of non-interacting spins is self-similar with correlation length $\xi = 0$. Qualitatively, it looks the same on all length scales. Quantitatively, the probability distribution of microstates is invariant under the renormalisation transformation.

Equation (2.246b) admits a second fixed point $K_1^\star = \infty$. This fixed point, commonly known as the low-temperature (strong coupling) fixed point associated with $T = 0$, corresponds to a system of fully aligned spins. A system of fully aligned spins is self-similar with correlation length $\xi = 0$.

Qualitatively, it looks the same on all length scales. Note, however, that the zero correlation length at the non-trivial fixed point is peculiar to one dimension. In higher dimensions, the correlation length is infinite at the non-trivial fixed point.

In summary, the renormalisation transformation has two fixed points in one dimension. The stable trivial fixed point $K_1^\star = 0$ corresponds to a system of non-interacting spins. The unstable non-trivial fixed point $K_1^\star = \infty$ corresponds to a system of fully aligned spins. Any finite initial value of K_1 flows away from the unstable fixed point towards the stable fixed point.

Therefore, the renormalisation transformation correctly predicts that there is no phase transition in zero external field for any finite temperature in the $d = 1$ Ising model. The unstable fixed point is associated with a phase transition at $(T_c, H_c) = (0, 0)$.

2.15.5 Renormalisation in $d = 2$ on a square lattice

Consider the $d = 2$ Ising model of N spins on a square lattice in zero external field [Maris and Kadanoff, 1978]. The partition function

$$Z(K_1, N) = \sum_{\{s_i\}} \exp(-\beta E_{\{s_i\}})$$

$$= \sum_{\{s_i\}} \exp\left(K_1 \sum_{\langle ij \rangle}^N s_i s_j\right), \qquad (2.250)$$

where $K_1 = J/(k_B T)$ is the reduced nearest-neighbour coupling constant and the sum in the exponential runs over all distinct nearest-neighbour pairs.

We apply a renormalisation transformation where the coarse graining is effected by summing out every second spin in the original lattice, which is a realisation of a decimation coarsening rule. In Figure 2.35(a), the decimated spins, that is, the spins to be summed over, have been shaded dark grey. Note that each pair of remaining spins, for example (s_1, s_2), has two common nearest neighbours of decimated spins. The remaining spins form a square lattice rotated by $45°$ with lattice constant $ba = \sqrt{2}a$, see Figure 2.35(b). To complete the renormalisation transformation, all length scales are reduced by the factor $b = \sqrt{2}$ to restore the original lattice spacing. After a $45°$ clockwise rotation, we are left with a renormalised version of the original system with $N' = N/b^2$ spins, see Figure 2.35(c).

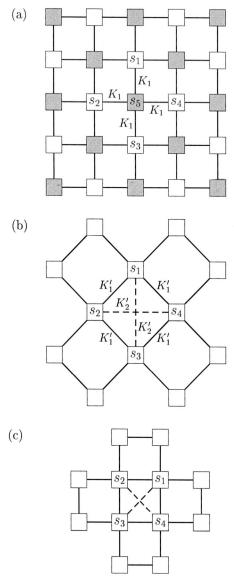

Fig. 2.35 Real-space renormalisation group transformation of the two-dimensional Ising model on a square lattice. (a) Every second spin shown in dark grey is marked for decimation. (b) The coarse graining is effected by summing out every second spin in the original lattice. This process generates renormalised nearest-neighbour couplings, K_1', next-nearest-neighbour couplings, K_2', and quadruple couplings, K_3' (not shown). The number of remaining spins, $N' = N/b^2$. (c) All length scales are reduced by the factor $b = \sqrt{2}$ and after a 45° rotation, a renormalised version of the original lattice is obtained.

To determine the partition function for the renormalised system, we have to perform the coarse graining explicitly by summing over every second spin. Since each spin has four nearest neighbours, each spin appears four times in the exponent of the exponential.

Collecting each decimated spin in a single term, we find

$$Z(K_1, N) = \sum_{\substack{\text{remaining} \\ \text{spins}}} \sum_{\substack{\text{decimated} \\ \text{spins}}} \exp\left(K_1 \sum_{\langle ij \rangle} s_i s_j\right)$$

$$= \sum_{\substack{\text{remaining} \\ \text{spins}}} \sum_{\substack{\text{decimated} \\ \text{spins}}} \cdots \exp\left(K_1 s_5[s_1 + s_2 + s_3 + s_4]\right) \cdots$$

$$= \sum_{\substack{\text{remaining} \\ \text{spins}}} \cdots 2\cosh\left(K_1[s_1 + s_2 + s_3 + s_4]\right) \cdots, \quad (2.251)$$

where the coarse graining over each decimated spins is readily performed. For example, for the spin s_5 that couples to spins s_1, s_2, s_3, s_4, see Figure 2.35(a), we find

$$\sum_{s_5 = \pm 1} \exp\left(K_1 s_5[s_1 + s_2 + s_3 + s_4]\right) = 2\cosh\left(K_1[s_1 + s_2 + s_3 + s_4]\right). \quad (2.252)$$

The quadruple of spins (s_1, s_2, s_3, s_4) can be in one of $2^4 = 16$ microstates. However, the right-hand side of this equation takes only three different values because of symmetry, and may be written with four appropriately defined renormalised coupling constants K'_0, K'_1, K'_2 and K'_3 in the form

$$2\cosh\left(K_1[s_1 + s_2 + s_3 + s_4]\right) = \quad (2.253)$$

$$\exp\left(K'_0 + \frac{K'_1}{2}[s_1 s_2 + s_1 s_4 + s_2 s_3 + s_3 s_4] + K'_2[s_1 s_3 + s_2 s_4] + K'_3 s_1 s_2 s_3 s_4\right).$$

Just as in the one-dimensional case, the coarse graining generates a constant coupling term K'_0 which only plays a role as a free energy offset that does not affect the calculation of probabilities and ensemble averages. Similarly, the nearest-neighbour coupling constant K_1 is renormalised to become K'_1.

This is not all, however. Contrary to one dimension, the coarse graining in two dimensions generates, in addition, renormalised coupling constants K'_2, representing next-nearest-neighbour (nnn) interactions and K'_3, representing quadruple ($\square$) interactions.

Physically, the introduction of these extra coupling constants K'_2 and K'_3 can be understood with reference to Figure 2.35(a). For example, spins s_1 and s_3 interact indirectly through spin s_5. Therefore, when summing out spin s_5, an effective coupling K'_2 between s_1 and s_3 must be introduced. Likewise, the quadruple of spins s_1, s_2, s_3, s_4 interact indirectly through spin s_5. Therefore, when summing out spin s_5, an effective coupling K'_3 between s_1, s_2, s_3, s_4 must also be introduced.

The four simultaneous equations that determine the renormalised coupling constants are

$$2\cosh 4K_1 = \exp\left(K'_0 + 2K'_1 + 2K'_2 + K'_3\right) \quad \{\; s_1 = s_2 = s_3 = s_4,$$

$$2\cosh 2K_1 = \exp\left(K'_0 - K'_3\right) \quad \begin{cases} -s_1 = s_2 = s_3 = s_4, \\ s_1 = -s_2 = s_3 = s_4, \\ s_1 = s_2 = -s_3 = s_4, \\ s_1 = s_2 = s_3 = -s_4, \end{cases}$$

$$2 = \exp\left(K'_0 - 2K'_2 + K'_3\right) \quad \begin{cases} s_1 = -s_2 = -s_3 = s_4, \\ s_1 = s_2 = -s_3 = -s_4, \end{cases}$$

$$2 = \exp\left(K'_0 - 2K'_1 + 2K'_2 + K'_3\right) \quad \{\; s_1 = -s_2 = s_3 = -s_4.$$

Each of the eight conditions specifies two microstates of the quadruple of spins. For example, $-s_1 = s_2 = s_3 = s_4$ specifies either $(-1, +1, +1, +1)$ or $(+1, -1, -1, -1)$.

Solving for K'_0, K'_1, K'_2 and K'_3 in terms of K_1, we find after some algebra

$$K'_0 = \ln\left(2\sqrt{\cosh 2K_1}\,(\cosh 4K_1)^{1/8}\right), \qquad (2.254\text{a})$$

$$K'_1 = \frac{1}{4}\ln(\cosh 4K_1), \qquad (2.254\text{b})$$

$$K'_2 = \frac{1}{8}\ln(\cosh 4K_1), \qquad (2.254\text{c})$$

$$K'_3 = \frac{1}{8}\ln(\cosh 4K_1) - \frac{1}{2}\ln(\cosh 2K_1). \qquad (2.254\text{d})$$

Therefore, the partition function can be expressed in terms of the renormalised coupling constants.

Note that an additional term of $\exp\left(K'_1/2\left[s_1s_2 + s_1s_4 + s_2s_3 + s_3s_4\right]\right)$ arises from the additional common nearest neighbour for each pair of the remaining spin $(s_1, s_2), (s_1 s_4), (s_2, s_3), (s_3, s_4)$, see Figure 2.35(a).

Relabelling the remaining spins from the original lattice as $s_I, s_J, \ldots$ and factoring out the constant $\exp(N'K_0')$, we have

$$Z(K_1, N) = \exp(N'K_0') \sum_{\{s_I\}} \exp\left(K_1' \sum_{\langle IJ \rangle} s_I s_J + K_2' \sum_{\text{nnn}} s_I s_J + K_3' \sum_{\square} s_I s_J s_K s_L\right)$$
$$= \exp(N'K_0')\, Z(K_1', K_2', K_3', N'), \tag{2.255}$$

where the sums in the exponential run over all distinct nearest-neighbour and next-nearest-neighbour pairs and quadruples, respectively. Without the couplings K_2', K_3', the sum over block spin configurations $\{s_I\}$ takes the same functional form as the original partition function but with a reduced number of spins $N' = N/b^2$ and a renormalised coupling constant K_1' given by Equation (2.254b). With the couplings K_2', K_3', however, the energy $E_{\{s_i\}}$ must be generalised to include next-nearest-neighbour and quadruple spin interactions for the sum over block spin configurations $\{s_I\}$ to be identified with a partition function $Z(K_1', K_2', K_3', N')$, see Equation (2.250).

In fact, upon successive applications of the renormalisation transformation, the coupling constants K_0', K_1', K_2' and K_3' are renormalised in turn and additional renormalised coupling constants are generated at each iteration. As a result, the energy must be generalised still further to include all possible spin interactions of which there are, in principle, infinitely many.

The possible spin interactions must respect the symmetry of the problem. In the Ising model in zero external field, for example, the energy must be invariant under the reversal of spins $s_i \mapsto -s_i$, thereby precluding interaction terms with an odd number of spins, such as $s_i s_j s_k$. In Section 2.16 we present a general theory of real-space renormalisation group transformations that will allow for an infinite number of couplings.

In order to calculate the partition function for the two-dimensional Ising model exactly, all generated couplings must be retained. However, let us investigate whether a truncated coupling space can yield a phase transition at non-zero temperature.

The most drastic approximation is to ignore all the generated couplings. After setting $K_2' = K_3' = \cdots = 0$, the flow of the remaining coupling constants is described by

$$K_0' = \ln\left(2\sqrt{\cosh 2K_1}\,(\cosh 4K_1)^{1/8}\right), \tag{2.256a}$$

$$K_1' = \frac{1}{4}\ln(\cosh 4K_1). \tag{2.256b}$$

Equation (2.256b) is similar to Equation (2.246b) describing the flow in the one-dimensional case and has only two fixed points: an unstable low-temperature fixed point $K_1^\star = \infty$ and a stable high-temperature fixed point $K_1^\star = 0$. Only the low- and high-temperature fixed points survive this crude truncation of the coupling space. Therefore, when ignoring all the generated couplings, the renormalisation transformation incorrectly predicts that there is no phase transition in zero external field for any finite temperature in the $d = 2$ Ising model.

A less drastic approximation is to ignore all the generated couplings, except K_2'. After setting $K_3' = \cdots = 0$, the flow of the remaining coupling constants is described by

$$K_0' = \ln\left(2\sqrt{\cosh 2K_1}\,(\cosh 4K_1)^{1/8}\right), \qquad (2.257a)$$

$$K_1' = \frac{1}{4}\ln(\cosh 4K_1), \qquad (2.257b)$$

$$K_2' = \frac{1}{8}\ln(\cosh 4K_1). \qquad (2.257c)$$

Both coupling constants K_1' and K_2' are positive and favour the alignment of spins. It is therefore reasonable to combine their effect into a single coupling constant $\tilde{K}_1'$. To estimate how much of a contribution K_2' makes in the effective nearest-neighbour coupling constant $\tilde{K}_1'$, consider a system of fully aligned spins. Since there are $N'z/2$ different nearest-neighbour and next-nearest-neighbour pairs, the renormalised reduced energy (without the constant offset $N'K_0'$)

$$E_{\{s_I\}} = K_1' \sum_{\langle IJ \rangle} s_I s_J + K_2' \sum_{\text{nnn}} s_I s_J$$

$$= (K_1' + K_2')\frac{N'z}{2}. \qquad (2.258)$$

Therefore, the effective nearest-neighbour coupling constant

$$\tilde{K}_1' = K_1' + K_2'$$

$$= \frac{3}{8}\ln\left(\cosh 4\tilde{K}_1\right). \qquad (2.259)$$

Solving the associated fixed point equation for $\tilde{K}_1$ graphically, we find three fixed points, see Figure 2.36(a). The trivial stable low-temperature fixed point $\tilde{K}_1^\star = \infty$ corresponds to a system of fully aligned spins with $\xi = 0$. The trivial stable high-temperature fixed point $\tilde{K}_1^\star = 0$ corresponds

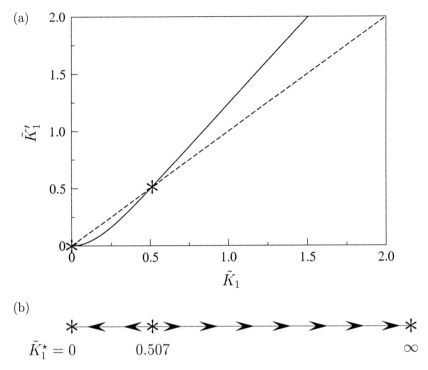

Fig. 2.36 (a) The renormalised coupling constant, $\tilde{K}'_1$, for the two-dimensional Ising model on a square lattice in zero external field versus the coupling constant in the original lattice, $\tilde{K}_1$. The fixed points ($\star$) lie at the intersections between the graph for $\tilde{K}'_1$ and the dashed line $\tilde{K}'_1 = \tilde{K}_1$. The fixed point $\tilde{K}^\star_1 = \infty$ implied by Equation (2.259) is not visible. (b) The associated renormalisation transformation flow in $\tilde{K}_1$-space. The fixed point $\tilde{K}^\star_1 = 0.507$ is unstable. For $0 < \tilde{K}_1 < 0.507$, applying the renormalisation transformation will induce a flow towards the stable fixed point $\tilde{K}^\star_1 = 0$. For $\tilde{K}_1 > 0.507$, applying the renormalisation transformation will induce a flow towards the stable fixed point $\tilde{K}^\star_1 = \infty$.

to a system of non-interacting spins with $\xi = 0$. The non-trivial unstable fixed point $\tilde{K}^\star_1 = 0.507$ corresponds to a system at a phase transition with $\xi = \infty$. Initial values of $\tilde{K}_1 > 0.507$ flow away from the unstable fixed point towards the stable low-temperature fixed point $\tilde{K}^\star_1 = \infty$. Initial values of $\tilde{K}_1 < 0.507$ flow away from the unstable fixed point towards the stable high-temperature fixed point $\tilde{K}^\star_1 = 0$, see Figure 2.36(b).

Therefore, when including the next-nearest-neighbour couplings K'_2 in an effective nearest-neighbour coupling constant $\tilde{K}'_1$, the renormalisation transformation correctly predicts that there is a phase transition in zero external field at a finite temperature in the $d = 2$ Ising model.

2.16 Wilson's Renormalisation Group Theory

Wilson was awarded the Nobel prize in physics 1982 'for his theory for critical phenomena in connection with phase transitions.' We are now in a position to outline his theory for the renormalisation group transformation, providing a beautiful and firm mathematical foundation for critical phenomena [Wilson, 1971a; Wilson, 1971b]. The theory explains the existence of universality classes and, furthermore, allows for the critical exponents to be calculated, at least in principle.

The formulation of the general framework for the renormalisation group transformation encapsulates the insight provided by the heuristic scaling picture based on the block spin transformation proposed by Kadanoff [Kadanoff, 1966]. It also, of course, encompasses the behaviour of the one- and two-dimensional Ising models under renormalisation presented in Section 2.15.4 and Section 2.15.5, respectively. For simplicity we restrict ourselves to zero external field in the following.

2.16.1 Coupling space and renormalisation group flow

First, we introduce the concept of an infinite-dimensional space of coupling constants and the so-called renormalisation group flow therein.

Applying the renormalisation group transformation once to the one-dimensional Ising model reduces the degrees of freedom from N to $N' = N/b$, and generates a renormalised nearest-neighbour coupling constant K_1' and a constant coupling term K_0' – but, importantly, no further coupling constants are generated. Therefore, in the renormalisation of the one-dimensional Ising model, the functional form of the partition function for the renormalised system is identical to that for the original system, apart from a constant prefactor $\exp(N'K_0')$, see Equation (2.247). The constant coupling term acts as a free energy offset, but does not play a role in the calculation of expectation values. Hence, successive applications of the renormalisation group transformation to the one-dimensional Ising model can be visualised as a flow in a one-dimensional coupling space:

$$K_1 \mapsto K_1' \mapsto K_1^{(2)} \mapsto \cdots \mapsto K_1^{(n-1)} \mapsto K_1^{(n)} \mapsto \cdots, \quad (2.260)$$

where $K_1^{(n)}$ denotes the renormalised nearest-neighbour coupling constant after n coarse-graining procedures.[10]

---

[10] For ease of notation, however, we will use a prime to indicate a variable that has been renormalised once, for example, $K_1' = K^{(1)}$.

Similarly, applying the renormalisation group transformation once to the two-dimensional Ising model reduces the degrees of freedom from N to $N' = N/b^2$, and generates a renormalised nearest-neighbour coupling constant K_1' and a constant coupling term K_0'. However, in contrast to one dimension, next-nearest neighbour interactions K_2' and quadruple spin interactions K_3' are generated in addition – even though $K_2 = K_3 = 0$ in the original system. In general for $d > 1$, a coupling constant that is zero in the original system may be non-zero in the renormalised system. On successive applications of the renormalisation group transformation, these coupling constants are themselves renormalised and, furthermore, additional renormalised coupling constants are generated. In fact, applying the renormalisation group transformation indefinitely generates an infinite number of renormalised coupling constants. Therefore, the renormalisation group transformation applied to the two-dimensional Ising model is associated with a flow in an infinite-dimensional coupling space:

$$\begin{aligned} K_1 &\mapsto K_1' \mapsto K_1^{(2)} \mapsto \cdots \mapsto K_1^{(n-1)} \mapsto K_1^{(n)} \mapsto \cdots \\ 0 = K_2 &\mapsto K_2' \mapsto K_2^{(2)} \mapsto \cdots \mapsto K_2^{(n-1)} \mapsto K_2^{(n)} \mapsto \cdots \\ 0 = K_3 &\mapsto K_3' \mapsto K_3^{(2)} \mapsto \cdots \mapsto K_3^{(n-1)} \mapsto K_3^{(n)} \mapsto \cdots \\ &\vdots \qquad \vdots \qquad \vdots \qquad \vdots \qquad \vdots \qquad \vdots \end{aligned} \qquad (2.261)$$

This motivates the introduction of an infinite-dimensional coupling space consisting of all possible coupling constants

$$\mathbf{K} = (K_1, K_2, K_3, \ldots), \qquad (2.262)$$

see Figure 2.37. Since the constant coupling term is not on the same footing as all the other coupling constants, it is not included in the coupling space. Physically, the constant coupling term represents a contribution to the free energy arising from summing out the degrees of freedom over the short length scale ba. Even though the constant coupling term neither affects expectation values nor is included in the coupling space, it plays a vital role of its own since its contribution to the free energy guarantees that the formalism is self-consistent.

For the following discussion, recall that the coupling constants are proportional to $1/(k_B T)$. Consider the 'original' Ising model at a given temperature T in zero external field with coupling constant $\mathbf{K} = (K_1, 0, 0, \ldots)$ represented by a point lying on the K_1-axis in the infinite-dimensional coupling space, see Figure 2.37. The temperature determines where the

224   *Complexity and Criticality*

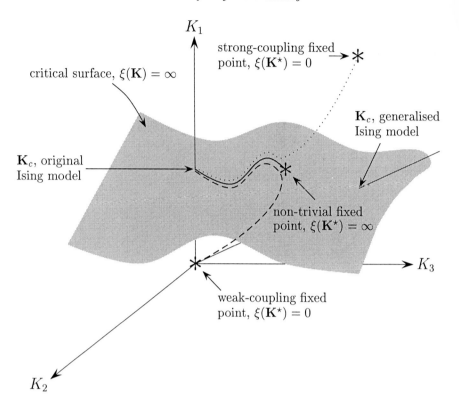

Fig. 2.37   A sketch of the infinite-dimensional coupling space where only K_1, K_2, K_3 of the infinitely many coupling constants are represented. To each point in the coupling space, $\mathbf{K} = (K_1, K_2, K_3, \ldots)$, is associated a generalised Ising model. The original Ising model is represented by a point $\mathbf{K} = (K_1, 0, 0, \ldots)$ on the K_1-axis. The three trajectories shown are associated with the renormalisation group transformation on the original Ising model with initial reduced temperature $t < 0$ (dotted line), $t = 0$ (solid line) and $t > 0$ (dashed line), respectively, previously visualised by the configurational flow shown in Figure 2.32. The critical surface, consisting of all coupling constants with $\xi(\mathbf{K}) = \infty$, is shaded dark grey. Critical coupling constants on the critical surface are shown for the original and a particular generalised Ising model. The three fixed points (∗) of the renormalisation group transformation are: the weak-coupling fixed point $\mathbf{K}^\star = (0, 0, 0, \ldots)$ with $\xi(\mathbf{K}^\star) = 0$, the strong-coupling fixed point $\mathbf{K}^\star = (\infty, \infty, \infty, \ldots)$ with $\xi(\mathbf{K}^\star) = 0$, and the non-trivial fixed point $\mathbf{K}^\star$ on the critical surface with $\xi(\mathbf{K}^\star) = \infty$. Strictly speaking, since the strong-coupling fixed point lies at infinity, it should not be visible in the figure.

original Ising model lies along the K_1-axis. For example, if $T = \infty$, the coupling constant $\mathbf{K} = (0, 0, 0, \ldots)$ and the associated correlation length $\xi(\mathbf{K}) = 0$. Therefore, the weak-coupling (high-temperature) Ising model lies at the origin. If $T = T_c$, the coupling constant takes its critical value $\mathbf{K}_c = (K_{1c}, 0, 0, \ldots)$ and the associated correlation length $\xi(\mathbf{K}_c) = \infty$. If $T = 0$, the coupling constant $\mathbf{K} = (\infty, 0, 0, \ldots)$ and the associated correlation length $\xi(\mathbf{K}) = 0$. Therefore, the strong-coupling (low temperature) Ising model lies at infinity on the K_1-axis.

Now, consider a 'generalised' Ising model at a given temperature T in zero external field with coupling constant $\mathbf{K} = (K_1, K_2, K_3, \ldots)$ represented by a point in the infinite-dimensional coupling space. The temperature determines where the particular generalised Ising model lies along the line from the origin in the direction given by $\mathbf{K}$, see Figure 2.37. If $T = \infty$, the coupling constant $\mathbf{K} = (0, 0, 0, \ldots)$ and the associated correlation length $\xi(\mathbf{K}) = 0$. Therefore, the weak-coupling (high-temperature) generalised Ising model lies at the origin. If $T = T_c$, the coupling constant takes its critical value $\mathbf{K}_c = (K_{1c}, K_{2c}, K_{3c}, \ldots)$ and the associated correlation length $\xi(\mathbf{K}_c) = \infty$. If $T = 0$, the coupling constant $\mathbf{K} = (\infty, \infty, \infty, \ldots)$ and the associated correlation length $\xi(\mathbf{K}) = 0$. The strong-coupling (low temperature) generalised Ising model lies at infinity.

Now that we have become familiar with the infinite-dimensional coupling space, we can show that the repeated application of the renormalisation group transformation on a generalised Ising model can be visualised as a discrete flow in this space.

As a starting point, consider a generalised Ising model in zero external field that allows for an infinite number of coupling constants representing nearest-neighbour interactions, next-nearest-neighbour interactions, quadruple spins interactions, etc. Such a generalised Ising model is defined by a coupling constant $\mathbf{K}$ and a constant coupling term K_0. The associated generalised reduced energy takes the form[11]

$$-\beta E_{\{s_i\}} = K_0 N + K_1 \sum_{\langle ij \rangle} s_i s_j + K_2 \sum_{\mathrm{nnn}} s_i s_j + K_3 \sum_{\square} s_i s_j s_k s_l + \cdots . \quad (2.263)$$

The generalised reduced energy contains all possible spin interactions, of which there are infinitely many, respecting the symmetry of the problem: in zero external field, for example, interaction terms with an odd number of spins, such as $s_i s_j s_k$, cannot be present since the energy must be invariant under the reversal of spins $s_i \mapsto -s_i$; however, all interaction terms with an even number of spins, such as $s_i s_j s_k s_l$, are present. Note that we may recover the original Ising model in zero external field by setting all coupling

---

[11] Since the external field $H = 0$, the term $h \sum_{i=1}^{N} s_i$ with $h = \beta H$ is absent.

constants except K_1 equal to zero.

Applying the renormalisation group transformation once to a generalised Ising model renormalises the associated reduced energy. The reduced energy in the renormalised system, $\beta E'_{\{s_I\}}$, is defined through the equation

$$-\beta E'_{\{s_I\}} = \ln[\sum_{\substack{\text{configurations } \{s_i\} \\ \text{consistent with } \{s_I\}}} \exp\left(-\beta E_{\{s_i\}}\right)] \qquad (2.264)$$

$$= K'_0 N' + K'_1 \sum_{\langle IJ \rangle} s_I s_J + K'_2 \sum_{\text{nnn}} s_I s_J + K'_3 \sum_{\square} s_I s_J s_K s_L + \cdots.$$

The renormalisation group transformation reduces the degrees of freedom from N to $N' = N/b^d$ block spin variables $\{S_I\}$, whose couplings are given by the renormalised coupling constants $\mathbf{K}' = (K'_1, K'_2, K'_3, \ldots)$ and, in addition, the spin-independent term renormalises to $K'_0 N'$. Note that, in contrast to the Kadanoff block spin transformation, this general formulation allows for 'new' coupling constants to be generated.

Let us now formally introduce the renormalisation group transformation R_b as a transformation acting on the members of the infinite-dimensional coupling space, that effects coarse-graining over blocks of size ba:

$$\mathbf{K}' = R_b(\mathbf{K}). \qquad (2.265)$$

Equation (2.265) expresses a recursion relation that can be applied indefinitely on a system in the thermodynamic limit. Since the renormalisation group transformation involves a coarse-graining procedure over a block with a finite number of b^d spins, the transformation is analytic. This analyticity will allow for a Taylor expansion of R_b which will prove important shortly. However, if the renormalisation group transformation is applied indefinitely to a system in the thermodynamic limit, singular behaviour may occur. In this respect the renormalisation group transformation is able to account for critical phenomena. Note also that renormalising twice, first with R_{b_1} and then with R_{b_2}, is equivalent to renormalising once with $R_{b_1 b_2}$. The renormalisation group transformation therefore satisfies $R_{b_2}(R_{b_1}(\mathbf{K})) = R_{b_1 b_2}(\mathbf{K})$ for all $\mathbf{K}$. Expressed as an operator identity:[12]

$$R_{b_2} R_{b_1} = R_{b_1 b_2}. \qquad (2.266)$$

---

[12] The renormalisation group transformation has a trivial identity element for $b = 1$. However, when $b > 1$ there is no inverse transformation. Thus, the renormalisation group transformation forms a semi-group rather than a group.

Note, however, that we have not specified the coarse-graining procedure and indeed 'the' renormalisation group transformation is by no means unique.

Successive applications of the renormalisation group transformation on a generalised Ising model induces a discrete flow in the infinite-dimensional coupling space that can be visualised as points along a trajectory originating at $\mathbf{K}$:

$$\mathbf{K} \mapsto R_b(\mathbf{K}) \mapsto R_b^2(\mathbf{K}) \mapsto \cdots \mapsto R_b^{n-1}(\mathbf{K}) \mapsto R_b^n(\mathbf{K}) \mapsto \cdots. \quad (2.267)$$

2.16.2 Self-similarity and fixed points

A fixed point of the renormalisation group transformation is a point $\mathbf{K}^\star$ in the infinite-dimensional coupling space that is invariant under renormalisation, that is,

$$R_b(\mathbf{K}^\star) = \mathbf{K}^\star. \quad (2.268)$$

Equation (2.268) is the fixed point equation for the renormalisation group transformation. For a fixed point $\mathbf{K}^\star$, the associated reduced energy is invariant under renormalisations; hence, the associated generalised Ising model is invariant under rescaling. We will show that the fixed points of the renormalisation group transformation are associated with zero or infinite correlation length. Hence, as in percolation, self-similarity and scale invariance are associated with the fixed points of the renormalisation group transformation.

The discrete flow in the infinite-dimensional space of coupling constants generated by applying the renormalisation group transformation is associated with a flow in the correlation length. Applying the renormalisation group transformation once reduces the correlation length from $\xi(\mathbf{K})$ to $\xi(\mathbf{K}')$ where

$$\xi(\mathbf{K}') = \frac{\xi(\mathbf{K})}{b}. \quad (2.269)$$

Since the correlation length after each application of the renormalisation group transformation is reduced by the rescaling factor b, after n successive transformations

$$\xi\left[R_b^n(\mathbf{K})\right] = \frac{\xi(\mathbf{K})}{b^n}, \quad \text{for } n = 1, 2, \ldots. \quad (2.270)$$

If the initial correlation length is finite, $\xi(\mathbf{K}) < \infty$, then the correlation length after n transformations is reduced by a factor b^n and the renormalised system moves further and further away from criticality, disclosing the large scale behaviour of the original system. The correlation length eventually shrinks to zero as $n \to \infty$. If, however, the initial correlation length is infinite, $\xi(\mathbf{K}) = \infty$, then so too is the correlation length in all the renormalised systems.

We will now show that a system with no characteristic scale is a fixed point of the renormalisation group transformation and vice versa.

Applying the renormalisation group transformation indefinitely,

$$\xi\left[\lim_{n\to\infty} R_b^n(\mathbf{K})\right] = \lim_{n\to\infty} \frac{\xi(\mathbf{K})}{b^n} = \begin{cases} 0 & \text{for } \xi(\mathbf{K}) < \infty \\ \infty & \text{for } \xi(\mathbf{K}) = \infty. \end{cases} \quad (2.271)$$

The correlation length associated with the generalised Ising model corresponding to the point $\lim_{n\to\infty} R_b^n(\mathbf{K})$ in coupling space is zero or infinite so there is no characteristic scale associated with $\lim_{n\to\infty} R_b^n(\mathbf{K})$. Therefore, $\lim_{n\to\infty} R_b^n(\mathbf{K})$ must be invariant under the renormalisation group transformation, that is,

$$R_b\left[\lim_{n\to\infty} R_b^n(\mathbf{K})\right] = \lim_{n\to\infty} R_b^n(\mathbf{K}). \quad (2.272)$$

Equation (2.272) demonstrates that $\lim_{n\to\infty} R_b^n(\mathbf{K})$ satisfies the fixed point equation for the renormalisation group transformation and we identify

$$\mathbf{K}^\star = \lim_{n\to\infty} R_b^n(\mathbf{K}), \quad (2.273)$$

with an associated correlation length that is $\xi(\mathbf{K}^\star) = 0$ or $\xi(\mathbf{K}^\star) = \infty$.

Now, assume that there exists a fixed point $\mathbf{K}^\star$ for the renormalisation group transformation. From Equation (2.269) we have $\xi(R_b(\mathbf{K}^\star)) = \xi(\mathbf{K}^\star)/b$, and after applying Equation (2.268) we find for the correlation length at a fixed point

$$\xi(\mathbf{K}^\star) = \frac{\xi(\mathbf{K}^\star)}{b} \quad \Leftrightarrow \quad \xi(\mathbf{K}^\star) = \begin{cases} 0 & \text{'trivial'} \\ \infty & \text{'non-trivial'}. \end{cases} \quad (2.274)$$

In summary, a fixed point of the renormalisation group transformation implies that there is no characteristic scale; scale invariance prevails. Likewise, if scale invariance prevails, it is associated with a fixed point of the renormalisation group transformation.

2.16.3 Basin of attraction of fixed points

In general, a renormalisation group transformation may have several fixed points. For simplicity, consider a renormalisation group transformation which has only three fixed points in the infinite-dimensional coupling space: the weak-coupling (high-temperature) fixed point $\mathbf{K}^\star = (0,0,0,\ldots)$ lying at the origin with $\xi(\mathbf{K}^\star) = 0$, the strong-coupling (low-temperature) fixed point $\mathbf{K}^\star = (\infty,\infty,\infty,\ldots)$ lying at infinity with $\xi(\mathbf{K}^\star) = 0$, and a non-trivial fixed point $\mathbf{K}^\star$ with $\xi(\mathbf{K}^\star) = \infty$, see Figure 2.37.

Each of the fixed points $\mathbf{K}^\star$ will have a so-called basin of attraction, consisting of all points in the coupling space that flow into the fixed point $\mathbf{K}^\star$ when the renormalisation group transformation is applied indefinitely.

The basin of attraction of the non-trivial fixed point $\mathbf{K}^\star$ with $\xi(\mathbf{K}^\star) = \infty$ is known as the 'critical surface' or, more generally the 'critical manifold', since its dimensionality need not be restricted to 2. We can show that for $\mathbf{K}$ to lie in the basin of attraction of the non-trivial fixed point, its associated correlation length must be infinite, $\xi(\mathbf{K}) = \infty$. A simple rearrangement of Equation (2.270) yields

$$\xi(\mathbf{K}) = b^n \xi\left[R_b^n(\mathbf{K})\right]. \tag{2.275}$$

The correlation length at an initial point $\mathbf{K}$ in coupling space is a factor b^n larger that the correlation length associated with the point after n transformations. The left-hand side of Equation (2.275) is independent of n. Therefore, if $\mathbf{K}$ lies in the basin of attraction of the non-trivial fixed point $\mathbf{K}^\star$ with $\xi(\mathbf{K}^\star) = \infty$, then taking the limit of $n \to \infty$ we find

$$\xi(\mathbf{K}) = \lim_{n\to\infty} b^n \xi\left[R_b^n(\mathbf{K})\right] = \lim_{n\to\infty} b^n \xi(\mathbf{K}^\star) = \infty. \tag{2.276}$$

Hence, for $\mathbf{K}$ to lie in the basin of attraction of the non-trivial fixed point, the associated correlation length must be infinite, that is, $\xi(\mathbf{K}) = \infty$. Equivalently, we may also define the critical surface as the set of all coupling constants $\mathbf{K}$ where $\xi(\mathbf{K}) = \infty$. In Figure 2.37, the surface that has been shaded dark grey is part of the infinite critical surface.

The critical surface divides the coupling space into the basin of attraction of the weak-coupling fixed point, lying at the origin, consisting of all points 'below' the critical surface, the basin of attraction of the strong-coupling fixed point, lying at infinity, consisting of all points 'above' the critical surface, and finally the basin of attraction of the non-trivial critical fixed point, lying on the critical surface, consisting of all points on the critical surface, see Figure 2.37.

2.16.4 RG flow in coupling and configurational space

Consider the original Ising model in zero external field at reduced temperature t. If t is in the neighbourhood of the critical temperature $t = 0$, the model lies along the K_1-axis in the neighbourhood of the critical surface in the infinite-dimensional coupling space, see Figure 2.37. Applying the renormalisation group transformation induces a discrete flow in coupling space. In Figure 2.32, we have seen the associated flow in configurational space for the two-dimensional Ising model with a particular choice of the renormalisation group transformation, namely the majority rule with $b = 3$.

Assume that the initial temperature is slightly below the critical temperature, that is, $t < 0$. Applying the renormalisation group transformation repeatedly induces the flow in configurational space displayed in the left-hand column of Figure 2.32 towards the trivial fixed point of all spins aligned associated with $T = 0$. This flow in configurational space is associated with the discrete flow in coupling space along the dotted line commencing on the K_1-axis just above the critical surface and 'terminating' at the strong-coupling fixed point $\mathbf{K}^\star = (\infty, \infty, \infty, \ldots)$ lying at infinity with correlation length $\xi(\mathbf{K}^\star) = 0$. Therefore, the two microstates associated with $\mathbf{K}^\star = (\infty, \infty, \infty, \ldots)$ are trivially self-similar.

Next assume that the initial temperature is slightly above the critical temperature, that is, $t > 0$. Applying the renormalisation group transformation repeatedly induces the flow in configurational space displayed in the right-hand column of Figure 2.32 towards the trivial fixed point of randomly orientated spins associated with $T = \infty$. This flow in configurational space is associated with the discrete flow in coupling space along the dashed line commencing on the K_1-axis just below the critical surface and terminating at the weak-coupling fixed point $\mathbf{K}^\star = (0, 0, 0, \ldots)$ lying at the origin with correlation length $\xi(\mathbf{K}^\star) = 0$. Therefore, the microstates associated with $\mathbf{K}^\star = (0, 0, 0, \ldots)$ are trivially self-similar.

Finally, assume that the initial temperature is critical, that is, $t = 0$. Applying the renormalisation group transformation repeatedly does not induce a flow in configurational space, see the middle column of Figure 2.32. Nevertheless, the absence of flow in configurational space is associated with the flow in coupling space commencing on the K_1-axis on the critical surface along the solid line terminating at the non-trivial fixed point $\mathbf{K}^\star$ with correlation length $\xi(\mathbf{K}^\star) = \infty$. Therefore, the microstates associated with $\mathbf{K}^\star$ are self-similar.

2.16.5 Universality and RG flow near fixed point

Assume that $\mathbf{K}_c$ lies on the critical surface, that is, $\xi(\mathbf{K}_c) = \infty$. As we have just seen, it is not the initial critical value of the coupling constant nor the flow in the neighbourhood of $\mathbf{K}_c$ that determines the critical behaviour of the associated generalised Ising model. Rather, the critical behaviour is determined by the flow of $R_b^n(\mathbf{K})$ for $n \to \infty$, which takes place in the neighbourhood of the non-trivial fixed point $\mathbf{K}^\star$ on the critical surface!

Each point $\mathbf{K}_c$ on the critical surface represents a particular generalised Ising model at its critical point. However, since they all flow into the same non-trivial fixed point when applying the renormalisation group transformation indefinitely, their critical behaviour is determined by the flow close to this fixed point. Indeed, universality refers to the identical behaviour shown by systems close to the fixed point $\mathbf{K}^\star$, rather than at the fixed point itself. Therefore, we need to investigate in more detail the flow in coupling space close to the non-trivial fixed point on the critical surface. In doing so, we will, as a by-product, also be able to demonstrate how the Widom scaling ansatz for the singular part of the free energy per spin can be derived using the general framework of the renormalisation group transformation.

Let $\mathbf{K} = \mathbf{K}^\star + \delta\mathbf{K}$ be close to the fixed point $\mathbf{K}^\star$, that is, the entries $\delta\mathbf{K}$ in the deviation from the fixed point are small. Applying the renormalisation group transformation once, $R_b(\mathbf{K}) = \mathbf{K}' = \mathbf{K}^\star + \delta\mathbf{K}'$ where $\delta\mathbf{K}'$, the deviation from the fixed point after renormalisation, is a function of $\mathbf{K}$. Since $\mathbf{K}^\star$ is a fixed point of the renormalisation group transformation, we have

$$R_b(\mathbf{K}) = \mathbf{K}^\star + \delta\mathbf{K}'$$
$$= R_b(\mathbf{K}^\star) + \delta\mathbf{K}'. \qquad (2.277)$$

However, since R_b is analytic, we can Taylor expand the renormalisation group transformation around the fixed point $\mathbf{K}^\star$, and to first order in $\delta\mathbf{K}$ we find

$$R_b(\mathbf{K}) = R_b(\mathbf{K}^\star) + \mathsf{M}(b)\delta\mathbf{K} + \mathcal{O}\left(\delta\mathbf{K}^2\right), \qquad (2.278)$$

where the ijth entry in the matrix $\mathsf{M}(b) = \partial\mathbf{K}'/\partial\mathbf{K}$

$$[\mathsf{M}(b)]_{ij} = \left.\frac{\partial K_i'}{\partial K_j}\right|_{\mathbf{K}^\star} \qquad (2.279)$$

is evaluated at the non-trivial fixed point $\mathbf{K}^\star$. The matrix $\mathsf{M}(b)$ is the so-called linearised renormalisation group transformation in the vicinity of the fixed point $\mathbf{K}^\star$.

Comparing Equation (2.277) with Equation (2.278), we identify that, to first order in $\delta\mathbf{K}$, the deviation from the fixed point after renormalisation $\delta\mathbf{K}'$ is related to the original deviation from the fixed point $\delta\mathbf{K}$ via the matrix $\mathbf{M}(b)$:

$$\delta\mathbf{K}' = \mathbf{M}(b)\delta\mathbf{K}. \qquad (2.280)$$

For the ith entry $\delta K'_i$, Equation (2.280) implies that

$$\begin{aligned}\delta K'_i &= \sum_j [\mathbf{M}(b)]_{ij}\, \delta K_j \\ &= \sum_j \left.\frac{\partial K'_i}{\partial K_j}\right|_{\mathbf{K}^\star} \delta K_j.\end{aligned} \qquad (2.281)$$

Hence, the matrix $\mathbf{M}(b)$ determines the speed of the flow towards or away from the fixed point. The real matrix $\mathbf{M}(b)$ is not necessarily symmetric. For simplicity, however, we assume that $\mathbf{M}(b)$ is symmetric to guarantee that its eigenvalues are real and that the associated set of eigenvectors are orthogonal and form a convenient basis in which to discuss the flow. The aim is to investigate the flow in the coupling space along the eigenvectors of $\mathbf{M}(b)$. This will allow us to make simple but far reaching conclusions about the nature of the flow.[13]

To investigate the flow in detail, we introduce the eigenvalues, $\lambda_i(b)$, and the eigenvectors $\mathbf{e}_i(b)$ of the matrix $\mathbf{M}(b)$:

$$\mathbf{M}(b)\mathbf{e}_i(b) = \lambda_i(b)\mathbf{e}_i(b). \qquad (2.282)$$

The eigenvectors $\{\mathbf{e}_i(b)\}$ are normalised and orthogonal

$$\mathbf{e}_i(b) \cdot \mathbf{e}_j(b) = \begin{cases} 1 & \text{for } i = j \\ 0 & \text{for } i \neq j \end{cases} \qquad (2.283)$$

and form a basis for the infinite-dimensional coupling space. We further assume that this basis is complete, that is, every point in the infinite-dimensional coupling space can be written as a linear combination of $\{\mathbf{e}_i(b)\}$.

---

[13] We refer the reader to [Goldenfeld, 1992; Cardy, 1996] for a discussion of the theory when $\mathbf{M}(b)$ is not symmetric.

We next define the (scalar) scaling field, u_i, as the length of the projection of the deviation from the fixed point along the direction of $\mathbf{e}_i(b)$:

$$u_i = \mathbf{e}_i(b) \cdot \delta\mathbf{K}, \qquad (2.284)$$

such that the component of $\delta\mathbf{K}$ in the direction of $\mathbf{e}_i(b)$ is $[\mathbf{e}_i(b) \cdot \delta\mathbf{K}]\,\mathbf{e}_i(b)$ and therefore we may write

$$\begin{aligned}
\delta\mathbf{K} &= \sum_{i=1}^{\infty} [\mathbf{e}_i(b) \cdot \delta\mathbf{K}]\,\mathbf{e}_i(b) \\
&= \sum_{i=1}^{\infty} u_i \mathbf{e}_i(b).
\end{aligned} \qquad (2.285)$$

We will shortly be able to identify the scaling fields with reduced control parameters, such as the reduced temperature, reduced external field and so on. Hence, as experimenters we can control the initial value of the scaling fields.

The renormalised scaling field, u_i', is the projection of the deviation from the fixed point along the direction of $\mathbf{e}_i(b)$ after renormalisation. Using the expansion in Equation (2.285) of $\delta\mathbf{K}$ along the directions of the eigenvectors, we can relate the transformed scaling field to the original scaling field:

$$\begin{aligned}
u_i' &= \mathbf{e}_i(b) \cdot \delta\mathbf{K}' \\
&= \mathbf{e}_i(b) \cdot \mathbf{M}(b)\delta\mathbf{K} \\
&= \mathbf{e}_i(b) \cdot \mathbf{M}(b) \sum_{j=1}^{\infty} u_j \mathbf{e}_j(b) \\
&= \mathbf{e}_i(b) \cdot \sum_{j=1}^{\infty} u_j \mathbf{M}(b)\mathbf{e}_j(b) \\
&= \mathbf{e}_i(b) \cdot \sum_{j=1}^{\infty} u_j \lambda_i(b)\mathbf{e}_j(b) \\
&= \lambda_i(b) u_i,
\end{aligned} \qquad (2.286)$$

where in the last step we have made use of the orthonormality of the eigenvectors, see Equation (2.283). Hence, the transformed scaling field is related to the original scaling field by the factor $\lambda_i(b)$, which is the eigenvalue of $\mathbf{M}(b)$ associated with the eigenvector in the direction $\mathbf{e}_i(b)$.

Now, from the semi-group property of the renormalisation group transformation in Equation (2.266) it follows that

$$\mathbf{M}(b_2)\mathbf{M}(b_1) = \mathbf{M}(b_2 b_1), \tag{2.287}$$

implying that the eigenvalues satisfy the condition

$$\lambda_i(b_2)\lambda_i(b_1) = \lambda_i(b_2 b_1). \tag{2.288}$$

The unique solution of the functional equation in Equation (2.288) is a power law, such that

$$\lambda_i(b) = b^{y_i}, \tag{2.289}$$

where y_i are the so-called renormalisation group eigenvalues. Substituting Equation (2.289) into Equation (2.286) we arrive at our principle result

$$u'_i = b^{y_i} u_i. \tag{2.290}$$

The derivation of the relation in Equation (2.290) between the renormalised and original scaling field puts the heuristic assumptions in Equation (2.218) of the Kadanoff block spin transition on a firm mathematical footing.

The factor b^{y_i} determines whether the scaling field increases, decreases or remains constant under renormalisation. We may distinguish three different cases:

- *Relevant scaling field.* If $y_i > 0$, then $\lambda_i(b) = b^{y_i} > 1$. The deviation from the fixed point along the direction of $\mathbf{e}_i(b)$ increases upon renormalisation. The scaling field u_i with $y_i > 0$ is said to be relevant, since the renormalisation group flow is driven away from the non-trivial fixed point. A relevant scaling field eventually explodes upon application of the renormalisation group transformation.
- *Irrelevant scaling field.* If $y_i < 0$, then $\lambda_i(b) = b^{y_i} < 1$. The deviation from the fixed point along the direction of $\mathbf{e}_i(b)$ decreases upon renormalisation. The scaling field u_i with $y_i < 0$ is said to be irrelevant, since the renormalisation group flow is driven towards the non-trivial fixed point. An irrelevant scaling field eventually vanishes upon application of the renormalisation group transformation.
- *Marginal scaling field.* If $y_i = 0$, then $\lambda_i(b) = b^{y_i} = 1$. The scaling field u_i with $y_i = 0$ is said to be marginal. A first order Taylor expansion in $\delta \mathbf{K}$ is insufficient to determine whether the flow is directed towards or away from the fixed point.

This observation allows us to make the following far reaching conclusion: When applying the renormalisation group transformation, the initial deviation $\delta\mathbf{K}$ from the fixed point $\mathbf{K}^\star$ transforms according to Equation (2.280) into $\delta\mathbf{K}' = \mathbf{M}(b)\delta\mathbf{K}$. However, only the components of $\delta\mathbf{K}$ along the eigenvectors of $\mathbf{M}(b)$ with eigenvalues $\lambda_i(b) = b^{y_i} > 1$ are relevant. The components along the eigenvectors of $\mathbf{M}(b)$ with eigenvalues $\lambda_i(b) = b^{y_i} < 1$ are irrelevant, since they will vanish when the renormalisation group transformation has been applied indefinitely.

Therefore, we see that the eigenvectors corresponding to the irrelevant eigenvalues span the critical surface. The dimension of the critical surface equals the number of irrelevant eigenvalues. In other words, the difference between the dimensionality of the coupling space itself and the critical surface is the number of relevant eigenvalues.

For the general Ising model, there are only two relevant scaling fields, and we can identify

$$u_1 = t \quad \text{with } y_1 = y_t, \tag{2.291a}$$
$$u_2 = h \quad \text{with } y_2 = y_h. \tag{2.291b}$$

All the remaining scaling fields $u_3, u_4, \ldots$ are irrelevant.

Let us therefore re-interpret the flow in coupling space discussed in Section 2.16.4. For a trajectory to lie on the critical surface, all its relevant scaling fields must be zero. The trajectory is thereby confined to remain on the critical surface, and flows towards the non-trivial fixed point as the irrelevant scaling fields vanish.

A trajectory that starts close to the critical surface initially flows towards the non-trivial fixed point as the irrelevant scaling fields decrease. Eventually, however, the trajectory is expelled from the neighbourhood of the critical surface as the relevant scaling fields increase, terminating at one of the two trivial fixed points.

2.16.6 Widom scaling form

We will show in general that the singular part of the free energy per spin f_s transforms homogeneously under a renormalisation group transformation. Together with the results above, this leads to the Widom scaling form for the singular part of the free energy.

Apart from the constant factor $\exp(N'K_0')$, the partition function remains invariant under a renormalisation group transformation:

$$\begin{aligned}
Z(\mathbf{K}, N) &= \sum_{\{s_i\}} \exp\left(-\beta E_{\{s_i\}}\right) \\
&= \sum_{\{s_I\}} \sum_{\substack{\text{configurations } \{s_i\} \\ \text{consistent with } \{s_I\}}} \exp\left(-\beta E_{\{s_i\}}\right) \\
&= \exp(N' K_0') \sum_{\{s_I\}} \exp\left(-\beta E'_{\{s_I\}}\right) \\
&= \exp(N' K_0') Z(\mathbf{K}', N').
\end{aligned} \qquad (2.292)$$

Meanwhile, since the number of spins is reduced to $N' = N/b^d$, the free energy per spin transforms according to

$$\begin{aligned}
f(\mathbf{K}) &= -\frac{1}{N} k_B T \ln Z(N, \mathbf{K}) \\
&= -b^{-d} \frac{1}{N'} k_B T \left[N' K_0' + \ln Z(N', \mathbf{K}')\right] \\
&= f_r(\mathbf{K}) + b^{-d} f_s(\mathbf{K}'). \qquad (2.293)
\end{aligned}$$

We identify the term arising from the renormalised constant coupling K_0' as the regular part of the free energy per spin[14] $f_r(\mathbf{K})$. Therefore, the free energy per spin transforms inhomogeneously. However, the singular part of the free energy per spin transforms as a generalised homogeneous function according to

$$f_s(\mathbf{K}) = b^{-d} f_s(\mathbf{K}'). \qquad (2.294)$$

Although coupling constants involving K_0, renormalised or otherwise, do not appear in the probability distribution of microstates, they are nevertheless required for the correct transformation of the free energy per spin. To see this, we argue that Equation (2.294) alone cannot be valid: Assume, for the moment, that (2.294) is a valid transformation for the free energy per spin. Then, at the non-trivial fixed point $f(\mathbf{K}^\star) = b^{-d} f(\mathbf{K}^\star)$, implying that $f(\mathbf{K}^\star) = 0$ or $f(\mathbf{K}^\star) = \infty$. The latter is not physical and the former cannot be true in general. Equation (2.294) only pertains to the singular part of the free energy, which indeed takes the value $f_s(\mathbf{K}^\star) = 0$, but with derivatives that are diverging [Wilson, 1971a].

---

[14]Since the term is proportional to N, it is common in the literature to identify $f_r(\mathbf{K}) = N g(\mathbf{K})$.

Close to the fixed point on the critical surface it is convenient to use the eigenvectors of **M** as the basis. Since the scaling fields are the deviations of **K** from the fixed point, we find that after applying the renormalisation group transformation n times

$$f_s(t, h, u_3, u_4, \ldots) = b^{-d} f_s(b^{y_t} t, b^{y_h} h, b^{y_3} u_3, b^{y_4} u_4, \ldots)$$
$$= b^{-nd} f_s(b^{ny_t} t, b^{ny_h} h, b^{ny_3} u_3, b^{ny_4} u_4, \ldots). \qquad (2.295)$$

In the limit of $n \to \infty$ all the irrelevant scaling fields $u_3, u_4, \ldots$ iterate to zero and

$$f_s(t, h, u_3, u_4, \ldots) = b^{-nd} f_s(b^{ny_t} t, b^{ny_h} h, 0, 0, \ldots). \qquad (2.296)$$

However, we would like to express the transformation in Equation (2.296) in a form that does not explicitly refer to the arbitrary scale factor b. Therefore, choosing $b^{ny_t} = |t|^{-1}$ we find

$$f_s(t, h, u_3, u_4, \ldots) = |t|^{d/y_t} f_s(t/|t|, h/|t|^{y_h/y_t}, 0, 0, \ldots)$$
$$= |t|^{d/y_t} f_s(\pm 1, h/|t|^{y_h/y_t}, 0, 0, \ldots)$$
$$= |t|^{d/y_t} \mathcal{F}_{\pm}\left(h/|t|^{y_h/y_t}\right). \qquad (2.297)$$

Hence, we have derived the Widom scaling ansatz in the general formalism developed by Wilson. In principle, the eigenvalues and eigenvectors of the renormalisation group transformation can be determined and hence critical exponents calculated, see Exercise 2.8.

2.17 Summary

In this chapter, we introduced the Ising model, one of the most widely-studied models in statistical mechanics. The model is defined through its configurational energy, so that the problem is to evaluate either the free energy or the partition function. To gain practice in such calculations, we examined a collection of non-interacting spins and introduced quantities such as the magnetisation, susceptibility per spin, and specific heat.

We first described the behaviour of the Ising model in purely qualitative terms. The free energy can be thought of as a competition between a tendency for spins to align and lower the internal energy of the system and a tendency for spins to point randomly and increase the entropy of the system. We speculated that in zero external field there was a temperature at which these two competing influences balance, thereby marking an order-disorder phase transition. We qualified these remarks by discussing the thermodynamic limit, which makes a phase transition mathematically precise, and symmetry breaking, which specifies how statistical mechanical averages should be performed.

We then applied this knowledge to the one-dimensional Ising model, which we could solve in full. We found that there was no phase transition at non-zero temperature, and we explained this result in terms of the energetics of creating disorder.

We next turned our attention to a mean-field version of the Ising model that gives a qualitatively more accurate description of the continuous phase transition. By neglecting fluctuations, we were able to solve this model to find a non-zero critical temperature in zero external field, at which point the behaviour of the thermodynamic quantities were characterised by critical exponents. We were able to visualise the behaviour of the mean-field Ising model by plotting the free energy surface as a function of temperature and external field. We were able to identify the location of the continuous phase transition as the point at which the free energy loses analyticity as the temperature is lowered in zero external field.

To generalise our findings, we introduced the Landau phenomenological theory of continuous phase transitions, which expands the free energy as a Taylor series in terms of the order parameter, respecting any symmetries enjoyed by the system. We found that the Landau theory reproduced the mean-field critical exponents. The Landau theory gave us an intuitive picture for the emergence of two ordered phases below the critical point. We analysed the validity of the mean-field approximation by deriving the so-called Ginzburg criterion for when fluctuations may be safely neglected.

In two dimensions, we did not derive the rather technical Onsager solution, but proceeded as in percolation by appealing to numerical results. These results were able to confirm what we had anticipated, that thermodynamic quantities such as the susceptibility and the specific heat diverge in the vicinity of the critical point.

We accounted for this behaviour by hypothesising that fluctuations in the Ising model have no characteristic scale at the critical point, and this led us to postulate that the singular part of the critical free energy transforms as a generalised homogeneous function. From this, we were able to derive scaling relations for various critical exponents, as well as scaling functions for the susceptibility per spin and specific heat.

To express the idea of scale invariance at the critical point in a general framework, we applied the renormalisation group approach to the Ising model, and we tracked fluctuations by using the renormalised correlation length, giving a measure of the size of clusters of correlated spins. We argued that the system could remain self-similar if and only if the partition function remained invariant under a global change of scale, and we used this to constrain the form of the free energy, confirming that its singular part indeed transforms as a generalised homogeneous function.

We applied these ideas to the one-dimensional Ising model, the solution of which is reproduced exactly in the renormalisation group approach. We found that the renormalisation flow could be described in terms of renormalised coupling constants. In two-dimensions, we found that with each renormalisation step we generally had to introduce new coupling constants describing non-nearest neighbour interactions if we were to exactly preserve all the information of the original system. But arguing that the model could be well approximated by just a few renormalised coupling constants, we dropped all higher order couplings but two, which allowed us to locate a non-trivial fixed point.

We presented Wilson's general framework for the renormalisation group transformation. We introduced a generalised infinite-dimensional coupling space that includes all possible spin-spin interactions obeying the symmetry of the system. A generalised Ising model, defined by arbitrary initial coupling constants, is similarly represented by a point in the infinite-dimensional coupling space. Upon application of the renormalisation group transformation, higher couplings are in general generated even if they are absent in the original system. Thus, the renormalisation group transformation induces a discrete flow in the infinite-dimensional coupling space.

We then studied the flow of the renormalisation group transformation in greater detail and argued for the existence of fixed points. We associated the fixed points of the renormalisation group transformation with self-similar microstates in configurational space. In this way we established a correspondence between zero correlation length and the weak- and strong-coupling fixed points, and infinite correlation length and the non-trivial critical coupling fixed point. We introduced the basin of attraction of a fixed point as the points in coupling space that flow into the fixed point under renormalisation. The critical surface is the basin of attraction of the non-trivial fixed point. Hence, all generalised Ising models initially on the critical surface flow towards the non-trivial fixed point. We argued that the existence of a unique non-trivial fixed point is the origin of universality. Thus, each universality class is characterised by a flow in the vicinity of the non-trivial fixed point particular to that universality class.

Since the renormalisation group transformation is analytic, we were able to perform a Taylor expansion to first order in the vicinity of the non-trivial fixed point. By linearising the flow in such a way, we described how nearby trajectories are attracted or repelled according to the eigenvalues and eigenvectors of the linearised renormalisation group transformation. Thus, the trajectory of a point in the space of couplings is governed by the action of so-called scaling fields which can be classed as relevant, irrelevant, or marginal, according to the value of the associated eigenvalues. For example, the Ising model has two relevant fields, one associated with the temperature and the other associated with the external field. An experimenter must therefore tune these two parameters to their critical values. This corresponds to positioning the Ising model on the critical surface such that the system flows towards the non-trivial fixed point under successive applications of the renormalisation group transformation.

Finally, we discussed the transformation of the free energy under the renormalisation group transformation. Although the transformation is inhomogeneous, we were able to identify regular and singular parts of the free energy, the latter part being responsible for the critical behaviour of thermodynamic quantities. The singular part of the free energy is a generalised homogeneous function. Combining this with our knowledge of the relevant scaling fields, we were able to rederive the Widom scaling ansatz for the free energy. In addition to explaining the concept of universality, the renormalisation group theory also allows for the calculation of the critical exponents.

Exercises

2.1 *The entropy and the free energy.*

The entropy S of a thermal system at equilibrium is defined as

$$S = -k_B \sum_r p_r \ln p_r, \tag{2.1.1}$$

where k_B is Boltzmann's constant and p_r is the probability of the system being in a microstate r.

(i) Use the Boltzmann's distribution for p_r to show that

$$S = k_B \ln Z + \langle E \rangle / T. \tag{2.1.2}$$

(ii) The total free energy

$$F = -k_B T \ln Z. \tag{2.1.3}$$

Show that

$$F = \langle E \rangle - TS. \tag{2.1.4}$$

2.2 *Fluctuation-dissipation theorem.*

Consider the Ising model with N spins. The susceptibility per spin, χ, is related to the variance of the total magnetisation by

$$k_B T \chi = \frac{1}{N} \left(\langle M^2 \rangle - \langle M \rangle^2 \right), \tag{2.2.1}$$

as proved in Section 2.1.2. Using a similar strategy, prove that the specific heat c is related to the variance of the total energy by

$$k_B T^2 c = \frac{1}{N} \left(\langle E^2 \rangle - \langle E \rangle^2 \right). \tag{2.2.2}$$

2.3 *Eigenvalues, eigenvectors and diagonalisation.*

Let f be a linear function from $\mathcal{R}^n$ into $\mathcal{R}^n$. If there is a non-zero vector $\mathbf{x} \in \mathcal{R}^n$ and a number λ such that $f(\mathbf{x}) = \lambda \mathbf{x}$, then $\mathbf{x}$ is called an eigenvector of the linear function f, and λ is called its associated eigenvalue.

(i) Show that if $\mathbf{x}$ is an eigenvector for f so is $\alpha \mathbf{x}$ for any number $\alpha \neq 0$.

(ii) Let **A** be the associated matrix for the linear function f. Argue that the eigenvalues for the function f are the solutions to the characteristic equation

$$\det(\mathbf{A} - \lambda \mathbf{I}) = 0. \qquad (2.3.1)$$

(iii) Assume that f is symmetric, that is,

$$f(\mathbf{x}) \cdot \mathbf{y} = \mathbf{x} \cdot f(\mathbf{y}) \quad \text{for all } \mathbf{x}, \mathbf{y} \in \mathcal{R}^n, \qquad (2.3.2)$$

or, if **A** is the associated symmetric matrix,

$$(\mathbf{A}\mathbf{x}) \cdot \mathbf{y} = \mathbf{x} \cdot (\mathbf{A}\mathbf{y}) \quad \text{for all } \mathbf{x}, \mathbf{y} \in \mathcal{R}^n. \qquad (2.3.3)$$

Show that if $\mathbf{x}_1$ and $\mathbf{x}_2$ are eigenvectors of f corresponding to distinct eigenvalues λ_1 and λ_2, the $\mathbf{x}_1$ and $\mathbf{x}_2$ are orthogonal, that is, $\mathbf{x}_1 \cdot \mathbf{x}_2 = 0$.

(iv) Consider the real and symmetric 2×2 matrix **T** in Section 2.4, Equation (2.61) on page 141.
  (a) Find the eigenvalues $\lambda_\pm$ for **T**.
  (b) Find explicitly two associated eigenvectors for **T** and check that they are orthogonal.
  (c) Determine explicitly the matrix **U** such that

$$\mathbf{U}^{-1}\mathbf{T}\mathbf{U} = \begin{pmatrix} \lambda_+ & 0 \\ 0 & \lambda_- \end{pmatrix}. \qquad (2.3.4)$$

2.4 *Critical exponents inequality.*

Given the thermodynamic relation

$$\chi(C_H - C_M) = T \left(\frac{\partial \langle M \rangle}{\partial T} \right)_H^2, \qquad (2.4.1)$$

show that as $C_M \geq 0$ it follows generally that the critical exponents satisfy the inequality

$$\alpha + 2\beta + \gamma \geq 2. \qquad (2.4.2)$$

2.5 *The spin-spin correlation function and scaling relations.*

The spin-spin correlation function

$$g(\mathbf{r}_i, \mathbf{r}_j) = \langle (s_i - \langle s_i \rangle)(s_j - \langle s_j \rangle) \rangle, \qquad (2.5.1)$$

measures the correlations in the fluctuations of spins s_i and s_j at positions $\mathbf{r}_i$ and $\mathbf{r}_j$ around their average values $\langle s_i \rangle$ and $\langle s_j \rangle$.

(i) Show that

$$g(\mathbf{r}_i, \mathbf{r}_j) = \langle s_i s_j \rangle - \langle s_i \rangle \langle s_j \rangle. \tag{2.5.2}$$

(ii) Assume the system is translationally invariant, that is, $\langle s_i \rangle = \langle s_j \rangle = m$. Discuss why this implies that the correlation function can only be a function of the relative distance, that is, $g(\mathbf{r}_i, \mathbf{r}_j) = g(|\mathbf{r}_i - \mathbf{r}_j|)$.

(iii) (a) Discuss the behaviour of the correlation function $g(\mathbf{r}_i, \mathbf{r}_j)$ in the limit $|\mathbf{r}_i - \mathbf{r}_j| \to \infty$ assuming $T \neq T_c$ and zero external field.

(b) Discuss the behaviour of the correlation function $g(\mathbf{r}_i, \mathbf{r}_i)$ of spin i with itself (i.e., in the limit $|\mathbf{r}_i - \mathbf{r}_j| \to 0$) as a function of temperature in zero external field.

(c) Discuss the behaviour of the correlation function $g(\mathbf{r}_i, \mathbf{r}_j)$ in the limits of $J/k_B T \ll 1$ and $J/k_B T \gg 1$.

The volume integral of the correlation function is related to the susceptibility per spin χ, by

$$k_B T \chi = \sum_j g(\mathbf{r}_i, \mathbf{r}_j) \propto \int_V g(\mathbf{r}_i, \mathbf{r}_j) d\mathbf{r}_j. \tag{2.5.3}$$

(iv) Convince yourself that the divergence of the susceptibility per spin at the critical point $(T, H) = (T_c, 0)$ is achieved with an algebraically decaying correlation function

$$g(\mathbf{r}) \propto |\mathbf{r}|^{-(d-2+\eta)} \quad \text{for } T = T_c, |\mathbf{r}| \gg 1, \tag{2.5.4}$$

defining a new critical exponent, η, and prove that $\eta \leq 2$.
Hint: Assume Equation (2.5.4) is valid for all r and recall that for a function depending only on the distance $r = |\mathbf{r}|$ but not the direction $\int_V f(\mathbf{r}) d^d\mathbf{r} \propto \int_0^\infty f(r) r^{d-1} dr$ in d dimensions.

For $T \neq T_c$, the correlation function will have a cutoff, defining implicitly the correlation length ξ, by

$$g(\mathbf{r}) \propto r^{-(d-2+\eta)} \mathcal{G}_\pm(r/\xi) \quad \text{for } T \to T_c^\pm, r \gg 1, \tag{2.5.5a}$$
$$\xi \propto |T_c - T|^{-\nu} \quad \text{for } T \to T_c, H = 0, \tag{2.5.5b}$$

and the scaling functions

$$\mathcal{G}_\pm(x) = \begin{cases} \text{constant} & \text{for } x \ll 1 \\ \text{decays rapidly} & \text{for } x \gg 1. \end{cases} \quad (2.5.6)$$

(v) (a) Sketch the correlation length as a function of temperature T, and discuss the physical interpretation of the correlation length ξ.

(b) Use Equation (2.5.3) and assume Equation (2.5.5) is valid for all r to show the scaling relation

$$\gamma = \nu(2 - \eta). \quad (2.5.7)$$

(c) Prove that

$$d - 2 + \eta = 2\beta/\nu. \quad (2.5.8)$$

Hint: Assume $T \leq T_c$. Define $\tilde{g}(r) = g(r) + m^2$ and consider the limit $r \to \infty$, that is, $r \gg \xi$.

2.6 *Diluted Ising model.*

Consider the diluted Ising model in zero external field with the energy

$$E_{\{s_i\}} = -\sum_{\langle i,j \rangle} J_{ij} s_i s_j, \quad (2.6.1)$$

where $s_i = \pm 1$ is the spin at lattice position i, the sum runs over different pairs of nearest neighbour sites, and the coupling constants

$$J_{ij} = \begin{cases} J > 0 & \text{with probability } p \\ 0 & \text{with probability } (1 - p). \end{cases} \quad (2.6.2)$$

(i) Discuss how this problem is related to the percolation.
(ii) In the following, assume the temperature $T = 0$.

(a) What is the ground state of the diluted Ising model?
(b) Show that

$$\langle s_i s_j \rangle = \begin{cases} 1 & i,j \text{ in the same percolation cluster} \\ 0 & \text{otherwise}. \end{cases} \quad (2.6.3)$$

(c) Argue why the average magnetisation per spin $m_0(p) = 0$ for $p \leq p_c$. Based on your knowledge of percolation, find an expression for $m_0(p)$ when $p > p_c$.

You may assume that for small non-zero external field H and low temperatures $k_B T \ll J$, the magnetisation per spin

$$m(p, H) = \pm P_\infty(p) + \sum_{s=1}^{\infty} sn(s,p) \tanh(sH/k_B T), \tag{2.6.4}$$

where $n(s, p)$ is the cluster number density.

(iii) (a) What does the term $P_\infty(p)$ represent? Find the magnetisation $m_0(p)$ in the limit of $H \to 0$.
(b) Define the susceptibility χ_T. Assuming $H \ll k_B T$, show that the susceptibility diverges when $p \to p_c$.

(iv) A version of the fluctuation-dissipation theorem states that

$$\chi_T = \frac{1}{k_B T} \sum_i \sum_j (\langle s_i s_j \rangle - \langle s_i \rangle \langle s_j \rangle) \tag{2.6.5}$$

where $\langle s_i \rangle = \langle s_j \rangle = m_0$. Assume p approaches p_c from below and $k_B T \ll J$. Calculate the susceptibility using this formula and show it is consistent with the result derived in (iii)(b).

2.7 *Second-order phase transition in a mass-spring system: Landau theory.*

A rigid massless rod of length a can rotate around a fixed point $\mathcal{O}$ in the vertical plane only. The orientation of the rod is given by its angle θ to be measured positive clockwise from the vertical. At the top of the rod is placed a *variable* mass m which is linked to a circular harmonic spring of radius a and spring constant k. When the rod is vertical, the length of the spring equals its natural length, $\pi a/2$ [Sivardière, 1997].

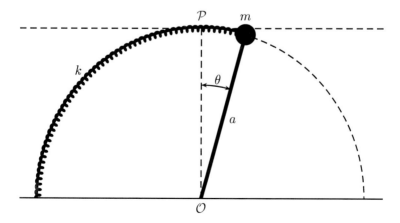

(i) Show that the total energy of the mass-spring system is

$$U(\theta) = \frac{1}{2}ka^2\theta^2 + mga(\cos\theta - 1), \quad (2.7.1)$$

where the zeroth-level of the gravitational potential energy is defined as the horizontal dashed line passing through the point $\mathcal{P}$, the position of the mass when the rod is vertical.

(ii) (a) Expand the function $U(\theta)$ in Equation (2.7.1) around $\theta = 0$ up to fourth order, to show that

$$U(\theta) = \frac{a}{2}(ka - mg)\theta^2 + \frac{mga}{24}\theta^4. \quad (2.7.2)$$

(b) Explain why only terms of even order appear in the expansion in Equation (2.7.2).

(c) Sketch the function $U(\theta)$ in Equation (2.7.2) for $ka > mg$, $ka = mg$, and $ka < mg$.

(d) Using Equation (2.7.2), find an explicit expression for the angle of equilibrium $\theta_0(m)$ when $ka > mg$ and $ka < mg$.

(e) Sketch the solution of the angle of equilibrium $\theta_0(m)$ as a function of the ratio ka/mg. Relate the graph to the sketches from question (c).

(f) Briefly outline the Landau theory of second-order phase transitions in general.

(g) What is the order parameter of the mass-spring system? What is the critical value m_c of the variable mass m? Explain your answers.

2.8 *Scaling ansatz of free energy and scaling relations.*

Consider the Ising model on a d-dimensional lattice in an external field H.

(i) (a) Write down the energy $E_{\{s_i\}}$ for the Ising model. Clearly identify all symbols.

(b) Identify the order parameter for the Ising model and discuss qualitatively its behaviour as a function of temperature T in zero external field.

Let $t = (T - T_c)/T_c$ and $h = H/k_BT$ denote the reduced temperature and external field, respectively. Assume that the singular part of free energy per spin is a generalised homogeneous function,

$$f(t,h) = b^{-d}f(b^{y_t}t, b^{y_h}h) \quad \text{for } t \to 0^\pm, h \to 0, b > 0, \quad (2.8.1)$$

where d is the dimension and y_t, y_h are positive exponents.

(ii) (a) Define the critical exponent α associated with the specific heat in zero external field and show that Equation (2.8.1) implies

$$\alpha = \frac{2y_t - d}{y_t}. \qquad (2.8.2)$$

(b) Define the critical exponent β associated with the order parameter in zero external field and show that Equation (2.8.1) implies

$$\beta = \frac{d - y_h}{y_t}. \qquad (2.8.3)$$

(c) Define the critical exponent γ associated with the susceptibility in zero external field and show that Equation (2.8.1) implies

$$\gamma = \frac{2y_h - d}{y_t}. \qquad (2.8.4)$$

(d) Define the critical exponent δ associated with the order parameter at the critical temperature and show that Equation (2.8.1) implies

$$\delta = \frac{y_h}{d - y_h}. \qquad (2.8.5)$$

(e) Hence confirm the two scaling relations

$$\alpha + 2\beta + \gamma = 2, \qquad (2.8.6\text{a})$$
$$\gamma = \beta(\delta - 1). \qquad (2.8.6\text{b})$$

Chapter 3

Self-Organised Criticality

3.1 Introduction

In the Ising model we can, in principle, calculate any thermodynamic quantity from the partition function or, equivalently, from the free energy using the relations in Table 2.2. This prescription applies to any system in equilibrium. For a system to reach equilibrium, it must be completely isolated from external influences (thermal, mechanical, chemical, etc.) such that no external flux of mass or energy can pass through it. Such contrived conditions can be met under laboratory conditions but almost never in Nature.

There are no truly isolated natural systems and most systems have an external flux of mass or energy passing through them. For example, the Earth system receives a constant flux of energy from the Sun in the form of solar radiation. This externally injected energy is harnessed by the Earth's biosphere and heats up the land, oceans and atmosphere. Some of the incoming flux of energy is intermediately stored, for example, in the form of chemical energy or the latent heat of condensation of evaporated water in the atmosphere, before eventually being dissipated and lost to space in the form of long wavelength radiation. Since the temperature of the Earth is approximately constant, the average amount of energy influx from the Sun to the Earth equals the average amount of energy outflux from the Earth to space. The Earth is therefore not in equilibrium but rather in a so-called non-equilibrium steady state.

We must take great care not to apply results from equilibrium systems outside their range of validity. For example, an equilibrium system that is not at a critical point is linear in the sense that its response is always proportional to the external perturbation. This may no longer hold for a non-equilibrium steady state system, as we shall soon discover.

There are many examples in Nature of systems in a non-equilibrium steady state that resemble, in some respects, equilibrium systems poised at a critical point. For example, non-equilibrium systems might respond to an external perturbation with events of all sizes and no apparent characteristic scale. The objective of the science of 'complexity' is to address the mechanisms that may give rise to scale-free behaviour in Nature. Are there any generic mechanisms for scale-free behaviour in non-equilibrium systems that, contrary to equilibrium systems, do not rely on an outside experimenter to carefully fine-tune any control parameters to a critical point?

3.1.1 Sandpile metaphor

One paradigm that has been proposed as a generic mechanism for scale invariance in a particular class of non-equilibrium systems is self-organised criticality, introduced in 1987 by Bak, Tang and Wiesenfeld [Bak *et al.*, 1987]. A sandpile metaphor illustrates conceptually the ideas of self-organised criticality, see Figure 3.1.

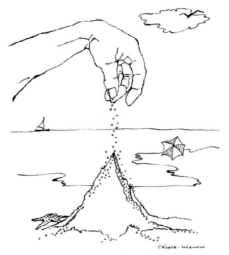

Fig. 3.1 The sandpile metaphor for self-organised criticality. When slowly driven, the pile spontaneously organises itself into a critical state in which the addition of a single grain may trigger an avalanche of any size. Drawing courtesy of Mrs. Elaine K. D. Wiesenfeld, U.S.A.

Imagine a table top onto which we drop grains of sand, one at a time, at random positions. Initially, the grains simply settle on the table in no particular pattern and the pile is approximately flat. Avalanches will be induced by the addition of a grain if the local slope of the pile exceeds a threshold slope determined by the 'stickiness' of the grains. Initially, any avalanches tend to be very small and localised. This is the response that we would expect of an equilibrium system away from a critical point: the response is simply proportional to the external perturbation. However, if we continue to drop grains slowly onto the table, the sandpile grows in size because the pile can maintain a finite slope thanks to the friction between grains. But the sandpile cannot grow indefinitely. The avalanches will ensure that the pile does not become too steep but, because of the friction between the grains, they do not collapse or flatten the pile either. Rather, the avalanches decrease the local slope just below the threshold slope by rearranging the internal configuration of the pile and, once in a while, transport sand off the table. Therefore, even after an avalanche, the pile is still in a highly susceptible state. The slowly driven pile eventually reaches a non-equilibrium steady state where the average influx of grains and energy equals the average outflux of grains and energy. The pile is certainly not in equilibrium.

If the slope of the pile becomes too shallow, the addition of grains will tend to increase it. On the other hand, if the slope becomes too steep, avalanches will tend to decrease it. Therefore, the pile 'self-organises' into a steady state in which its slope fluctuates about a constant angle of repose. The finite size of the pile will necessarily introduce an upper limit to the avalanche size. This so-called cutoff avalanche size will diverge with system size and each added grain may intermittently induce an avalanche of any size, with no typical avalanche size other than that set by the size of the pile. There are many small avalanches, fewer medium avalanches, and very few large avalanches, all obeying a well-defined statistical law spanning many orders of magnitude. Therefore, the response of the pile is no longer restricted to being small and localised. In this sense, the pile is said to be 'critical'.

The avalanche size is a measure of the response of the pile to an external perturbation, and the average avalanche size can be thought of as the susceptibility of the pile. For a pile resting on a finite table, the susceptibility is finite. However, the average distance to the boundary will increase with system size. To maintain a steady state, the added grains must be transported over larger and larger distances to reach the boundaries and therefore the average avalanche size must diverge with system size.

In summary, the slowly driven sandpile spontaneously organises itself into a highly susceptible state where the average response diverges with system size. The slowly driven pile displays self-organised criticality.

The applications of self-organised criticality go well beyond granular piles, but the basic picture remains the same: many slowly driven non-equilibrium systems organise themselves, without any external fine-tuning of control parameters, into a poised state – the critical state – where the susceptibility is very large, diverging with system size.

The seismic system is probably the best candidate for a self-organised critical system in Nature. The slow relative movement of tectonic plates causes stress to build up along the plate boundary. The friction between the plates is finite. Therefore, the strain energy is intermediately stored in the crust of the Earth rather than being relaxed continuously. When the stress exceeds the static friction threshold, an earthquake is triggered that propagates through the crust – much like an avalanche, triggered by a slope exceeding the threshold slope, propagates through the pile. The quakes come in all sizes and obey the same statistical law, the Gutenberg-Richter law, which states that the number of quakes with an energy release larger than $s = 10^m$, where m is the magnitude of an earthquake, decays approximately as $1/s$. That earthquakes obey one and the same statistical law, irrespective of size, strongly suggests that large earthquakes and small earthquakes have one and the same physical origin. Compared to the number of earthquakes larger than, say, magnitude $m = 6$, there are ten times as many earthquakes larger than magnitude $m = 5$, and one hundred times as many earthquakes larger than magnitude $m = 4$, and so on. There are many small earthquakes, fewer medium earthquakes and very few large quakes, but there is no typical size of an earthquake: the response of the crust of the Earth is highly non-linear. Even though larger earthquakes are much less frequent than smaller earthquakes, the average energy release is, in fact, mainly determined by the rare large quakes. Therefore, the crust of the Earth can be viewed as a slowly driven non-equilibrium steady state system with intermittent behaviour that is scale free. The crust of the Earth may be an example of a system displaying self-organised criticality.

If the crust of the Earth could be described by equilibrium statistical mechanics, the response would be proportional to the cause and one would expect to see a Gaussian distribution of earthquake sizes, unless the system is deliberately tuned to a critical point. A Gaussian distribution would imply that a typical size of events exists and that large fluctuations away from this typical event size would happen with exponentially small probability. However, the crust of the Earth is inherently non-equilibrium and cannot be described using established statistical physics for equilibrium systems.

The avalanches of various sizes can be literal, as in the granular pile or the seismic system, or they can be metaphorical, like rainfall in the atmosphere, mass extinctions in biology, or stock market crashes in economics. In all cases, the large events are intrinsic to the dynamics and need no special explanation. The underlying mechanism that caused the small earthquake in San Francisco yesterday also caused the 1906 earthquake of magnitude $m = 8$. Similarly, the mechanism which made the Dow Jones average rise five points yesterday also caused the crash of 1987.

The science of complexity is highly interdisciplinary. It deals with dynamical systems composed of many interacting units, for example grains in granular media, rocks in the crust of the Earth, water droplets in the atmosphere, networks of organisms in biology or agents in economics, etc. We address the question as to whether self-organised criticality is a very general paradigm for the emergence of complexity in Nature. Therefore, in this chapter, we devote our attention to the particular class of non-equilibrium steady state systems that are slowly driven, store mass or energy intermediately, and relax intermittently. Crucial to these systems is the existence of a local threshold, without which the systems could neither store mass nor energy intermediately, nor relax intermittently.

In order to make the concept of self-organised criticality more tangible, Bak, Tang and Wiesenfeld studied a so-called cellular automaton which is discrete in space, time and in its dynamical variables [Bak *et al.*, 1987; Bak *et al.*, 1988]. The configuration of the cellular automaton at a given time is fully specified by the values of the dynamical variables at each site. For example, the dynamical variables could represent the local slopes of a sandpile. An imposed set of local dynamical rules, in the form of an algorithm, defines the model and its evolution. This algorithm is applied *ad infinitum*. Such simple models might be capable of displaying complexity.

We stress that the complexity is not contained in the algorithm itself, but rather it emerges as a result of the repeated local interactions among the dynamical variables in the extended system. Since self-organised criticality attempts to account for complexity in the form of scale invariance, we must not put complexity into the algorithm by hand. Such an approach is in line with 'developing complexity out of simplicity', to use the words of Anderson [Anderson, 1991]. After all, it would be no feat to create a model which displays scale invariance if it is encoded in the algorithm to begin with.

For a lively non-technical overview of self-organised criticality, we can recommend [Bak, 1996], written by one of the key pioneers of the field. A mathematical, succinct and lucid presentation is given by [Jensen, 1998].

We begin with the one-dimensional Bak-Tang-Wiesenfeld (BTW) model of a sandpile which displays self-organised criticality. This allows us to introduce all the main ideas of self-organised criticality in a simple setting. This model has the advantage of being exactly solvable and we can identify analogies with percolation and the Ising model in one dimension. Further insight into self-organised criticality is gained by considering a BTW-like mean-field model where we can reveal the mechanism of self-organisation and the emergence of scale invariance in the avalanche-size probability. By visualising an avalanche as a branching process, we can determine the avalanche-size probability explicitly and, at the same time, make connections to percolation on a Bethe lattice. Armed with the exact solutions for the one-dimensional and mean-field BTW model, we propose a general scaling ansatz for the avalanche-size probability and derive scaling relations among the critical exponents.

We then begin a dialogue between modelling and the real world, starting gently with an experiment of avalanches in a slowly driven one-dimensional ricepile [Frette et al., 1996]. The outcome of the experiment indicates that scale invariance in the avalanche-size probability density can be observed in real granular media. We can find an appropriate algorithm to describe the evolution of the slowly driven pile by adapting the one-dimensional BTW model into the Oslo model [Christensen et al., 1996]. The Oslo model is rich and non-trivial in its behaviour, and its avalanche-size probability is consistent with the proposed general framework for scaling and data collapse.

Next, we show that the frequency-size distribution of earthquakes provides evidence that the crust of the Earth is in a non-equilibrium critical state. The essential features of the earthquake mechanism, the slow continuous drive of the tectonic plates and their stick-slip behaviour, is captured in the Olami-Feder-Christensen (OFC) model, which is discrete in space, but continuous in time and its dynamical variables [Olami et al., 1992]. Moreover, contrary to the BTW model, the OFC model is inherently non-conservative and driven globally rather than locally. We show that the OFC model is able to account for the observed Gutenberg-Richter law.

Finally, we consider the atmosphere as another example of a slowly driven non-equilibrium system that naturally enters into a steady state, and we show that rain 'events' display scale invariance [Peters et al., 2002].

We close the chapter by summarising the analogies between avalanches in the ricepile, earthquakes in the crust of the Earth, and rain events in the atmosphere.

3.2 BTW Model in $d = 1$

The dynamical algorithm for the one-dimensional BTW model, at least intuitively, resembles the dynamics of a slowly driven sandpile: a grain topples from a site to its downward nearest neighbour when the dynamical variable, representing the local slope, exceeds a threshold slope [Bak et al., 1987; Bak et al., 1988]. Consider a one-dimensional lattice composed of L sites, $i = 1, 2, \ldots, L$. The system is sealed at the left boundary next to site $i = 1$, preventing grains from escaping. However, the system remains open at the right boundary next to site $i = L$, where grains can drop out. The number of grains at site i is referred to as the height, h_i, while the slope at site i is defined by $z_i = h_i - h_{i+1}$, with the convention that $h_{L+1} = 0$. Each site is assigned the same threshold slope, z^{th}. The system is driven by adding a grain to a random site, $h_i \to h_i + 1$. If the slope at site i is greater than the threshold slope, $z_i > z^{\text{th}}$, a grain topples from site i to site $i + 1$, that is, $h_i \to h_i - 1$ and $h_{i+1} \to h_{i+1} + 1$. As a result of the initial toppling, some sites may have slopes greater than the threshold slope and topple in turn, causing an avalanche to propagate until a stable configuration with $z_i \leq z^{\text{th}}$ for all i is reached, see Figure 3.2(a).

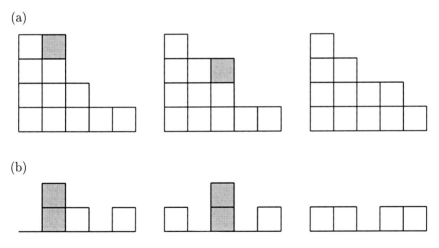

Fig. 3.2 A realisation of an avalanche of size $s = 2$ in the one-dimensional BTW model on a lattice of size $L = 5$. A grain at site i topples to site $i + 1$ when the slope is greater than the threshold slope $z^{\text{th}} = 1$. A grain is added at site $i = 2$. The slope of site $i = 2$ exceeds the threshold slope and an avalanche is initiated. The avalanche eventually terminates when the toppling grain reaches site $i = 4$. (a) The drive and relaxation in terms of grains. Toppling grains are shown in grey. (b) The drive and relaxation in terms of slope units. Toppling slope units are shown in grey.

When a grain topples from site i to site $i+1$, one slope unit moves to each of the two nearest neighbours, that is, $z_i \to z_i - 2$ and $z_{i\pm 1} \to z_{i\pm 1} + 1$, see Figure 3.2(b).

Special care is required at the boundaries where sites only have one neighbour. When site $i = 1$ topples, $z_1 \to z_1 - 2$ and $z_2 \to z_2 + 1$, and when site $i = L$ topples, $z_L \to z_L - 1$ and $z_{L-1} \to z_{L-1} + 1$. In this perspective, the left boundary is open to slope units while the right boundary is sealed.

3.2.1 Algorithm of the BTW model in $d = 1$

We summarise the dynamics of the one-dimensional BTW model in terms of the slope units with the following algorithm:

1. *Initialisation.* Prepare the system in an arbitrary stable configuration with $z_i \leq z^{\text{th}}$ for all i.
2. *Drive.* Add a grain at random site i.
   For $i = 1$:
   $$z_1 \to z_1 + 1. \tag{3.1a}$$
   For $i = 2, \ldots, L$:
   $$z_i \to z_i + 1,$$
   $$z_{i-1} \to z_{i-1} - 1. \tag{3.1b}$$
3. *Relaxation.* If $z_i > z^{\text{th}}$, relax site i.
   For $i = 1$:
   $$z_1 \to z_1 - 2,$$
   $$z_2 \to z_2 + 1. \tag{3.2a}$$
   For $i = 2, \ldots, L - 1$:
   $$z_i \to z_i - 2,$$
   $$z_{i\pm 1} \to z_{i\pm 1} + 1. \tag{3.2b}$$
   For $i = L$:
   $$z_L \to z_L - 1,$$
   $$z_{L-1} \to z_{L-1} + 1. \tag{3.2c}$$
   Continue relaxing sites until $z_i \leq z^{\text{th}}$ for all i.
4. *Iteration.* Return to **2**.

The value of the threshold slope is irrelevant for the behaviour of the model and we may choose $z^{\text{th}} = 1$ for simplicity.

Grains are only added when the system is in a stable configuration where no sites have slope greater than the threshold slope. This amounts to a separation of time scales. The response time of the system (the duration of the avalanche) is instantaneous compared to the time between additions.

Physically, the rules of the one-dimensional BTW model describe some of the features of a slowly driven friction-dominated granular pile: adding a grain to the pile is equivalent to adding potential energy. When the friction between grains is overcome, an avalanche is triggered. During an avalanche, potential energy is converted into kinetic energy which is eventually dissipated as heat and sound because of the friction between grains.

The 'stickiness' of grains is captured in the one-dimensional BTW model by using a non-zero threshold slope, which allows a pile to build up. With zero threshold slope, it is impossible to build up a pile because the grains would flow out of the system like a fluid. Therefore, the non-zero threshold slope affords the existences of stable configurations and avalanches.

3.2.2 Transient and recurrent configurations

The number of stable configurations in a system with L sites is finite, but not all of them are accessible after a sufficiently large number of iterations. For example, a system initialised in the empty configuration with $z_i = 0$ for all i will, after one iteration, never return to this stable configuration. We can disjointly divide the set of stable configurations, $\mathcal{S}$, into a set of transient configurations, $\mathcal{T}$, and a set of recurrent configurations, $\mathcal{R}$. By definition, transient configurations are encountered at most once and are not accessible once the system has entered the set $\mathcal{R}$ of recurrent configurations. By contrast, recurrent configurations are revisited indefinitely. The set of recurrent configurations may be thought of as the attractor of the dynamics in the L-dimensional space of all configurations.

When adding a grain to a stable configuration $\mathcal{S}_j$, the system evolves into another stable configuration $\mathcal{S}_{j+1}$. Symbolically, we write $\mathcal{S}_j \mapsto \mathcal{S}_{j+1}$, where the arrow is a shorthand notation for the operation of adding a grain and, if necessary, relaxing the system until it reaches a stable configuration. Therefore, the index j counts the number of grains added and is associated with the slow timescale of the system. For example, after preparing the system in the transient empty configuration, the system passes through a series of transient configurations before entering the set of recurrent configurations after n additions,

$$\underbrace{\mathcal{T}_0 \mapsto \mathcal{T}_1 \mapsto \cdots \mapsto \mathcal{T}_{n-1}}_{\text{transient configurations}} \mapsto \underbrace{\mathcal{R}_n \mapsto \mathcal{R}_{n+1} \mapsto \cdots}_{\text{recurrent configurations}}. \qquad (3.3)$$

For the one-dimensional BTW model, the set of recurrent configurations is particularly simple, since it contains only one configuration, namely the stable configuration, $z_i = z^{\text{th}}$ for all i, see Figure 3.3(a).

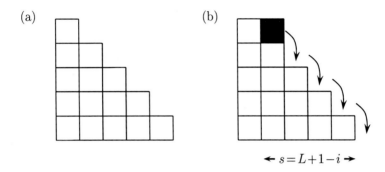

Fig. 3.3 (a) The unique recurrent configuration for the one-dimensional BTW model on a lattice of size $L = 5$ with threshold slope $z^{\text{th}} = 1$. All sites have slope $z_i = z^{\text{th}}$. (b) Adding a grain at site $i = 2$ triggers an avalanche. The grain will tumble down the 'staircase' and drop out of the system after $s = L + 1 - i = 4$ topplings. The avalanche size s is a measure of the dissipated potential energy.

Any grains added to the system in this 'staircase' recurrent configuration will simply topple to the right and drop out of the system at the open boundary, see Figure 3.3(b). Therefore, when the system has entered the set of recurrent configurations, the average number of grains added to the system equals the average number of grains leaving the system. We say that the system is in a steady state where

$$\langle \text{influx} \rangle = \langle \text{outflux} \rangle. \qquad (3.4)$$

In terms of energy, the average energy added equals the average energy dissipated. A measure of the energy dissipated by an avalanche is the total number of topplings, which is defined as the avalanche size, s.

Grains are, on average, added in the middle of the pile. When the system has entered the 'staircase' recurrent configuration, grains will, on average, take $(L + 1)/2$ steps before leaving the system. Therefore, the average avalanche size, $\langle s \rangle$, is proportional to the system size.

3.2.3 Avalanche time series

When the system is in the recurrent configuration, a grain added at site i will trigger an avalanche of size $s = L + 1 - i$, see Figure 3.3(b). Therefore, avalanche sizes range from $s = L$, when adding a grain at site $i = 1$, to $s = 1$, when adding a grain at site $i = L$. The largest possible avalanche size $s_{\max} = L$ scales linearly with system size. However, the finite size of the system introduces an upper cutoff of the avalanche size. Since the grains are added at random positions, all avalanche sizes $1 \leq s \leq s_{\max}$ occur with equal probability.

Figure 3.4 displays a time series of avalanches in the one-dimensional BTW model on a lattice of size $L = 256$. The time, t, is measured in the number of additions after the system has reached the recurrent configuration. An avalanche of size s is rescaled by the size of the largest observed avalanche in the time series, $s_{\max}$, and appears as a spike of height, $s/s_{\max}$. We will focus mainly on the statistics of avalanche sizes derived from such time series.

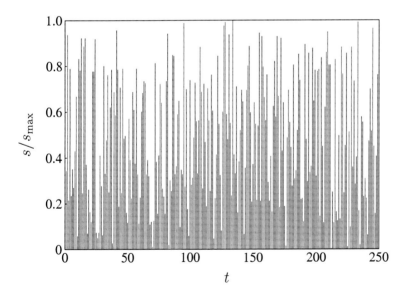

Fig. 3.4 Numerical results for the rescaled avalanche size, $s/s_{\max}$, versus the time, t, measured in number of additions for the one-dimensional BTW model in the recurrent configuration on a lattice of size $L = 256$. The largest observed avalanche is of size $s_{\max} = L$. The avalanche sizes are equally probable and span over approximately two orders of magnitude.

3.2.4 Avalanche-size probability

The probability of observing an avalanche of size s in a system of size L is denoted by $P(s; L)$. In the recurrent configuration, all avalanche sizes $s = 1, \ldots, L$ are equally probable so that

$$P(s; L) = \begin{cases} 1/L & \text{for } 1 \leq s \leq L \\ 0 & \text{otherwise,} \end{cases} \quad (3.5)$$

where the avalanche-size probability is normalised:

$$\sum_{s=1}^{\infty} P(s; L) = 1. \quad (3.6)$$

Figure 3.5 displays the avalanche-size probability $P(s; L)$ versus the avalanche size s for the one-dimensional BTW model on lattices of size $L = 64, 128, 256$.

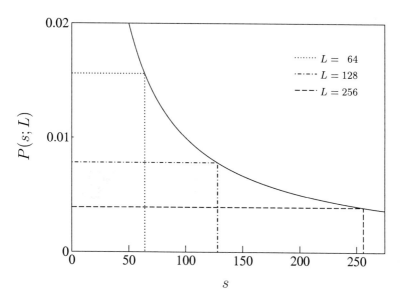

Fig. 3.5 Exact solution of the avalanche-size probability, $P(s; L)$, versus the avalanche size, s, for the one-dimensional BTW model on lattices of size $L = 64, 128, 256$ marked with lines of increasing dash length. All avalanche sizes $1 \leq s \leq s_{\max}$ are equally probable. The cutoff avalanche size $s_c = L$, which, in this case, is the largest avalanche possible. The probability of the distinctive cutoff avalanche size decays like s^{-1} (solid line).

It is possible to rewrite the avalanche-size probability in a scaling form. Introducing the Heaviside step function

$$\Theta(x) = \begin{cases} 1 & \text{for } x \geq 0 \\ 0 & \text{otherwise,} \end{cases} \quad (3.7)$$

we may write for $s = 1, 2, \ldots$

$$\begin{aligned} P(s; L) &= \frac{1}{L} \Theta(1 - s/L) \\ &= s^{-1} \frac{s}{L} \Theta(1 - s/L) \\ &= s^{-1} \mathcal{G}_{\text{1d}}^{\text{BTW}}(s/s_c), \end{aligned} \quad (3.8)$$

where, in the last step, we have introduced the cutoff avalanche size, $s_c(L) = L$, and the scaling function for the avalanche-size probability

$$\mathcal{G}_{\text{1d}}^{\text{BTW}}(x) = x\Theta(1 - x) \quad (3.9)$$

which is a function of the rescaled avalanche size $x = s/s_c$.

Therefore, the avalanche-size probability for the one-dimensional BTW model on a lattice of size L satisfies the scaling form

$$P(s; L) = s^{-1} \mathcal{G}_{\text{1d}}^{\text{BTW}}(s/s_c), \quad (3.10\text{a})$$
$$s_c(L) = L, \quad (3.10\text{b})$$

where the cutoff avalanche size $s_c(L)$ diverges linearly with system size.

The scaling form in Equation (3.10) for the avalanche-size probability allows for a data collapse since it can be rearranged as

$$sP(s; L) = \mathcal{G}_{\text{1d}}^{\text{BTW}}(s/s_c). \quad (3.11)$$

Although the left-hand side of Equation (3.11) is a function of two variables, s and L, the right-hand side is only a function of one rescaled variable, s/s_c. A data collapse for different system sizes is achieved by plotting the rescaled avalanche-size probability $sP(s; L)$ versus the rescaled avalanche size s/s_c with the cutoff avalanche size $s_c(L) = L$, see Figure 3.6.

The reader may recognise Equation (3.10) as a finite-size scaling ansatz. However, unlike percolation and the Ising model, the BTW model organises itself into a critical state where the avalanche-size probability obeys finite-size scaling, without the fine-tuning of any control parameters. The model is said to display self-organised criticality. The behaviour of the

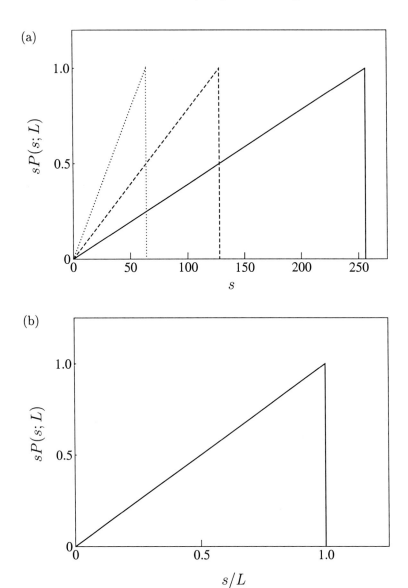

Fig. 3.6 Data collapse of the exact solution of the avalanche-size probabilities for the one-dimensional BTW model on lattices of size $L = 64, 128, 256$ marked with lines of increasing dash length. (a) The transformed avalanche-size probability, $sP(s;L)$, versus the avalanche size, s. (b) The transformed avalanche-size probability, $sP(s;L)$, versus the rescaled avalanche size, s/s_c, where the cutoff avalanche size $s_c(L) = L$. The curves collapse onto the graph for the scaling function $\mathcal{G}_{\text{1d}}^{\text{BTW}}(x) = x\Theta(1-x)$.

one-dimensional BTW model is similar, in some respects, to that of equilibrium critical systems in the lower critical dimension. For example, in one-dimensional percolation the critical occupation probability $p_c = 1$, and there is only one configuration where all sites are occupied. In the one-dimensional Ising model the critical point $(T_c, H_c) = (0, 0)$, and all spins are aligned. Thus all three models have trivial critical configurations. Despite their simplicity, these one-dimensional models also fit into the general framework of scaling and data collapse applicable in higher dimensions.

The scaling form in Equation (3.10) for the avalanche-size probability provides complete statistical information about the avalanche behaviour in the one-dimensional BTW model. We may regard avalanches as the system's response to external perturbations. Therefore, the average avalanche size affords a measure of the system's susceptibility to the external perturbation of adding a single grain. The average avalanche size $\langle s \rangle$ is the first moment of the avalanche-size probability and can be calculated explicitly:

$$\begin{aligned} \langle s \rangle &= \sum_{s=1}^{\infty} s P(s; L) \\ &= \frac{1}{L} \sum_{s=1}^{L} s \\ &= \frac{L+1}{2}. \end{aligned} \quad (3.12)$$

The average avalanche size is linear in L and diverges as the system size tends to infinity. We can generalise this calculation to the kth moment of the avalanche-size probability:

$$\begin{aligned} \langle s^k \rangle &= \sum_{s=1}^{\infty} s^k P(s; L) \\ &= \frac{1}{L} \sum_{s=1}^{L} s^k \\ &\approx \frac{1}{L} \int_1^L s^k ds \quad \text{for } L \gg 1 \\ &\propto \frac{L^k}{1+k} \quad \text{for } L \gg 1. \end{aligned} \quad (3.13)$$

The kth moment of the avalanche-size probability scales like L^k for lattices of size $L \gg 1$. Indeed, if the sum in Equation (3.13) is calculated explicitly, we find that the leading order has a correction term that vanishes like $1/L$.

3.3 Mean-Field Theory of the BTW Model

3.3.1 Random neighbour BTW model

It is instructive to examine a BTW-like mean-field model of spatially uncorrelated sites in which we are able to pinpoint the process of self-organisation and the emergence of scale invariance [Christensen and Olami, 1993]. The simplicity of the model allows us to calculate various quantities analytically such as the average avalanche size and the so-called branching ratio.

Consider N sites labelled $i = 1, 2, \ldots, N$, each assigned an integer variable z_i with a threshold of zth. The system is driven by adding a unit to a random site, $z_i \to z_i + 1$. If the dynamical variable at site i is greater than the threshold, $z_i > z^{\text{th}}$, the site relaxes by an amount z^{th}, that is, $z_i \to z_i - z^{\text{th}}$. We introduce a discrete parameter $\alpha_n = n/z^{\text{th}}, n = 0, 1, \ldots, z^{\text{th}}$ such that one unit is redistributed to each of $n = \alpha_n z^{\text{th}}$ different random 'neighbour' sites, with the remainder $z^{\text{th}} - n = (1 - \alpha_n)z^{\text{th}}$ being dissipated. As a result, some of the random 'neighbour' sites that have received a unit may exceed the threshold and topple in turn, causing an avalanche to propagate until a stable configuration with $z_i \leq z^{\text{th}}$ for all i is reached. For simplicity, new random 'neighbour' sites are chosen each time a site relaxes as opposed to being fixed throughout the dynamics.

3.3.2 Algorithm of the random neighbour BTW model

The algorithm of the random neighbour BTW model is defined by:

1. *Initialisation.* Prepare the system in an arbitrary stable configuration[1] with $1 \leq z_i \leq z^{\text{th}}$ for all i.
2. *Drive.* Add a unit at random site i.

$$z_i \to z_i + 1. \qquad (3.14)$$

3. *Relaxation.* If $z_i > z^{\text{th}}$, relax site i.

$$\begin{aligned} z_i &\to z_i - z^{\text{th}}, \\ z_{j_k} &\to z_{j_k} + 1 \quad \text{for } k = 1, \ldots, \alpha_n z^{\text{th}}, \end{aligned} \qquad (3.15)$$

where the sites j_k are chosen randomly. Continue relaxing sites until $z_i \leq z^{\text{th}}$ for all i.
4. *Iteration.* Return to **2**.

---
[1] For ease of notation in the following, the minimum dynamical variable is set to one.

3.3.3 Steady state and the average avalanche size

The number of stable configurations $N_S = \left(z^{\text{th}}\right)^N$, since each of the N sites can take z^{th} possible stable values. However, not all of these configurations are recurrent. After a transient period, the system self-organises into the set of recurrent configurations where the average dynamical variable fluctuates around a constant value. We can identify this constant as well as the mechanism of self-organisation by considering the dynamics of P_z, the probability that a site contains z units.

The drive and the relaxation steps add one unit of the dynamical variable to randomly chosen sites. Therefore, the flow of the dynamical variable into z is proportional to the probability P_{z-1} that a randomly chosen site contains $z-1$ units. Similarly, the flow of the dynamical variable out of z is proportional to P_z.

Assume that $P_{z-1} > P_z$ for some stable configuration. The flow into z exceeds the flow out of z, thereby increasing P_z until $P_{z-1} = P_z$. Assume that $P_{z-1} < P_z$ for some stable configuration. The flow out of z exceeds the flow into z, thereby decreasing P_z until $P_{z-1} = P_z$. Therefore, the system will settle into a steady state where, on average, $P_{z-1} = P_z$, in which the flow into z equals the flow out of z.

Since the probability that a site contains z units is normalised,

$$P_z = \frac{1}{z^{\text{th}}} \quad \text{for } z = 1, \ldots, z^{\text{th}}, \tag{3.16}$$

and we find that the dynamical variable fluctuates about the average value

$$\langle z \rangle = \sum_{z=1}^{z^{\text{th}}} z P_z = \frac{z^{\text{th}} + 1}{2}. \tag{3.17}$$

The drive increases the dynamical variable of randomly chosen sites by one unit until an avalanche is initiated when a site is chosen with z^{th} units. The probability of initiating an avalanche is therefore $P_{z^{\text{th}}}$, so that the total average influx in between avalanches is

$$\langle \text{influx} \rangle = \frac{1}{P_{z^{\text{th}}}} = z^{\text{th}}. \tag{3.18}$$

A relaxing site decreases by an amount z^{th}, while $\alpha_n z^{\text{th}}$ receiving sites each increase by one unit. The net amount dissipated by a single relaxation is $(1 - \alpha_n) z^{\text{th}}$. Therefore, the total average outflux during avalanches is

$$\langle \text{outflux} \rangle = (1 - \alpha_n) z^{\text{th}} \langle s \rangle. \tag{3.19}$$

Equating the average influx with the average outflux and rearranging for the average avalanche size, we find

$$\langle s \rangle = \frac{1}{1 - \alpha_n} \quad \text{for } \alpha_n < 1. \tag{3.20}$$

The value of $(1 - \alpha_n)$ controls the amount of dissipation when a site relaxes. When the dissipation is non-zero, criticality, in the sense of a diverging average avalanche size, is destroyed, and a cutoff avalanche size exists. Only in the limit of zero dissipation does the average avalanche size diverge, and no cutoff avalanche size exists. When $\alpha_n = 1$, the system is incapable of dissipation and the average value of the dynamical variable can only increase as the system is driven. If $\langle z \rangle$ reaches z^{th}, that is, $z_i = z^{\text{th}}$ for all sites i, the addition of a single unit would initiate an infinite avalanche.

The average avalanche size can be related to the so-called average branching ratio of relaxing sites. A relaxing site adds one unit to randomly chosen sites which may relax in turn. The probability that a relaxation induces b of the $\alpha_n z^{\text{th}}$ randomly chosen sites to relax is given by the binomial distribution

$$\binom{\alpha_n z^{\text{th}}}{b} P_{z^{\text{th}}}^b (1 - P_{z^{\text{th}}})^{\alpha_n z^{\text{th}} - b}, \tag{3.21}$$

since $P_{z^{\text{th}}}$ is the probability that a randomly chosen site relaxes and $(1 - P_{z^{\text{th}}})$ is the probability that it does not. The average number of induced relaxations, also known as the average branching ratio, is therefore

$$\langle b \rangle = \sum_{b=0}^{\alpha_n z^{\text{th}}} b \binom{\alpha_n z^{\text{th}}}{b} P_{z^{\text{th}}}^b (1 - P_{z^{\text{th}}})^{\alpha_n z^{\text{th}} - b} = \alpha_n z^{\text{th}} P_{z^{\text{th}}} = \alpha_n, \tag{3.22}$$

where we have used Equation (3.16) in the final step.

A 'sub-critical' average branching ratio, $\langle b \rangle < 1$, corresponds to a dissipative system with only finite avalanche sizes. A 'critical' average branching ratio, $\langle b \rangle = 1$, corresponds to a non-dissipative system with an infinite average avalanche size. Some avalanches propagate indefinitely where each relaxing site induces, on average, one further site to relax.

Substituting Equation (3.22) into Equation (3.20), we re-express the average avalanche size in terms of the average branching ratio

$$\langle s \rangle = \frac{1}{1 - \langle b \rangle} \quad \text{for } \langle b \rangle < 1. \tag{3.23}$$

We will now re-derive the relation between the average avalanche size and the average branching ratio using the more general framework of a branching process.

3.4 Branching Process

It is instructive to visualise an avalanche as a rooted tree consisting of nodes (relaxing sites) and branches (induced relaxations). Figure 3.7 shows a rooted tree for an avalanche of size $s = 13$. The initial relaxing site forms the root of the tree and has no incoming branch. Sites that do not induce further relaxations have no outgoing branches. The size of the avalanche is the total number of nodes in the tree.

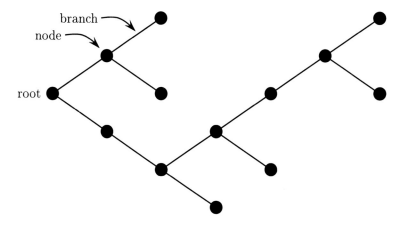

Fig. 3.7 A finite avalanche of size $s = 13$ represented as a rooted tree. The nodes represent relaxing sites. The number of branches leaving a node is the number of induced relaxations. Nodes that do not induce further relaxation have zero outgoing branches. All nodes have exactly one incoming branch except the root of the tree, representing the site where the avalanche is initiated. Therefore, there is a total of $s - 1 = 12$ branches.

3.4.1 *Branching ratio*

Consider a particular finite tree, labelled k, of size s_k. Each of the s_k nodes has exactly one incoming branch, except for the root which has none, so that the number of branches is $(s_k - 1)$. The total number of nodes in an ensemble of finite trees is $\sum_k s_k$, while the total number of branches is $\sum_k (s_k - 1)$, where the sums run over all finite trees. Therefore, the average branching ratio for an ensemble of finite trees is

$$\langle b \rangle = \frac{\sum_k (s_k - 1)}{\sum_k s_k}$$
$$= 1 - \frac{\sum_k 1}{\sum_k s_k}$$
$$= 1 - \frac{1}{\langle s \rangle} \quad \text{for } \langle b \rangle < 1, \tag{3.24}$$

which is equivalent to Equation (3.23) but derived under more general considerations. We have made no reference to the probability with which relaxing sites induce further relaxations, nor whether sites are correlated during an avalanche.

3.4.2 Avalanche-size probability – exact

If we restrict ourselves to an uncorrelated branching process where all nodes induce further relaxations with the same probability, we can deduce a scaling form of the avalanche-size probability.

Given that an avalanche has been initiated, let us calculate the probability of an avalanche of size s. Assume that a node induces a relaxation with probability p, and that it can induce up to $\alpha_n z^{\text{th}}$ relaxations. A tree of size s contains $(s-1)$ out of a possible $s\alpha_n z^{\text{th}}$ branches. Therefore, the number of absent branches (the perimeter) is $t = s(\alpha_n z^{\text{th}} - 1) + 1$, which, by definition, is equal to the number of sites that were not induced to relax. The probability that the root, present with probability one, gives rise to a particular finite tree of size s is then

$$p^{s-1} (1-p)^t. \tag{3.25}$$

The number of trees $g(s,t)$ of size s and perimeter $t = s(\alpha_n z^{\text{th}} - 1) + 1$ depends only on s for a given $\alpha_n z^{\text{th}}$ and can be enumerated exactly [Grimmett, 1999]. In the following, we will use the exact expression for $g(s,t)$ to directly derive the scaling form of the avalanche-size probability. This approach is complementary to that in Section 1.3.5, where we avoided $g(s,t)$ altogether by considering the ratio of the cluster number density at and away from p_c.

For simplicity, we choose the maximum number of outgoing branches $\alpha_n z^{\text{th}} = 2$, implying that the perimeter $t = s+1$. Therefore, the avalanche-size probability

$$P(s,p) = g(s,s+1)\,p^{s-1}\,(1-p)^{s+1}$$
$$= \frac{1}{s+1}\binom{2s}{s} p^{s-1}\,(1-p)^{s+1}$$
$$= \frac{1}{s+1}\binom{2s}{s}\frac{1-p}{p}[p(1-p)]^s \quad \text{for } s \geq 1, \qquad (3.26)$$

where we have used the exact result for the number of trees of size s and perimeter $t = s + 1$. Figure 3.8 shows the graphs of the avalanche-size probability for various values of p approaching p_c from below. Qualitatively, the graphs are reminiscent of a power-law decay with an exponential cutoff that increases with p and diverges at $p = p_c = 1/2$.

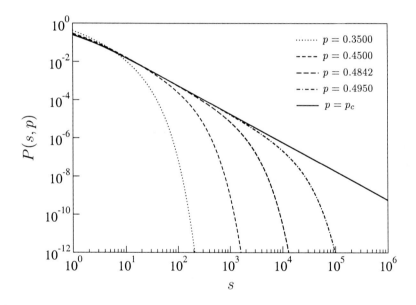

Fig. 3.8 Exact solution of the avalanche-size probability, $P(s,p)$, versus the avalanche size, s, for the branching process with $\alpha_n z^{\text{th}} = 2$ where $p_c = 1/2$. The five curves correspond to $p = 0.35, 0.45, 0.4842, 0.495, p_c$ with cutoff avalanche sizes $s_c(p) \approx 10, 100, 1000, 10\,000, \infty$, respectively. At $p = p_c$, the cutoff avalanche size is infinite and for large avalanche sizes, the avalanche-size probability is well approximated by a power-law decay, $P(s,p_c) \propto s^{-\tau_s}$ for $s \gg 1$, with $\tau_s = 3/2$.

3.4.3 Avalanche-size probability – scaling form

In order to expose the scaling form for the avalanche-size probability for large s, we make use of Stirling's formula $n! \approx \sqrt{2\pi n}\, n^n \exp(-n)$ valid for large n, on the factorials in the binomial coefficient.

$$g(s, s+1) = \frac{1}{s+1} \frac{(2s)!}{(s!)^2}$$

$$\approx \frac{1}{s+1} \frac{\sqrt{2\pi 2s}(2s)^{2s} \exp(-2s)}{2\pi s s^{2s} \exp(-2s)} \quad \text{for } s \gg 1$$

$$\approx \frac{1}{\sqrt{\pi}} s^{-3/2}\, 4^s \quad \text{for } s \gg 1, \quad (3.27)$$

where, in the last step, we have collected terms and approximated $s+1$ by s. For large s, the term 4^s dominates the expression and the number of trees increases exponentially with s. In the avalanche-size probability, this exponential increase is counterbalanced by the exponential decrease $[p(1-p)]^s$, ensuring that it is less than or equal to one. Substituting the approximation in Equation (3.27) back into Equation (3.26) and collecting terms raised to the power of s, the avalanche-size probability

$$P(s, p) \approx \frac{1}{\sqrt{\pi}} \frac{(1-p)}{p} s^{-3/2} [4p(1-p)]^s \quad \text{for } s \gg 1. \quad (3.28)$$

The avalanche-size probability has the form of a power-law decay, $s^{-3/2}$, multiplied by $[4p(1-p)]^s$. When $p = p_c = 1/2$, the latter term is absent and the avalanche-size probability is a pure power law. However, when $p < p_c = 1/2$, the latter term represents an exponential decay with avalanche size s and the avalanche-size probability has the form of a power-law decay multiplied by an exponential decay with an associated cutoff avalanche size, $s_c(p)$, see Figure 3.8.

When p tends to p_c, the ratio $(1-p)/p$ tends to one and the scaling form for the avalanche-size probability for an uncorrelated branching process is of the form

$$P(s, p) \propto s^{-\tau_s} \exp(-s/s_c) \quad \text{for } p \to p_c, s \gg 1, \quad (3.29a)$$

$$s_c(p) = -\frac{1}{\ln[4p(1-p)]}, \quad (3.29b)$$

where the avalanche-size exponent $\tau_s = 3/2$ and the cutoff avalanche size $s_c(p)$ diverges when p tends to p_c. We can test the validity of Equation (3.29) by performing a data collapse.

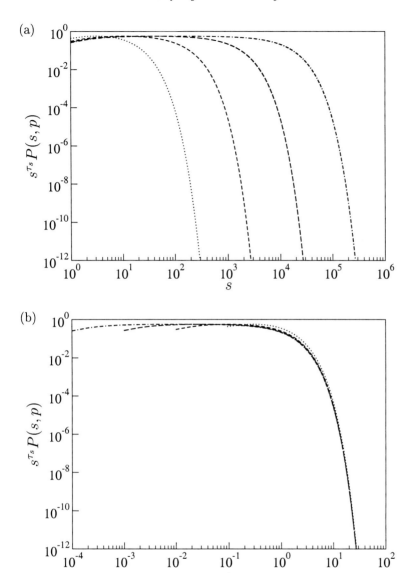

Fig. 3.9 Data collapse of the exact solution of the avalanche-size probabilities for the branching process with $\alpha_n z^{\text{th}} = 2$ where $p_c = 1/2$ for $p = 0.35, 0.45, 0.4842, 0.495$. (a) The transformed avalanche-size probability, $s^{\tau_s} P(s,p)$, versus the avalanche size, s, using the critical exponent $\tau_s = 3/2$. (b) The transformed avalanche-size probability, $s^{\tau_s} P(s,p)$, versus the rescaled argument, s/s_c, where the cutoff avalanche size $s_c(p) = -1/\ln[4p(1-p)]$. For $s \gg 1, p \to p_c$, the curves collapse onto the graph for the scaling function $\mathcal{G}^{\text{BP}}(x) = \exp(-x)$.

Figure 3.9(a) displays the transformed avalanche-size probability $s^{\tau_s} P(s,p)$ versus the avalanche size using the critical exponent $\tau_s = 3/2$. When rescaling the avalanche size s with the cutoff avalanche size s_c, the graphs collapse onto the graph for the scaling function $\mathcal{G}^{\text{BP}}(s/s_c) = \exp(-s/s_c)$, see Figure 3.9(b). The collapse is excellent for large s and p close to p_c. The deviations for small s are a result of Stirling's approximation, which overestimates the avalanche-size probability for small s. Consequently, the transformed avalanche-size probability $s^{\tau_s} P(s,p)$ deviates below the graph for the scaling function for small s. The deviations away from the graph for the scaling function for p far from p_c arise from neglecting the factor $(1-p)/p$ in Equation (3.28).[2]

In spite of these deviations, we should be able to recover the scaling of the average avalanche size with p. Using the scaling form in Equation (3.29) for the avalanche-size probability, the average avalanche size

$$\langle s \rangle = \sum_{s=1}^{\infty} s P(s,p)$$

$$\propto \sum_{s=1}^{\infty} s^{1-\tau_s} \exp(-s/s_c)$$

$$\approx \int_{1}^{\infty} s^{1-\tau_s} \exp(-s/s_c) \, ds$$

$$= \int_{1/s_c}^{\infty} (u s_c)^{1-\tau_s} \exp(-u) s_c \, du \quad \text{with } u = s/s_c$$

$$= s_c^{2-\tau_s} \int_{1/s_c}^{\infty} u^{1-\tau_s} \exp(-u) \, du$$

$$\propto s_c^{2-\tau_s} \qquad \qquad \text{for } p \to p_c. \qquad (3.30)$$

When $p \to p_c$, the cutoff avalanche size diverges and the lower limit of the integral tends to zero so that the integral is just a constant. Recalling that $\tau_s = 3/2$, we find that the average avalanche size scales with the square root of the cutoff avalanche size. Taylor expanding the logarithm in the denominator of Equation (3.29b), we indeed recover

$$\langle s \rangle = \left(\frac{-1}{\ln[4p(1-p)]} \right)^{1/2} = \frac{1}{2(p_c - p)} = \frac{1}{1-2p} \quad \text{for } p \to p_c^-, \qquad (3.31)$$

---

[2] Had we used the exact avalanche-size probability in Equation (3.26), we would have achieved a perfect data collapse by plotting $\frac{(1-p_c)p}{(1-p)p_c} \frac{P(s,p)}{P(s,p_c)}$ versus s/s_c. See Figure 1.20 on page 45 in Section 1.5.3 for a similar example.

in agreement with Equation (3.23) since, for $\alpha_n z^{\text{th}} = 2$, the average branching ratio

$$\langle b \rangle = \sum_{b=0}^{2} b \binom{2}{b} p^b (1-p)^{2-b} = 2p. \quad (3.32)$$

Equation (3.32) is consistent with $p_c = 1/2$ for a critical branching process.

In summary, for avalanches viewed as an uncorrelated branching process, the avalanche-size probability satisfies a scaling form for $s \gg 1, p \to p_c$ similar to the scaling form satisfied by the cluster number density for percolation on a Bethe lattice: a power-law decay with an exponential cutoff controlled by a cutoff avalanche size that diverges as p tends to p_c. The exponents describing the divergence of the cutoff avalanche size as $p \to p_c$ are identical for avalanches and clusters. However, the power-law decay exponent for avalanches is one less than for clusters, because the average avalanche size is the first moment of the avalanche-size probability, while the average cluster size is related to the second moment of the cluster number density.

Finally, we note that Equation (3.24) relating the average branching ratio to the average avalanche size is valid in general. In a system displaying self-organised criticality, the average avalanche size diverges with system size. In a mean-field picture, the avalanches may be viewed as an uncorrelated branching process with a branching ratio tending to one. However, in a real system, avalanches propagate on a lattice rather than a loopless tree and correlations must be taken into account. When correlations are important, the mean-field exponents will no longer apply.

3.5 Avalanche-Size Probability – Scaling Ansatz

We now set up a general framework for scaling and data collapse. Building on our experience of scaling in equilibrium critical phenomena, we propose a simple finite-size scaling ansatz for the avalanche-size probability valid for large avalanches and system sizes

$$P(s; L) \propto s^{-\tau_s} \mathcal{G}(s/s_c) \quad \text{for } L \gg 1, s \gg 1, \quad (3.33\text{a})$$
$$s_c(L) \propto L^D \quad \text{for } L \gg 1, \quad (3.33\text{b})$$

where the critical exponent D is the so-called avalanche dimension and the critical exponent τ_s is the avalanche-size exponent.

The scaling function $\mathcal{G}$ must fall off sufficiently fast such that all moments of the avalanche-size probability exist in a finite system. Assuming that the scaling function is well behaved for small arguments, we can Taylor expand it around zero. Therefore we expect

$$\mathcal{G}(x) \propto \begin{cases} \mathcal{G}(0) + \mathcal{G}'(0)x + \tfrac{1}{2}\mathcal{G}''(0)x^2 + \cdots & \text{for } x \ll 1 \\ \text{decay rapidly} & \text{for } x \gg 1. \end{cases} \quad (3.34)$$

Since the avalanche-size probability must be normalisable and have a diverging first moment, it must satisfy the constraints

$$\sum_{s=1}^{\infty} P(s;L) = 1 \quad \text{for all } L, \quad (3.35a)$$

$$\sum_{s=1}^{\infty} sP(s;L) \to \infty \quad \text{for } L \to \infty. \quad (3.35b)$$

Assuming, for now, that $\mathcal{G}(0) \neq 0$, then in the limit of infinite system size the avalanche-size probability takes the form

$$\lim_{L \to \infty} P(s;L) \propto s^{-\tau_s} \mathcal{G}(0). \quad (3.36)$$

Therefore, the two constraints in Equation (3.35) are satisfied if $1 < \tau_s \leq 2$.

The avalanche-size probability ansatz in Equation (3.33) is verified numerically by performing a data collapse. First, the avalanche-size probabilities are multiplied by s^{τ_s} to vertically align the distinctive feature of each graph, with only the horizontal positions of this feature distinguishing the graphs. Second, the avalanche size is rescaled by the factor L^D to horizontally align the distinctive feature of each graph.

We have already performed a data collapse by this procedure on the avalanche-size probability for the one-dimensional BTW model in Figure 3.6. The critical exponents are $D = 1$ and $\tau_s = 1$ and the scaling function $\mathcal{G}_{1d}^{BTW}(x) = x\Theta(1-x)$. Note that, just as in one-dimensional percolation, the scaling function is zero when $x = 0$. The one-dimensional BTW model highlights that the avalanche-size exponent τ_s is determined from the vertical alignment of the distinctive feature of the avalanche-size probabilities. Therefore, τ_s does not relate to the decay of the avalanche-size probability, but rather to the decay of its distinctive feature, namely the upper right corner of the avalanche-size probability signifying the cutoff, as traced by the dotted curve in Figure 3.5.

3.6 Scaling Relations

The avalanche-size probability is normalised so that the zeroth moment equals one, see Equation (3.35a). We can use the finite-size scaling ansatz to calculate higher moments $k \geq 1$ of the avalanche-size probability. Assuming Equation (3.33) is valid for all avalanche sizes s and approximating the sum with an integral, we find that the kth moment of the the avalanche-size probability

$$\begin{aligned}\langle s^k \rangle &= \sum_{s=1}^{\infty} s^k P(s;L) \\ &= \sum_{s=1}^{\infty} s^{k-\tau_s} \mathcal{G}\left(s/L^D\right) \\ &\propto \int_1^{\infty} s^{k-\tau_s} \mathcal{G}\left(s/L^D\right) ds \\ &= \int_{1/L^D}^{\infty} (uL^D)^{k-\tau_s} \mathcal{G}(u) L^D \, du \qquad u = s/L^D \\ &= L^{D(1+k-\tau_s)} \int_{1/L^D}^{\infty} u^{k-\tau_s} \mathcal{G}(u) \, du. \end{aligned} \qquad (3.37)$$

If the avalanche-size exponent satisfies $\tau_s < 1+k$, the integral converges in the lower limit. The scaling function decays sufficiently rapidly to ensure convergence in the upper limit. Therefore, the integral tends to a constant for L tending to infinity and

$$\langle s^k \rangle \propto L^{D(1+k-\tau_s)} \quad \text{for } L \gg 1, k \geq 1. \qquad (3.38)$$

The scaling of the kth moment of the avalanche-size probability with system size in Equation (3.38) is general and only relies on the avalanche-size probability satisfying the finite-size scaling ansatz in Equation (3.33). This 'simple-scaling' ansatz is responsible for the constant separation D between successive exponents of higher moments.

When the system has reached the set of recurrent configurations, it is often possible, using the notion of a steady state, to argue how the average avalanche size scale with system size without any prior knowledge of the avalanche-size probability. This will impose a scaling relation between D and τ_s.

Let us apply such a steady state argument to the 'staircase' configuration of the one-dimensional BTW model. In a steady state, every grain that is added will drop out at the open boundary. Therefore, every added grain must topple through of the order of L sites before reaching the open boundary, implying that the average avalanche size scales linearly with system size: $\langle s \rangle \propto L$. By Equation (3.38) we can infer the following scaling relation for the one-dimensional BTW model

$$D(2 - \tau_s) = 1, \tag{3.39}$$

which is indeed satisfied with $D = 1$ and $\tau_s = 1$.

3.7 Moment Analysis of Avalanche-Size Probability

The critical exponents D and τ_s can be determined by performing a data collapse of the avalanche-size probabilities for different system sizes. However, Equation (3.38) provides an alternative way to determine the critical exponents by analysing the scaling of the moments of the avalanche-size probability [De Menech et al., 1998]. Taking the logarithm on both sides of Equation (3.38) we have

$$\log\langle s^k \rangle = D(1 + k - \tau_s) \log L + \log(\text{constant}) \quad \text{for } L \gg 1, k \geq 1. \tag{3.40}$$

First, we measure the moments $k = 1, 2, \ldots$ of the avalanche-size probability as a function of system size. Plotting the logarithm of the kth moment versus the logarithm of system size for $k = 1, 2, \ldots$, the slope of each graph gives an estimate for the exponent $D(1 + k - \tau_s)$, see Equation (3.40).

Second, by plotting the estimates of $D(1 + k - \tau_s)$ versus the moment k, we obtain a straight line with slope D, intersecting the k-axis at $k = \tau_s - 1$.

We illustrate this procedure by examining the scaling of the moments in the one-dimensional BTW model. According to Equation (3.13)

$$\log\langle s^k \rangle \approx k \log L - \log(1 + k) \quad \text{for } L \gg 1. \tag{3.41}$$

Therefore, in a plot of $\log\langle s^k \rangle$ versus $\log L$ for $k = 1, 2, \ldots$, each graph has slope $k = 1, 2, \ldots$, see Figure 3.10(a). Plotting each of these slopes against k, the data lie on a straight line with slope $D = 1$, intersecting the k-axis at $k = 0$, implying that $\tau_s = 1$, see Figure 3.10(b). This is, of course, consistent with the critical exponents determined by data collapse of the avalanche-size probabilities.

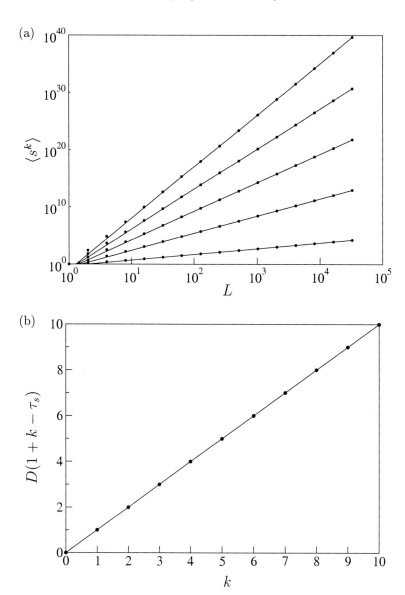

Fig. 3.10 Moment analysis of the avalanche-size probability for the one-dimensional BTW model on lattices of sizes $L = 1, 2, 4, \ldots, 32\,768$. (a) Exact solution of the kth moment $\langle s^k \rangle$, $k = 1, 3, \ldots, 9$ versus the system size, L (solid circles). Each straight k-line is the graph of $\langle s^k \rangle = L^k/(1+k)$ valid for $L \gg 1$, see Equation (3.13). (b) The exact value of the exponent, $D(1 + k - \tau_s)$, extracted as the slope $d\log\langle s^k \rangle / d\log L$ for $L \gg 1$ in (a) versus the moment, k (solid circles). The slope of the straight line is $D = 1$, intersecting the k-axis at $k = 0$, implying $\tau_s = 1$.

3.8 BTW Model in $d = 2$

Consider a two-dimensional square lattice composed of $N = L \times L$ sites, $i = 1, 2, \ldots, N$. We assign an integer variable z_i at each site i, commonly referred to as a height rather than a slope. It might, for example, represent the height of a column of sand on site i. We drive the system by choosing a site at random and increasing the height by one: $z_i \to z_i + 1$. Whenever the height at site i exceeds a threshold value, z^{th}, the site topples, $z_i \to z_i - 4$, and adds one height unit to each of its nearest neighbours, $z_{\text{nn}} \to z_{\text{nn}} + 1$. As a result of the initial toppling, some of the nearest neighbours may exceed the threshold value and topple in turn. The avalanche propagates in the system until a stable configuration with $z_i \leq z^{\text{th}}$ is reached. The avalanche size s is the total number of topplings during an avalanche.

3.8.1 Algorithm of the BTW model in $d = 2$

The algorithm of the two-dimensional BTW model on a square lattice is:

1. *Initialisation.* Prepare the system in an arbitrary stable configuration with $z_i \leq z^{\text{th}}$ for all i.
2. *Drive.* Add a grain at random site i.

$$z_i \to z_i + 1. \tag{3.42}$$

3. *Relaxation.* If $z_i > z^{\text{th}}$, relax site i.

$$\begin{aligned} z_i &\to z_i - 4, \\ z_{\text{nn}} &\to z_{\text{nn}} + 1. \end{aligned} \tag{3.43}$$

Continue relaxing sites until $z_i \leq z^{\text{th}}$ for all i.
4. *Iteration.* Return to **2**.

The value of the threshold is irrelevant for the behaviour of the model and we choose $z^{\text{th}} = 3$ for simplicity such that sites with $z_i = 0, 1, 2, 3$ are stable.

Figure 3.11 displays an avalanche propagating in the two-dimensional BTW model on a square lattice of size $L = 5$, triggered by the addition of a grain to the site in the middle of the second row. The avalanche stops after two topplings. Note that bulk sites have four nearest neighbours, edge sites three and corner sites two. Therefore, the amount of the dynamical variable z is conserved whenever a bulk site topples. Dissipation of the dynamical variable only occurs when an edge or corner site topples, where one and two units of height are lost, respectively.

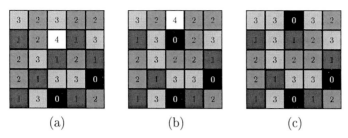

Fig. 3.11 The two-dimensional BTW model on a square lattice of size $L = 5$ with $z^{\text{th}} = 3$. (a) The white bulk site topples by losing four height units and adding one height unit to all of its four nearest neighbours. (b) As a consequence, the nearest neighbour at the edge exceeds the threshold and topples. Note that when an edge site relaxes, one unit is lost. (c) The avalanche has come to an end and the system is in a new stable configuration.

3.8.2 Steady state and the average avalanche size

The system will, eventually, enter the set of recurrent configurations, where the average number of grains added to the system equals the average number of grains leaving the system at the boundary. The system is in a steady state where the average height $\langle z \rangle = \frac{1}{N} \sum_{i=1}^{N} z_i$ fluctuates around a constant value. If the average height is too large, avalanches will transfer grains to the boundary, where they are lost. If the average height is too small, it will increase because of the driving.

In a steady state, added grains are eventually transported to the boundary. Consider a grain added somewhere in the bulk. Since the relaxation rule has no preferred direction, the 'trace' left by this particular grain is a random walk. How many relaxations (steps) does the grain need to reach the boundary of the system if the distance to the boundary is proportional to L? Equivalently, how many steps does a random walker need, on average, to cover a distance proportional to L?

Let us calculate the root-mean-square distance $\sqrt{\langle R^2 \rangle}$ after s steps, where $\mathbf{R} = \sum_{i=1}^{s} \mathbf{e}_i$, with $\mathbf{e}_i \in \{(\pm 1, 0), (0, \pm 1)\}$ all with equal probability:

$$R^2 = \sum_{i=1}^{s} \mathbf{e}_i \cdot \sum_{j=1}^{s} \mathbf{e}_j = \sum_{i=1}^{s} \mathbf{e}_i \cdot \mathbf{e}_i + \sum_{i=1}^{s} \sum_{\substack{j=1 \\ j \neq i}}^{s} \mathbf{e}_i \cdot \mathbf{e}_j = s + \sum_{i=1}^{s} \sum_{\substack{j=1 \\ j \neq i}}^{s} \mathbf{e}_i \cdot \mathbf{e}_j. \quad (3.44)$$

Since $\langle \mathbf{e}_i \cdot \mathbf{e}_j \rangle = 0$ because there is no preferred direction, we obtain that

$$\langle R^2 \rangle = \langle s \rangle. \quad (3.45)$$

In order for a grain to reach the boundary $\sqrt{\langle R^2 \rangle} \propto L$. The average avalanche size is equal to the average number of topplings an added grain suffers before reaching the boundary and therefore

$$\langle s \rangle \propto L^2. \tag{3.46}$$

The divergence of the average avalanche size with system size can only be accomplished if the associated avalanche-size probability is broad with a cutoff avalanche size diverging with system size.

3.8.3 Avalanche time series

Figure 3.12 displays a time series of avalanches for the two-dimensional BTW model on a square lattice of size $L = 128$. The time, t, is measured in the number of additions after the system has reached the set of recurrent configurations. An avalanche of size s is rescaled by the size of the largest observed avalanche in the time series, $s_{\max}$, and appears as a spike of height, $s/s_{\max}$. There is a large variability in avalanche sizes with the frequency of avalanches decaying with size.

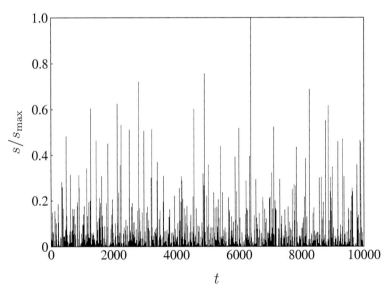

Fig. 3.12 Numerical results for the rescaled avalanche size, $s/s_{\max}$, versus the time, t, measured in number of additions for the two-dimensional BTW model in the set of recurrent configurations on a lattice of size $L = 128$. The largest avalanche recorded during a sequence of 10 000 additions is of size $s_{\max} = 76\,598$. The avalanche sizes span over nearly five orders of magnitude, with the frequency of avalanches decaying with size.

3.8.4 Avalanche-size probability

In order to sample the avalanche-size probability, we collect statistics of avalanche sizes when the system is in the set of recurrent configurations. We repeat simulations for a range of system sizes to measure the avalanche-size probability $P(s; L)$. Figure 3.13 displays $P(s; L)$ versus the avalanche size s for the two-dimensional BTW model on square lattices of size $L = 32, 128, 512, 2\,048$. The frequency of avalanches decays with size. The avalanche-size probability is indeed broad with a cutoff avalanche size that increases with system size.

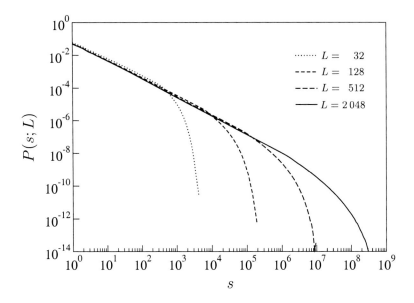

Fig. 3.13 Numerical results of the avalanche-size probability, $P(s; L)$, versus the avalanche size, s, for the two-dimensional BTW model on square lattices of size $L = 32, 128, 512, 2\,048$ marked with lines of increasing dash length. The frequency of avalanches decays with size. There is no typical size of an avalanche except for a cutoff avalanche size which increases with system size.

In order to investigate whether the avalanche-size probability satisfies the simple finite-size scaling ansatz in Equation (3.33) we attempt to extract the associated critical exponents by applying the moment scaling analysis outlined in Section 3.7.

Figure 3.14(a) displays the numerically estimated moments $\langle s^k \rangle$ versus system size L. For each moment k we use the numerical results for the

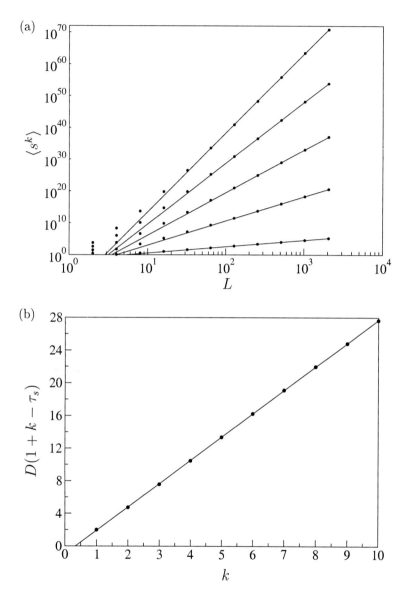

Fig. 3.14 Numerical moment analysis of the avalanche-size probability for the two-dimensional BTW model on square lattices of size $L = 1, 2, 4, \ldots, 2\,048$. (a) The kth moment $\langle s^k \rangle$, $k = 1, 3, \ldots, 9$ versus the system size, L (solid circles). The slope of the each straight k-line is the estimate of $d\log\langle s^k\rangle/d\log L$ for $L \gg 1$. (b) The estimated exponent, $D(1 + k - \tau_s)$, versus the moment, k (solid circles). The slope of the straight line $D = 2.86$, intersecting the k-axis at $k = 0.33$, implying $\tau_s = 1.33$.

largest lattices to estimate the slope of the graph $\log\langle s^k \rangle$ versus $\log L$, which is an estimate of the exponent $D(1 + k - \tau_s)$, see Equation (3.40). The straight lines in Figure 3.14(a) have slope $D(1 + k - \tau_s)$ and we notice that the moments measured on the smaller lattices fall outside their respective lines. But since the scaling is only valid asymptotically when $L \to \infty$ one would not expect the scaling to apply for small system sizes.

The estimated exponents $D(1 + k - \tau_s)$ are displayed in Figure 3.14(b) versus the moment, k. A linear fit has slope $D = 2.86$ and intersects the k-axis at $k = 0.33$, implying $\tau_s = 1.33$. However, we caution the reader against blindly quoting or trusting the results of the moment analysis without testing whether the estimated critical exponents D and τ_s produce a data collapse of the numerical results for the avalanche-size probabilities.

Assuming that the avalanche-size probability for the two-dimensional BTW model satisfies the simple finite-size scaling ansatz in Equation (3.33), we should be able to perform a data collapse.

Figure 3.15(a) displays the transformed avalanche-size probability $s^{\tau_s} P(s; L)$ versus the avalanche size s, using the moment analysis estimate of the critical exponent $\tau_s = 1.33$. Contrary to expectation, the distinctive feature of each graph does not align vertically.

After rescaling the avalanche size with the cutoff avalanche size $s_c(L) \propto L^D$, the moment analysis estimate of the critical exponent $D = 2.86$ horizontally aligns the onset of rapid decay of the transformed avalanche-size probabilities, see Figure 3.15(b).

However, it is obvious that the attempted data collapse of the numerical results for the avalanche-size probabilities for the two-dimensional BTW model seems to have failed since the data do not collapse convincingly onto a well-defined graph for a scaling function $\mathcal{G}^{\mathrm{BTW}}$.

There are two possible explanations for the absence of data collapse. The existence of a data collapse when plotting the transformed avalanche-size probability versus a rescaled avalanche size relies on the assumption that the avalanche-size probability satisfies the finite-size scaling ansatz in Equation (3.33). Therefore, one conclusion might be that finite-size scaling for the two-dimensional BTW model breaks down [De Menech et al., 1998]. This explanation is supported by the fact that the avalanches can be disjointly divided into non-dissipative avalanches that do not reach the boundary of the system, and dissipative avalanches that do. The two subsets of avalanches do not scale in the same way and hence finite-size scaling breaks down when considering the mixed set of avalanches [Drossel, 2000].

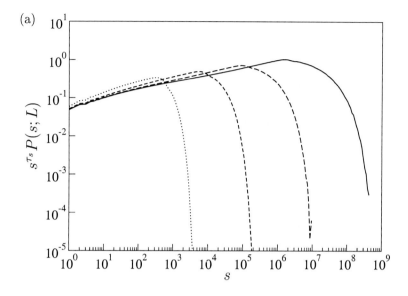
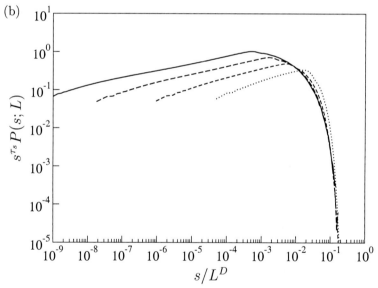

Fig. 3.15 Attempted data collapse of the numerical results for the avalanche-size probabilities for the two-dimensional BTW model on square lattices of size $L = 32, 128, 512, 2048$ marked with lines of increasing dash length using the critical exponents $D = 2.86$ and $\tau_s = 1.33$ estimated from moment analysis. (a) The transformed avalanche-size probability, $s^{\tau_s} P(s; L)$, versus the avalanche size, s. (b) The transformed avalanche-size probability, $s^{\tau_s} P(s; L)$, versus the rescaled avalanche size, s/L^D. The data do not collapse onto the graph for a scaling function $\mathcal{G}^{\mathrm{BTW}}$.

The simple finite-size scaling ansatz in Equation (3.33) is only valid in the limit of large avalanches and diverging system sizes. Inspecting Figure 3.15(b) reveals that the slope of the avalanche-size probability is increasing with system size. This indicates another explanation for the lack of data collapse: the numerical results available are not yet in the asymptotic scaling regime. Such a scaling regime may exist if one of the two subsets of avalanches were to dominate. The scaling regime would only be visible in system sizes not yet accessible to numerical simulations [Drossel, 2000].

Clearly, the scaling ansatz proposed in Equation (3.33) is not fulfilled for the lattice sizes we have investigated for the two-dimensional BTW model. But it remains an open question whether simple finite-size scaling breaks down for all system sizes or whether it will be recovered in the limit of $L \to \infty$. One might argue that since we have derived explicitly simple scaling forms for the avalanche-size probabilities for the one-dimensional BTW model and the mean-field theory of the BTW model, it would be rather surprising if the avalanche-size probability for the two-dimensional BTW model did not obey the scaling ansatz proposed in Equation (3.33) in the limit of $L \to \infty$.

3.9 Ricepile Experiment and the Oslo Model

3.9.1 *Ricepile experiment*

A slowly driven sandpile self-organising into a highly susceptible state provides a compelling metaphor for the critical behaviour of non-equilibrium systems with stick-slip dynamics. We are naturally led to ask whether this is more than just a metaphor. Do real granular media display self-organised criticality when driven slowly?

To test a physical system for self-organised criticality we must be able to measure the system's response to the small perturbation of adding a grain. In granular media, the response takes the form of an avalanche, the size of which is the (potential) energy dissipated. Practically all of this energy is dissipated in the bulk of the pile. Therefore an experimenter must have access to the internal rearrangements of a pile, rather than simply counting the number of grains dropping out of the system.

In the mid-nineties an experiment on a slowly driven pile of rice was performed at the University of Oslo, Norway [Frette et al., 1996]. We describe the experimental setup with the help of Figure 3.16.

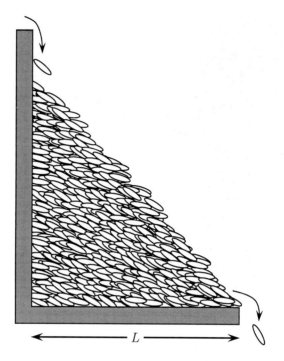

Fig. 3.16 A sketch of the setup of the ricepile experiment. The system size $L = 16$ in terms of the average grain length. The ricepile is narrowly confined between two glass plates. The pile ultimately reaches a steady state in which the average number of grains added at the left boundary equals the average number of grains dropping out of the system at the right boundary.

A ricepile was narrowly confined between two glass plates and sealed at the left boundary by a vertical wall, while the right boundary remained open. The separation between the glass plates was smaller than the length of the grains, such that the grains were aligned along the glass plates and restricted to move in practically one direction. Such a pile is said to be quasi one-dimensional. Grains were added at a slow rate and velocity into the system at the top of the pile next to the vertical wall at the left boundary. The pile was allowed to reach a steady state before the internal avalanches were recorded.

An example of a stable configuration is shown in Figure 3.17(a). Adding a single grain next to the vertical wall may trigger an avalanche which propagates down the slope of the ricepile as shown in Figure 3.17(b). The rearrangement of the pile as a result of an avalanche is characterised by a change in its profile.

Fig. 3.17 (a) A close-up photograph of a ricepile in a stable configuration. Most of the grains are aligned along the glass plates. However, there is a significant variation in the angular orientation of the grains, resulting in various packing arrangements and slopes along the profile. (b) An avalanche, triggered by a single grain at the top of the ricepile exceeding its threshold slope, propagates down the ricepile dissipating energy. Photos courtesy of Anders Malthe-Sørenssen, PGP, University of Oslo, Norway.

3.9.2 Ricepile avalanche time series

The dynamics of the pile were carefully recorded using a high-resolution digital camera. Profiles of the pile before and after each addition were compared, in order to identify the internal reorganisation of the pile, see Figure 3.18(a) and Figure 3.18(b).

The change in potential energy is a measure for the energy dissipated by an avalanche. Consider a profile with initial potential energy E^i_{pot}. Neglecting the kinetic energy of the grains added in and dropping out (if any), the energy E dissipated by an avalanche is

$$E = E^i_{\text{pot}} + E^{\text{add}}_{\text{pot}} - E^f_{\text{pot}}, \qquad (3.47)$$

where $E^{\text{add}}_{\text{pot}}$ is the potential energy added to the profile and E^f_{pot} is the final potential energy of the profile after the reorganisations caused by the

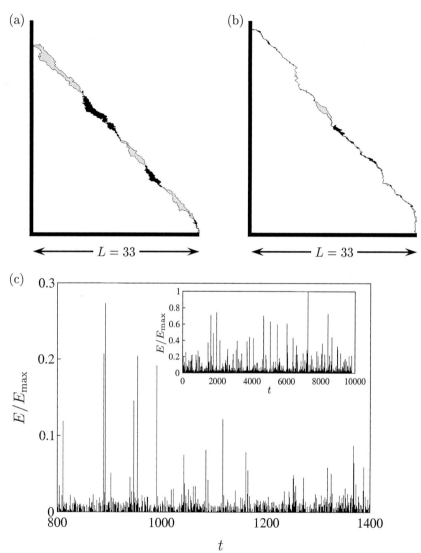

Fig. 3.18 Experimental results for the avalanche sizes in the quasi one-dimensional ricepile in a system of size $L = 33$. The energy is measured in units of $\langle m \rangle g \langle \ell \rangle = 1.54 \mu J$. (a) and (b) The area enclosed by the profiles of the ricepile before and after each addition reveal the internal reorganisations of mass. Loss of mass is shown by grey areas and gain of mass is shown by black areas. The energy dissipated is (a) $E = 2\,868$ and (b) $E = 294$. (c) The rescaled avalanche sizes, $E/E_{\max}$, versus the time, t, measured in number of avalanches. The largest avalanche recorded in the complete sequence of $N = 9\,817$ avalanches shown in the inset is of size $E_{\max} = 14\,981$. The avalanche sizes span over many orders of magnitude, with the frequency of avalanches decaying with size.

avalanche. The energy is measured in units of $\langle m \rangle g \langle \ell \rangle = 1.54\mu\text{J}$, where $\langle m \rangle = 0.02066$ g is the average grain mass, $g = 9.82$ m/s^2 is the acceleration due to gravity and $\langle \ell \rangle = 7.6$ mm is the average grain length.

Figure 3.18(c) shows a time series of 600 avalanches out of a total of $N = 9\,817$ avalanches shown in the inset. An avalanche appears as a spike because its duration is much smaller than the time in between additions. The height of a spike is $E/E_{\max}$, where the energy E released by an avalanche has been rescaled by the energy of the largest avalanche $E_{\max}$ that occurred during the entire experiment. Qualitatively, there is a huge variability in the energy of avalanches, with many small, some medium, and few large avalanches.

3.9.3 Ricepile avalanche-size probability density

Since E is a continuous variable, we are working with an avalanche-size probability density $P(E; L)\, dE$, that quantifies the relative number of avalanches in the interval $[E, E + dE]$, see Figure 3.19.

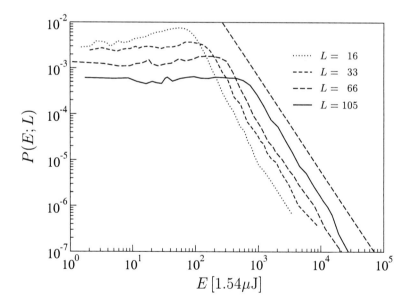

Fig. 3.19 Experimental results of the avalanche-size probability density, $P(E; L)$, versus the energy dissipation, E, for the quasi one-dimensional ricepile in systems of size $L = 16, 33, 66, 105$ marked with lines of increasing dash length. For large energy dissipation, the avalanche-size probability density is well approximated by a power-law decay, $P(E; L) \propto E^{-\tau_E}$ for $E \gg 1$, with $\tau_E = 2.06$. The dashed straight line has slope -2.06.

The probability densities decay approximately as E^{-2} for large avalanches without any visible cutoff for systems of size $L = 16, 33, 66, 105$ in terms of the grain length,

$$P(E; L) \propto E^{-\tau_E} \quad \text{for } E \gg 1 \tag{3.48}$$

where the ricepile avalanche-size exponent $\tau_E \approx 2.06$. Therefore, there is no cutoff size for the largest avalanches and they appear to be scale invariant.

The absence of a cutoff avalanche size is related to insufficient statistics. The expected number of avalanches with an energy release larger than E', $N(E > E') = N \int_{E'}^{\infty} P(E; L) \, dE$. This number must be much larger than one in order to probe the functional form of the avalanche-size probability for $E < E'$. In these experiments, the total number of avalanches, N, was too small to imply $N(E > E') \gg 1$. Without a cutoff avalanche size it is not possible to test whether the avalanche-size probability obeys the finite-size scaling ansatz in Equation (3.33). Nevertheless, the pile organises itself into a steady state where the probability of avalanche sizes is scale invariant. In this sense, the pile displays self-organised criticality.

The deficit of small avalanches is related to the practical difficulties of adding a single grain in between recorded profiles and has no bearing on the critical features of the pile.

3.9.4 Ricepile modelling

Let us investigate whether a boundary-driven one-dimensional BTW model, in which grains are only added at site $i = 1$, is a reasonable model for the ricepile experiment. First, it is apparent that modifying the driving does not give rise to any new recurrent configurations. The 'staircase' configuration shown in Figure 3.3 remains the unique recurrent configuration. This does not capture the variability in the configurations in the ricepile experiment. As a consequence of the uniqueness of the recurrent configuration, the boundary-driven one-dimensional BTW model predicts avalanches of size $s = L$ only. Clearly, this is entirely dissimilar to the time series of avalanches and the corresponding avalanche-size probability density observed in the experiment. In conclusion, the boundary-driven one-dimensional BTW model is too simple.

Let us try to identify the essential feature missing in the model. Looking at Figure 3.17(a), the spatial fluctuations in the profile can be attributed to the angular degree of freedom of the grains. Furthermore, it was observed that the slope at a given site also varied in time. However, the constant threshold slope in the one-dimensional BTW model does not allow any such spatial or temporal fluctuations.

The boundary-driven Oslo model is identical to the boundary-driven one-dimensional BTW model, except that the threshold slope of a site may change when the site topples [Christensen et al., 1996]. Specifically, the threshold slope z_i^{th} is chosen randomly to be 1 or 2 every time site i topples. This extra internal degree of freedom in the Oslo model is sufficient to capture, at least qualitatively, the spatial or temporal fluctuations in the slope along the profile.

3.9.5 Algorithm of the Oslo model

We summarise the dynamics of the boundary-driven Oslo model with the following algorithm:

1. *Initialisation.* Prepare the system in an arbitrary stable configuration with $z_i \leq z_i^{\text{th}}$ for all i.
2. *Drive.* Add a grain at the left-most site $i = 1$:

$$z_1 \to z_1 + 1. \tag{3.49}$$

3. *Relaxation.* If $z_i > z_i^{\text{th}}$, relax site i.
   For $i = 1$:

$$z_1 \to z_1 - 2,$$
$$z_2 \to z_2 + 1. \tag{3.50a}$$

   For $i = 2, \ldots, L - 1$:

$$z_i \to z_i - 2,$$
$$z_{i \pm 1} \to z_{i \pm 1} + 1. \tag{3.50b}$$

   For $i = L$:

$$z_L \to z_L - 1,$$
$$z_{L-1} \to z_{L-1} + 1. \tag{3.50c}$$

   Choose a new threshold slope $z_i^{\text{th}} \in \{1, 2\}$ for the relaxed site. Continue relaxing sites until $z_i \leq z_i^{\text{th}}$ for all i.
4. *Iteration.* Return to **2**.

3.9.6 Transient and recurrent configurations

In a stable configuration, $z_i \leq z_i^{\text{th}}$ for all i. Since $z_i^{\text{th}} \in \{1,2\}$, the possible slope values for site i are $z_i = 0, 1, 2$ and the total number of stable configurations for the entire system is $N_S = 3^L$. Not all of the stable configurations are recurrent. The number of recurrent configurations has been calculated to be

$$N_{\mathcal{R}} = \frac{1}{\sqrt{5}} \left(\phi(1+\phi)^L + \frac{1}{\phi(1+\phi)^L} \right) \propto (1+\phi)^L \quad \text{for } L \gg 1, \quad (3.51)$$

where $\phi = (1+\sqrt{5})/2 = 1.618033\ldots$ is the golden mean [Chua and Christensen, 2002]. Thus the number of recurrent configurations increases exponentially fast with system size. However, the fraction of stable configurations that are recurrent, $N_{\mathcal{R}}/N_S$, decreases exponentially fast with system size. Table 3.1 gives the number of recurrent configurations for various system sizes and the fraction of stable configurations that are recurrent.

Table 3.1 The number of recurrent configurations, $N_{\mathcal{R}}$, and the fraction of stable configurations that are recurrent, $N_{\mathcal{R}}/N_S$, as a function of system size, L, in the Oslo model with threshold slopes $z_i^{\text{th}} \in \{1,2\}$. Four leading digits are given for numbers written in decimal form.

L	$N_{\mathcal{R}}$	$N_{\mathcal{R}}/N_S$
1	2	0.6666
2	5	0.5555
4	34	0.4197
8	1 597	0.2434
16	3 524 578	0.08187
32	17 167 680 177 565	0.009264
64	407 305 795 904 080 553 832 073 954	0.0001186
128	2.292×10^{53}	1.944×10^{-8}
256	7.263×10^{106}	5.225×10^{-16}
512	7.291×10^{213}	3.773×10^{-31}
1 024	7.348×10^{427}	1.967×10^{-61}

When the number of recurrent configurations is sufficiently small, that is for system sizes up to $L = 7$ where $N_{\mathcal{R}} = 610$, it is possible to infer the exact avalanche behaviour of the Oslo model by brute force [Corral, 2004; Dhar, 2004]. However, we are ultimately interested in the behaviour of large systems in a steady state where the astronomical number of recurrent configurations rules out such an approach. Therefore, we must resort to numerical simulations.

3.9.7 Avalanche time series

Figure 3.20 displays a time series of avalanches in the Oslo model on a lattice of size $L = 128$. The time t is measured in the number of additions after the system has reached the set of recurrent configurations. An avalanche of size s is rescaled by the size of the largest observed avalanche in the time series $s_{\max}$ and appears as a spike of height $s/s_{\max}$.

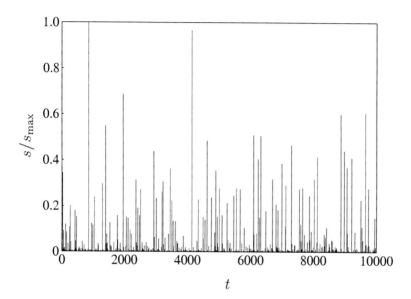

Fig. 3.20 Numerical results for the rescaled avalanche size, $s/s_{\max}$, versus the time, t, measured in number of additions for the Oslo model on a lattice of size $L = 128$. The largest avalanche recorded during a sequence of 10 000 additions is of size $s_{\max} = 45\,841$. The avalanche sizes span over four orders of magnitude, with the frequency of avalanches decaying with size.

The Oslo model avalanche time series resembles, at least qualitatively, the ricepile avalanche time series in Figure 3.18(c). There is a large variability in avalanche sizes with the frequency of avalanches decaying with size. Note that every avalanche is initiated by adding a single grain at site $i = 1$. For the system of size $L = 128$, this single mechanism causes avalanches with sizes spanning more than four orders of magnitude. Hence, the response of the ricepile is highly non-linear and the size of the response, measured by the size of the resulting avalanche, is not related in a trivial way to the size of the external perturbation.

3.9.8 Avalanche-size probability

In order to sample the avalanche-size probability, we collect statistics of avalanche sizes when the system is in the set of recurrent configurations. We repeat simulations for a range of system sizes to measure the avalanche-size probability $P(s; L)$ displayed in Figure 3.21, versus the avalanche size s for the Oslo model on lattices of size $L = 64, 256, 1\,024, 4\,096$.

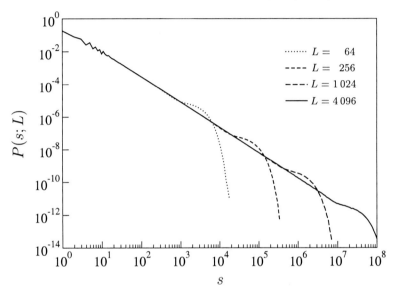

Fig. 3.21 Numerical results for the avalanche-size probability, $P(s; L)$, versus the avalanche size, s, for the Oslo model on lattices of size $L = 64, 256, 1\,024, 4\,096$ marked with lines of increasing dash length. There is no typical size of an avalanche except for a cutoff avalanche size, s_c, which increases with system size. For large avalanche sizes, the avalanche-size probability is well approximated by a power-law decay, $P(s; L) \propto s^{-\tau_s}$ for $1 \ll s \ll s_c$.

To test whether the avalanche-size probability satisfies the finite-size scaling ansatz given in Equation (3.33), we attempt to collapse data for different system sizes onto a single curve.

The distinctive feature of each graph is a pronounced bump[3] that reach the just before the avalanche-size probability decays rapidly. The bumps are aligned vertically by multiplying each avalanche-size probability with s^{τ_s}, see Figure 3.22(a), and then aligned horizontally by rescaling the avalanche size with the cutoff avalanche size $s_c(L) \propto L^D$, see Figure 3.22(b).

---
[3]The bump is related to system spanning avalanches where grains are dropping out at the right boundary next to $i = L$.

(a)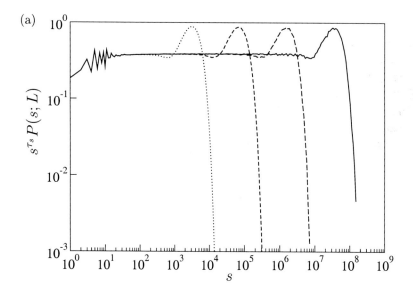

(b)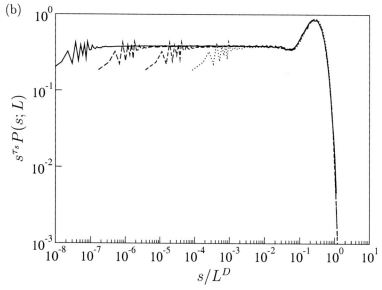

Fig. 3.22 Data collapse of the numerical results of the avalanche-size probabilities for the Oslo model on lattices of size $L = 64, 256, 1\,024, 4\,096$ marked with lines of increasing dash length. (a) The transformed avalanche-size probability, $s^{\tau_s} P(s; L)$, versus the avalanche size, s, using the avalanche-size exponent $\tau_s = 1.55$. (b) The transformed avalanche-size probability, $s^{\tau_s} P(s; L)$, versus the rescaled avalanche size, s/s_c, where the cutoff avalanche size, $s_c(L) \propto L^D$, with $D = 2.25$. For $s \gg 1, L \gg 1$, the curves collapse onto the graph for the scaling function $\mathcal{G}^{\text{Oslo}}$.

The quality of the data collapse is such that we can be reasonably confident that the avalanche-size probability satisfies the simple finite-size scaling ansatz in Equation (3.33). The critical exponents extracted from the data collapse are $D = 2.25$ and $\tau_s = 1.55$.

In addition to data collapse, one should also extract the numerical estimates of the critical exponents D and τ_s by applying the moment analysis of the avalanche-size probability.

Figure 3.23(a) displays the numerically estimated moments $\langle s^k \rangle$ for the Oslo model on lattices of size $L = 1, 2, 4, \ldots, 4\,096$. For each moment k, we estimate the slope of the graph $\log \langle s^k \rangle$ versus $\log L$ for $L \gg 1$.

In Figure 3.23(b) we plot the estimated slope $D(1 + k - \tau_s)$ versus the moment k. A linear fit shown with a straight line has slope $D = 2.252$ and intersects the k-axis at $k = 0.557$, implying $\tau_s = 1.557$. These estimates for the critical exponents are consistent with the estimates derived from the finite-size scaling data collapse of the avalanche-size probabilities in Figure 3.22.

When the system evolves within the set of recurrent configurations, the number of grains added at the left boundary equals, on average, the number of grains dropping out at the right boundary. This notion of a steady state will impose a scaling relation between D and τ_s just like for the one-dimensional BTW model, see Equation (3.39).

Indeed, the average avalanche size equals the system size L and therefore, according to Equation (3.38) with $k = 1$ we obtain the scaling relation for the boundary driven Oslo model:

$$D(2 - \tau_s) = 1. \qquad (3.52)$$

These measured exponents fulfil this scaling relation. In fact, the critical exponents of the Oslo model have been conjectured to be $D = 9/4 = 2.25$ and $\tau_s = 14/9 = 1.5555\ldots$.

We can find an upper bound on the value of the avalanche dimension D. The quantity L^D describes the scaling of the avalanche-size cutoff with system size L. The most extreme avalanche size would occur if it were possible for all grains to leave the pile. In a steady state, the height of the pile is proportional to the base so that there are of the order of L^2 grains in the pile. Each grain can topple at most L times before dropping out of the system, with a maximum avalanche size of $s_c \leq L^3$ such that $D \leq 3$. Since the average avalanche size scales linearly with system size, Equation (3.52) implies that $\tau_s \leq 5/3$, which is a tighter upper bound than that derived by imposing general constraints on the avalanche-size probability.

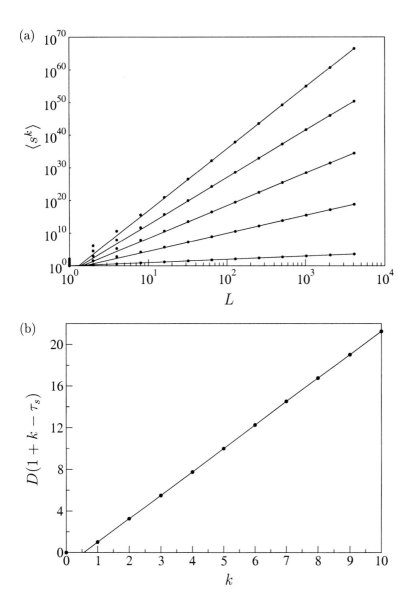

Fig. 3.23 Moment analysis of the avalanche-size probability for the Oslo model on lattices of size $L = 1, 2, 4, \ldots, 4\,096$. (a) The kth moment $\langle s^k \rangle$, $k = 1, 3, \ldots, 9$ versus the system size, L (solid circles). The slope of each straight k-line is the estimate of $d \log \langle s^k \rangle / d \log L$ based on $L = 256, 512, \ldots, 4096$. (b) The estimated exponent, $D(1 + k - \tau_s)$, versus the moment, k (solid circles). The slope of the straight line $D = 2.252$, intersecting the k-axis at $k = 0.557$, implying $\tau_s = 1.557$.

It is enlightening to consider the Oslo model when changing the drive from adding a grain to the boundary to adding a slope unit to a randomly chosen site i, that is, $z_i \to z_i + 1$. Figure 3.24 displays the numerical results for the avalanche-size probability.

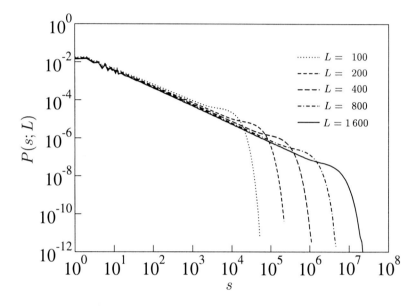

Fig. 3.24 Numerical results for the avalanche-size probability, $P(s;L)$, versus the avalanche size, s, for the bulk-driven Oslo model on lattices of size $L = 100, 200, 400, 800, 1\,600$ marked with lines of increasing dash length. There is no typical size of an avalanche except for a cutoff avalanche size, s_c, which increases with system size. For large avalanche sizes, the avalanche-size probability is well approximated by a power-law decay, $P(s;L) \propto s^{-\tau_s}$ for $1 \ll s \ll s_c$. Note, however, the slope is changing with system size.

The procedure of determining the critical exponents by data collapse of the avalanche-size probabilities appear to be difficult to apply since the slope of the avalanche-size probability is apparently increasing with system size, see Figure 3.24, much like the situation encountered for the two-dimensional BTW model. However, choosing $\tau_s = 1.11$ and $D = 2.25$ produces a reasonable collapse for large values of s, see Figure 3.25. Note again that the avalanche size exponent $\tau_s = 1.11$ does not reflect the decay of the avalanche-size probability. It is determined purely by making sure that the distinctive feature of each graph, namely where the rapid decay sets in, all lie at the same vertical position.

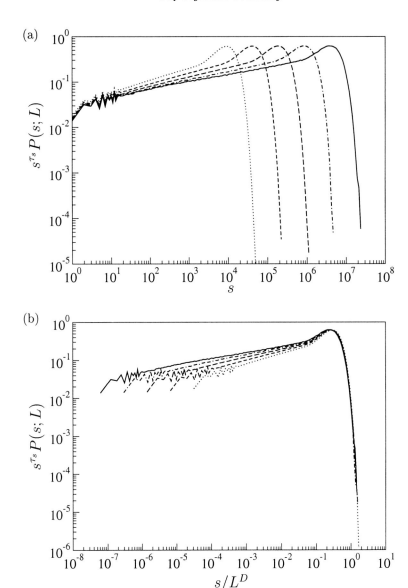

Fig. 3.25 Data collapse of the numerical results of the avalanche-size probabilities for the bulk-driven Oslo model on lattices of size $L = 100, 200, 400, 800, 1\,600$ marked with lines of increasing dash length. (a) The transformed avalanche-size probability, $s^{\tau_s} P(s; L)$, versus the avalanche size, s, using the avalanche-size exponent $\tau_s = 1.11$. (b) The transformed avalanche-size probability, $s^{\tau_s} P(s; L)$, versus the rescaled avalanche size, s/s_c, where the cutoff avalanche size, $s_c(L) \propto L^D$, with $D = 2.25$. For $s \gg 1, L \gg 1$, the curves collapse onto the graph for the scaling function $\mathcal{G}_{\text{bulk}}^{\text{Oslo}}$.

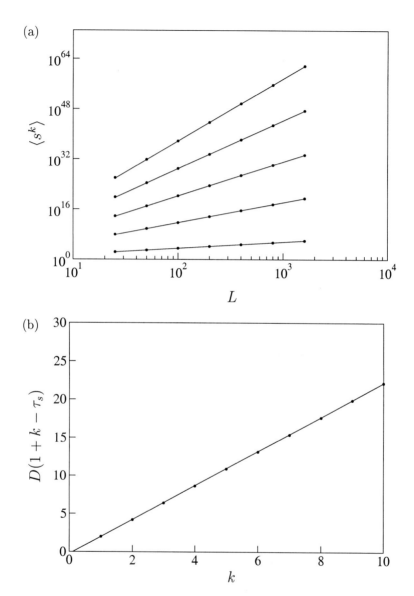

Fig. 3.26 Moment analysis of the avalanche-size probability for the bulk-driven Oslo model on lattices of size $L = 25, 50, 100, \ldots, 1\,600$. (a) The kth moment $\langle s^k \rangle$, $k = 1, 3, \ldots, 9$ versus the system size, L (solid circles). The slope of each straight k-line is the estimate of $d\log\langle s^k \rangle / d\log L$ based on $L = 100, 200, \ldots, 1\,600$. (b) The estimated exponent, $D(1 + k - \tau_s)$, versus the moment, k (solid circles). The slope of the straight line $D = 2.24$, intersecting the k-axis at $k = 0.11$, implying $\tau_s = 1.11$.

Applying the moment analysis of the avalanche-size probability with system size, we find $D = 2.24$ and $\tau_s = 1.11$, see Figure 3.26. These estimates for the critical exponents are consistent with the estimates derived from data collapse.

For the bulk-driven Oslo model, it has been shown analytically [Stapleton, 2004] that the average avalanche size

$$\langle s \rangle = L^2/3 + L/2 + 1/6. \qquad (3.53)$$

Therefore, the critical exponents D and τ_s must satisfy the scaling relation

$$D(2 - \tau_s) = \begin{cases} 1 & \text{boundary driven} \\ 2 & \text{bulk driven.} \end{cases} \qquad (3.54)$$

In fact, the critical exponents of the bulk-driven Oslo model have been conjectured to be $D = 9/4 = 2.25$ and $\tau_s = 10/9 = 1.1111\ldots$.

It is intriguing to note that the avalanche dimension, D, is identical for the boundary- and bulk-driven Oslo models while the avalanche size exponent τ_s changes in order to satisfy the relevant scaling relation in Equation (3.54). One might speculate whether D is a more fundamental exponent in a hierarchy of critical exponents associated with self-organised critical behaviour.

We stress that the observed scale-free behaviour in the Oslo model emerges without the need to fine-tune any external parameters. The algorithm defining the Oslo model contains only very simple local dynamical rules. However, when applying the algorithm *ad infinitum*, the system self-organises into a steady state that is highly susceptible and characterised by a scale-free behaviour in the avalanche-size probability. Remarkably, for a moderate system of size $L = 4\,096$ the avalanche-size probability spans seven orders of magnitude in s and twelve orders of magnitude in $P(s; L)$ for the boundary-driven Oslo model, see Figure 3.21. A single mechanism, in the form of a single grain dropped onto the apex of the pile, underlies all avalanche sizes, spanning seven orders of magnitude.

Because of the simplicity of the Oslo model, it may be considered the analogue of the Ising model for slowly driven threshold dominated models displaying self-organised criticality. Exact results are known for the Oslo model [Chua and Christensen, 2002; Corral, 2004; Dhar, 2004; Chua and Christensen, 2002; Stapleton, 2004]. However, despite its simplicity, it is fascinating that at the time of writing, no complete analytical solution for the Oslo model exists.

3.10 Earthquakes and the OFC Model

3.10.1 *Earthquake mechanism*

Since 1967 [McKenzie and Parker, 1967] it has been widely accepted that the outermost layer of the Earth down to about 100 km, called the lithosphere, consists of a number of rigid tectonic plates floating on a viscous medium, called the asthenosphere, as originally proposed by Wegener in 1915 [Wegener, 1924]. The convective flow in the asthenosphere, induced by the heat generated from fission processes in the centre of the Earth, slowly drives the plates relative to one another, resulting in the build up of stress at the plate boundaries. Because of friction between the plates, the strain energy is released intermittently rather than continuously in time, that is, the plates stick and slip rather than slide. These sudden stick-slip events are commonly known as earthquakes.

The San Andreas Fault in California, where the Pacific and the North American Plates make up the boundary, is an example of a so-called transform fault where the boundaries move parallel to each other with an average rate of approximately 4 cm per year. When the friction between the tectonic plates cannot sustain the stress any more an earthquake is 'nucleated'.

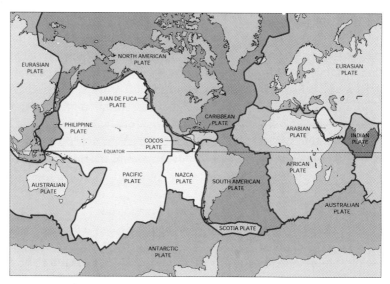

Fig. 3.27 The major tectonic plates. Due to the relative movement of the tectonic plates, stress builds up along the boundary of the plates. When the stress exceeds a friction-related threshold, the strain energy is released through an earthquake. Figure courtesy of U.S. Geological Survey.

The duration of an earthquake varies from fractions of a second (very small earthquake) to minutes (very large earthquake) and the relative displacement of the tectonic plates is of the order of centimetres to metres. For example, the relative displacement was in the range 0.6 − 6 m for the 1906 San Francisco earthquake on the San Andreas Fault.

3.10.2 Earthquake time series

There are numerous earthquake catalogues that record magnitude, time of occurrence, epicentre location, etc. The size s of an earthquake is related to its magnitude by $m = \log s$, and gives a measure of the energy released. Figure 3.28 displays a time series of earthquakes recorded in the Southern California Seismographic Network (SCSN) catalogue from 1984 to 2001. The time t is measured in years. An earthquake of size s is rescaled by the size of the largest observed earthquake in the time series $s_{\max} = 10^{7.3}$ and appears as a spike of height $s/s_{\max}$. There is a large variability in earthquake sizes, with the frequency of earthquakes decaying with size.

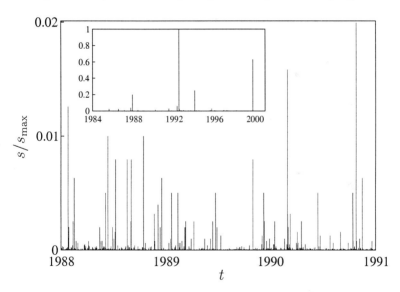

Fig. 3.28 Earthquake data for the rescaled earthquake size, $s/s_{\max}$, versus the time of occurrence, t, measured in years for part of Southern California according to the SCSN catalogue. The sequence of recorded earthquakes from 1998 to 1991 where $s/s_{\max} \in [0, 0.02]$. The largest earthquake recorded in the complete sequence of $N = 312\,973$ earthquakes during the period 1984−2001 (inset) is of size $s_{\max} = 10^{7.3}$. The earthquake sizes span over seven orders of magnitude, with the frequency of earthquakes decaying with size.

3.10.3 Earthquake-size frequency

Figure 3.29 displays the earthquake-size frequency derived from the time series in Figure 3.28. The annual number of earthquakes with a size S larger than s, $N(S > s)$, is plotted on a logarithmic scale versus $\log s$.

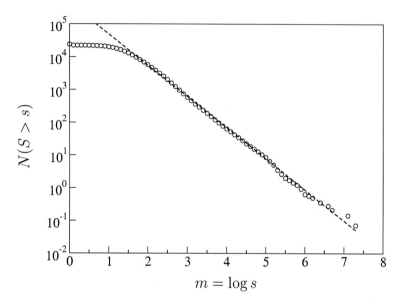

Fig. 3.29 Data for the annual number of earthquakes with a size S larger than s, $N(S > s)$, versus $\log s$, based on the SCSN catalogue 1984 − 2001. The size is a measure of the energy released during an earthquake. There is no typical size of an earthquake. For large earthquake sizes, the annual number of earthquakes with a size S larger than s is well approximated by a power-law decay, $N(S > s) \propto s^{-B}$ for $\log s \gtrsim 1.5$, with $B \approx 0.95$. The dashed straight line has slope -0.95.

For earthquakes with size $\log s \gtrsim 1.5$, the data are consistent with a power-law decay

$$N(S > s) \propto s^{-B} \quad \text{for } s \gtrsim 10^{1.5}, \tag{3.55}$$

where $B \approx 0.95$. This is the famous Gutenberg-Richter law for the earthquake-size frequency [Gutenberg and Richter, 1944].

Roughly speaking, earthquakes ten times larger in size are ten times less frequent. Since all earthquakes with size $\log s \gtrsim 1.5$ follow the same statistical distribution, we may conclude that a single mechanism underlies all earthquakes. Moreover, since the earthquake-size frequency is scale free, the stick-slip response of the Earth's crust is highly non-linear and gives rise to earthquakes of no characteristic size. Indeed, large earthquakes, although rare, may not need a separate description from small earthquakes.

The absence of a characteristic largest earthquake is related to insufficient statistics. In Figure 3.29 we can see that the expected annual number of earthquakes with size $\log s \geq 7.5$ is of the order of 10^{-1}. Therefore, to observe, say, 10 earthquakes with $\log s \geq 7.5$, we would need statistics covering roughly 100 years. Earthquakes with size $\log s \lesssim 2$ are usually called micro-earthquakes; they are rarely noticed by people and are often recorded only on local seismographs. The apparent deficit of earthquakes with size $\log s \lesssim 1.5$ might therefore be attributed to incomplete records and the practical difficulties of distinguishing micro-earthquakes from noise.

3.10.4 Earthquake modelling

Earthquakes may be thought of as avalanches in the crust of the Earth. The observed scale-free behaviour of earthquakes must be a consequence of a cooperative rather than a random process. One possible explanation is that the seismic system has self-organised into a highly susceptible state.

Having identified the stick-slip mechanism as the paradigm for earthquakes, we now attempt to capture the essential features of this mechanism in a simple model of a single fault. Although the dynamics of earthquakes are very complex, there are some basic components which must be included in any model of earthquakes:

– Earthquakes are caused by the slow continuous driving of the tectonic plates relative to one another and are triggered when the fault can no longer sustain the stress.
– Strain energy is built up over a long time scale and released over a short time scale.
– Earthquakes occur intermittently.

A simple picture of the internal dynamics of a single fault that includes these basic components was proposed by Burridge and Knopoff soon after the acceptance of plate tectonics in 1967 [Burridge and Knopoff, 1967]. They represented a fault by a two-dimensional network of blocks interconnected by springs. The blocks rest on a rigid plate and are connected by another set of springs to a rigid plate above, see Figure 3.30(a). The blocks are driven by the relative movement of the two plates. For convenience the lower plate is held fixed while the upper plate moves at a very slow speed.

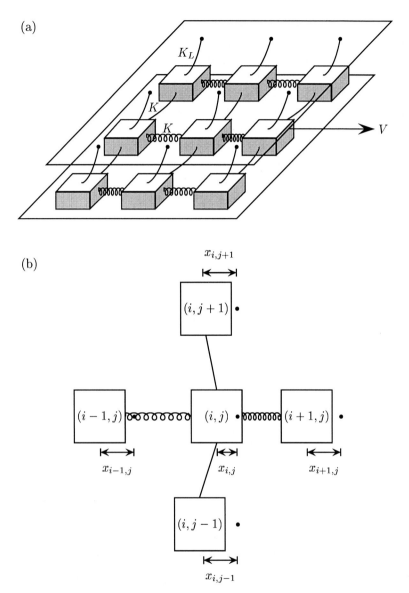

Fig. 3.30 The two-dimensional Burridge-Knopoff spring-block model. (a) The two-dimensional network of blocks is interconnected by springs with elastic constants K, and each block is connected to the upper plate by a leaf spring with elastic constant K_L. Because the plates move with a relative velocity V, the forces on the blocks increase at the same rate, $K_L V$. (b) The block (i,j) and its nearest neighbours. The solid circles mark the position of each leaf spring connecting the blocks to the upper plate. The displacements, $x_{i,j}$, of each block are measured relative to these positions.

As the plates move relative to one another, stress builds up on all the blocks at the same rate due to the springs connecting them to the upper plate. An 'earthquake' is triggered when the force on a block exceeds the frictional force resisting the block's motion. The slipping block interacts with its nearest neighbours via the springs, thereby increasing the forces on its nearest neighbours. This may induce neighbouring blocks to slip in turn so that an earthquake propagates and grows in size. The earthquake comes to come a halt when the force on all the blocks are less than or equal to the static frictional threshold force. The total number of blocks slipping, s, during an event gives a measure of the size of the earthquake.

This simple picture was transformed by Olami, Feder, and Christensen (OFC) into a so-called coupled map lattice of $L \times L$ blocks labelled (i, j), where $i, j = 1, \ldots, L$ [Olami et al., 1992]. The dynamical variable for the OFC model is given by the total force, $F_{i,j}$, exerted by the springs on the block (i, j).

The displacement, $x_{i,j}$, of each block is defined relative to the position of the leaf spring connecting the block to the upper plate. Figure 3.30(b) shows block (i, j) and its nearest neighbours together with their relative displacements.

Let $f_{i,j}^{i-1,j}, f_{i,j}^{i+1,j}, f_{i,j}^{i,j-1}, f_{i,j}^{i,j+1}$ denote the forces exerted by the four springs in the plane connected on the block (i, j), and let $f_{i,j}^L$ denote the driving force exerted by the leaf spring springs on the block (i, j). Using Hooke's law with elastic constants K and K_L for the springs in the plane and the leaf spring connecting the block to the upper plate, respectively, we can express the total force $F_{i,j}$ on block (i, j) in terms of the relative displacements:

$$F_{i,j} = f_{i,j}^{i-1,j} + f_{i,j}^{i+1,j} + f_{i,j}^{i,j-1} + f_{i,j}^{i,j+1} + f_{i,j}^L$$
$$= K(x_{i-1,j} - x_{i,j}) + K(x_{i+1,j} - x_{i,j}) +$$
$$K(x_{i,j-1} - x_{i,j}) + K(x_{i,j+1} - x_{i,j}) - K_L x_{i,j}$$
$$= K(x_{i-1,j} + x_{i+1,j} + x_{i,j-1} + x_{i,j+1} - 4x_{i,j}) - K_L x_{i,j}. \quad (3.56)$$

Because the two rigid plates move with a relative velocity V, the total force on each block increases with a rate $K_L V$. An earthquake is triggered when the first block slips. The condition for block (i, j) to slip is that the force exerted by the springs exceeds the frictional threshold force, $F_{i,j} > F^{\text{th}}$. In this way, the static frictional threshold force is incorporated in the model as the threshold F^{th}. For simplicity, we assume that the threshold is a constant, independent of the position of the block on the lower plate.

Let us calculate the change in the force, $\delta F_{i+1,j}$, on the nearest neighbour block $(i+1, j)$ when block (i, j) in the bulk slips. Assuming that the block (i, j) comes to rest at the zero force position $\tilde{x}_{i,j}$ after slipping,

$$0 = K(x_{i-1,j} + x_{i+1,j} + x_{i,j-1} + x_{i,j+1} - 4\tilde{x}_{i,j}) - K_L \tilde{x}_{i,j}. \quad (3.57)$$

By subtracting Equation (3.57) from Equation (3.56), the decrease in force on block (i, j) is given by

$$F_{i,j} = (4K + K_L)(\tilde{x}_{i,j} - x_{i,j}), \quad (3.58)$$

which is proportional to its displacement $(\tilde{x}_{i,j} - x_{i,j})$, as expected.

Since the force on the nearest neighbour block $(i+1, j)$ is

$$F_{i+1,j} = K(x_{i,j} + x_{i+2,j} + x_{i+1,j-1} + x_{i+1,j+1} - 4x_{i+1,j}) - K_L x_{i+1,j} \quad (3.59)$$

the change in the force $\delta F_{i+1,j}$ caused by the slip of block (i, j) is

$$\delta F_{i+1,j} = K(\tilde{x}_{i,j} - x_{i,j}) = \frac{K}{4K + K_L} F_{i,j}, \quad (3.60)$$

where we have used Equation (3.58) to relate the change in forces in terms of the elastic constants K and K_L.

Identical results follow for all other nearest neighbour blocks. Therefore, when a bulk block (i, j) slips, it relaxes to zero force position and the increase in the forces F_{nn} on the nearest neighbour blocks is proportional to the force on the block (i, j) before slipping, that is,

$$F_{nn} \to F_{nn} + \alpha F_{i,j},$$
$$F_{i,j} \to 0, \quad (3.61)$$

where the proportionality constant is given by the dimensionless ratio

$$\alpha = \frac{K}{4K + K_L}. \quad (3.62)$$

When a block at the edge or the corner slips, the factor multiplying the displacement in Equation (3.58) should read $(3K + K_L)$ or $(2K + K_L)$, respectively. Therefore, strictly speaking, the dimensionless ratio should be generalised to $\alpha_{i,j} = K/(q_{i,j}K + K_L)$, where $q_{i,j}$ is the coordination number of block (i, j). However, the dimensionless ratio in Equation (3.62) is commonly used for all relaxing blocks.

3.10.5 Algorithm of the OFC model

We summarise the dynamics of the two-dimensional OFC model with $0 \leq \alpha \leq 1/4$ on a square lattice with the following algorithm, where we identify blocks with sites:

1. *Initialisation.* Prepare the system in an arbitrary stable configuration with $F_{i,j} \leq F^{\text{th}}$ for all sites (i,j).
2. *Drive.* Increase the force on all sites by an infinitesimal amount δF:

$$F_{i,j} \to F_{i,j} + \delta F. \qquad (3.63)$$

3. *Relaxation.* If $F_{i,j} > F^{\text{th}}$, relax site (i,j):

$$F_{nn} \to F_{nn} + \alpha F_{i,j},$$
$$F_{i,j} \to 0. \qquad (3.64)$$

Continue relaxing sites until $F_{i,j} \leq F^{\text{th}}$ for all sites (i,j).

4. *Iteration.* Return to 2.

The value of the threshold is irrelevant for the behaviour of the model and we may choose $F^{\text{th}} = 1$ for simplicity.

At a first glance, the algorithm for the OFC model may appear similar to those of the BTW and Oslo models. However, there are important differences. First, the force is a continuous variable, whereas the slope is a discrete variable. Second, the drive is continuous and global rather than discrete and local. Third, the force on the relaxing site decreases to zero rather than by a fixed amount. Fourth, the force redistributed to the nearest neighbours is proportional to the force on the relaxing site rather than being a fixed amount. Finally, when $\alpha < 1/4$, not all of the force $F_{i,j}$ is redistributed when bulk sites relax, but only the fraction 4α. Therefore, when $0 \leq \alpha < 1/4$, the model is said to be 'non-conservative', while for $\alpha = 1/4$, the model is 'conservative' like the BTW and Oslo models, and the dissipation of the dynamical variable only takes place when edge or corner sites relax.

Note that when the springs interconnecting the blocks are absent, $K = 0$ in Equation (3.62) and hence $\alpha = 0$: the blocks are decoupled and completely independent of each other. The reader will have noticed that the conservative OFC model, where $\alpha = 1/4$, corresponds to $K_L = 0$ in Equation (3.62), implying that the leaf springs connecting the blocks to the upper plate are absent. While this is clearly unphysical, it represents no difficulties as far as the algorithm for the OFC model is concerned.

3.10.6 Steady state and the average avalanche size

In the BTW and Oslo models, the dynamical variables are discrete and the number of stable configurations is finite, implying the existence of recurrent configurations. In the OFC model, the dynamical variable is continuous and the number of stable configurations is infinite. Little is known about the recurrent configurations, if indeed they exist at all. Nevertheless, because a system of size L has a finite capacity of $F^{\mathrm{th}}L^2$, the concept of a steady state of the dynamical variable still applies in the sense that its average influx equals its average outflux.

The drive increases the dynamical variable of each site by δF before the system relaxes. Therefore, the total average influx in between avalanches

$$\langle \mathrm{influx} \rangle = \langle \delta F \rangle L^2. \tag{3.65}$$

The relaxation, meanwhile, decreases the dynamical variable by $(1 - 4\alpha)F_{\mathrm{b}}^{\mathrm{r}}$ for each slip in the bulk, where $F_{\mathrm{b}}^{\mathrm{r}} > F^{\mathrm{th}}$ is the value of the dynamical bulk variable just before relaxing. If an edge or corner site slips, the corresponding decrease in the dynamical variable is $(1-3\alpha)F_{\mathrm{e}}^{\mathrm{r}}$ and $(1-2\alpha)F_{\mathrm{c}}^{\mathrm{r}}$, respectively. Therefore, the total average outflux during avalanches is

$$\langle \mathrm{outflux} \rangle = [(1-4\alpha)\langle P_{\mathrm{b}}F_{\mathrm{b}}^{\mathrm{r}}\rangle + (1-3\alpha)\langle P_{\mathrm{e}}F_{\mathrm{e}}^{\mathrm{r}}\rangle + (1-2\alpha)\langle P_{\mathrm{c}}F_{\mathrm{c}}^{\mathrm{r}}\rangle]\langle s \rangle, \tag{3.66}$$

where $P_{\mathrm{b}}, P_{\mathrm{e}}, P_{\mathrm{c}}$ are the fractions of bulk, edge, and corner sites slipping, respectively, and $\langle s \rangle$ is the average avalanche size.

Equating the average influx with the average outflux and rearranging for the average avalanche size we find, that in the steady state

$$\langle s \rangle = \frac{\langle \delta F \rangle L^2}{(1 - 4\alpha)\langle P_{\mathrm{b}}F_{\mathrm{b}}^{\mathrm{r}}\rangle + (1 - 3\alpha)\langle P_{\mathrm{e}}F_{\mathrm{e}}^{\mathrm{r}}\rangle + (1 - 2\alpha)\langle P_{\mathrm{c}}F_{\mathrm{c}}^{\mathrm{r}}\rangle}. \tag{3.67}$$

Let us consider the average avalanche size in the three regimes, $\alpha = 0$ (decoupled), $0 < \alpha < 1/4$ (non-conservative), and $\alpha = 1/4$ (conservative).

When $\alpha = 0$, the sites are decoupled from each other and a relaxing site cannot cause other sites to exceed the threshold. Avalanches cannot propagate and they are all of size $s = 1$. We can check that Equation (3.67) is consistent with the average avalanche size being $\langle s \rangle = 1$. Sites relax with $F^{\mathrm{r}} = 1$ as a result of the drive, because all sites are decoupled. Therefore,

$$\langle s \rangle = \frac{\langle \delta F \rangle L^2}{\langle P_{\mathrm{b}} \rangle + \langle P_{\mathrm{e}} \rangle + \langle P_{\mathrm{c}} \rangle} = \langle \delta F \rangle L^2, \tag{3.68}$$

since the average fractions add up to one. Indeed, because the sites are decoupled, the fraction of bulk, edge and corner sites slipping is given by

$$P_{\rm b} = (L-2)^2/L^2, \tag{3.69a}$$
$$P_{\rm e} = 4(L-2)/L^2, \quad \text{for } \alpha = 0 \tag{3.69b}$$
$$P_{\rm c} = 4/L^2. \tag{3.69c}$$

The drive is determined by the distance of the maximal dynamical variable to the threshold, $\delta F = F^{\rm th} - \max(F_{i,j})$. The values of the L^2 dynamical variables in the initial configuration are arbitrarily distributed over the interval $[0,1]$ when $F^{\rm th} = 1$. The dynamics of the decoupled model reproduce the initial configuration periodically after a total of $\delta F = 1$ has been added to the system over L^2 steps. Therefore the average drive in between avalanches is $\langle \delta F \rangle = 1/L^2$. As a result, the average avalanche size

$$\langle s \rangle = \langle \delta F \rangle L^2 = 1 \quad \text{for } \alpha = 0. \tag{3.70}$$

Note that we have assumed that no sites are initialised with the same value of F. Otherwise, the initial configuration is reproduced after N_F steps, where N_F is the number of different values of F. The average drive in between avalanches generalises to $\langle \delta F \rangle = 1/N_F$ and the average avalanche size is L^2/N_F. Such synchronisation of the dynamical variable among sites may allow the average drive $\langle \delta F \rangle$ to deviate from the trivial $1/L^2$ scaling.

When $0 < \alpha < 1/4$, the sites are coupled to each other and a relaxing site may cause other sites to exceed the threshold. Therefore avalanches may propagate in the system. If all sites are equally likely to slip then $P_{\rm b}, P_{\rm e}$, and $P_{\rm c}$ would be identical to the fractions of bulk, edge, and corner sites, respectively, as is the case for $\alpha = 0$. Although this is not strictly true in practice for $0 < \alpha < 1/4$ (e.g., a given corner site, having only two nearest neighbours, is less likely to slip than a given bulk site, having four nearest neighbours), the implied relative scalings still hold:

$$P_{\rm e}/P_{\rm b} \propto P_{\rm c}/P_{\rm e} \propto 1/L. \tag{3.71}$$

When $0 < \alpha < 1/4$, the system is non-conservative. For a system with $L \to \infty$ all dissipation effectively takes place in the bulk since both $P_{\rm e}/P_{\rm b} \to 0$ and $P_{\rm c}/P_{\rm e} \to 0$. Furthermore, the average value of the dynamical variable just before a site relaxes is constant, so that the denominator in Equation (3.67) is constant and

$$\langle s \rangle \propto \langle \delta F \rangle L^2 \quad \text{for } 0 < \alpha < 1/4. \tag{3.72}$$

When $\alpha = 1/4$, the system is conservative in the bulk and the first term in the denominator in Equation (3.67) vanishes. Therefore, as $L \to \infty$, all dissipation effectively takes place at the edge since $P_c/P_e \to 0$ and only the second term in the denominator in Equation (3.67) remains. The average value of the dynamical variable just before a site relaxes is constant. However, numerical simulations show that for $\alpha = 1/4$, the fraction of edge sites slipping $P_e \propto 1/L^2$ unlike the case $0 \leq \alpha < 1/4$ where $P_e \propto 1/L$, see Figure 3.31. We will interpret this finding later. Equation (3.67) therefore reduces to

$$\langle s \rangle \propto \langle \delta F \rangle L^4 \quad \text{for } \alpha = 1/4. \tag{3.73}$$

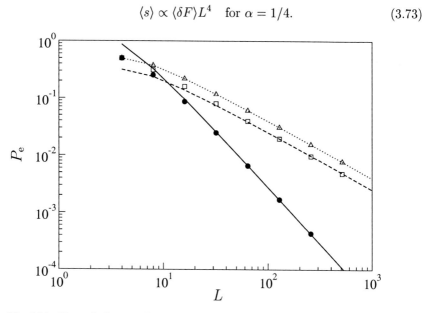

Fig. 3.31 Numerical results for the fractions of edge sites slipping, P_e, versus the system size, L, for the OFC model on square lattices of size $L = 1, 2, 4, \ldots, 512$. The data displayed are for the decoupled model where $\alpha = 0$ (open triangles), the non-conservative model with $\alpha = 0.20$ (open squares), and the conservative model where $\alpha = 1/4$ (solid circles). The dotted and dashed lines are the graph of $P_e \propto 4(L-2)/L^2 \propto 1/L$ for $L \gg 1$. The solid line is the graph of $P_e \propto (L-2)/L^3 \propto 1/L^2$ for $L \gg 1$.

Whether or not the average avalanche size diverges with system size depends on how the average drive in between avalanches, $\langle \delta F \rangle$, scales with system size. Assuming that there exists a function $\delta(\alpha)$ of the dimensionless ratio α such that

$$\langle \delta F \rangle \propto L^{-\delta(\alpha)}, \tag{3.74}$$

we can summarise the above results for the decoupled, non-conservative and conservative regimes as

$$\langle s \rangle \propto \begin{cases} L^{2-\delta(0)} & \text{for } \alpha = 0 & (3.75\text{a}) \\ L^{2-\delta(\alpha)} & \text{for } 0 < \alpha < 1/4 & (3.75\text{b}) \\ L^{4-\delta(1/4)} & \text{for } \alpha = 1/4. & (3.75\text{c}) \end{cases}$$

For $\alpha = 0$, we have argued that $\delta(0) = 2$ for all random initial configurations. The average avalanche size remains constant with system size.

For $\alpha = 1/4$, any excess dynamical variable can only be dissipated at the boundary. The avalanches must 'transport' excess dynamical variable added in the bulk through the system to the boundaries. Therefore, the average avalanche size diverges with system size. Hence $\delta(1/4) < 4$.

In fact, the numerical results for the average drive in between avalanches displayed in Figure 3.32 confirms that for the decoupled model, $\delta(0) = 2$ and indicates that for the conservative model, $\delta(1/4) \approx 1.75$.

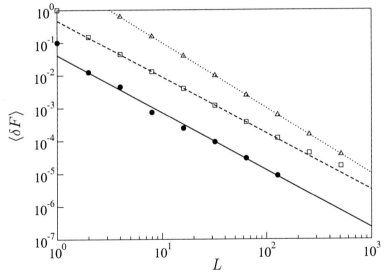

Fig. 3.32 Numerical results for the average drive in between avalanches, $\langle \delta F \rangle$, versus the system size, L, for the OFC model on lattices of size $L = 1, 2, 4, \ldots, 512$. The first 10^8 avalanches have been discarded and the data shown are averaged over further 10^8 avalanches. The data displayed are for the decoupled model where $\alpha = 0$ (open triangles), the non-conservative model with $\alpha = 0.20$ (open squares), and the conservative model where $\alpha = 1/4$ (solid circles). For clarity, the data have been shifted along the y-axis by multiplying with $10^1, 10^0$, and 10^{-1}, respectively. The average drive decay with system size, $\langle \delta F \rangle \propto L^{-\delta(\alpha)}$, with $\delta(0) = 2, \delta(0.20) \approx 1.72$, and $\delta(1/4) \approx 1.75$. The dotted, dashed, and solid straight lines have slopes $-2, -1.72$, and -1.75, respectively.

Since the average avalanche size is finite in the OFC model for $\alpha = 0$ and infinite for $\alpha = 1/4$ when $L \to \infty$, there exists a critical value, α_c, pinpointing a transition from finite average avalanche size when $0 \leq \alpha < \alpha_c$ to diverging average avalanche size when $\alpha \geq \alpha_c$.

For $0 < \alpha < 1/4$, dissipation takes place in the bulk and the average avalanche size diverges with system size only if the scaling exponent $\delta(\alpha) < 2$. Currently there are no theoretical predictions for $\delta(\alpha)$ in the range $0 < \alpha < 1/4$ and we must rely on numerical simulations for further investigations. We refer to the data in Figure 3.32 of the measured average drive in between avalanches for the OFC model with $\alpha = 0.20$ on lattices of size $L = 1, 2, 4, \ldots, 512$. For small system sizes, the data fall on the dashed line corresponding to $\delta(0.20) \approx 1.72$. For large system sizes, the data even appear to decay less rapidly, suggesting that the asymptotic scaling behaviour, assuming it exists, has $\delta(0.20) < 1.72$. Therefore, according to Equation (3.75b), the data for $\alpha = 0.20$ is consistent with an average avalanche size that diverges with system size.

However, it is still an open question at what value of α_c the transition from non-diverging to diverging average avalanche behaviour takes place.

3.10.7 Avalanche time series

Figure 3.33 displays two time series of avalanches in the OFC model. The time t is the accumulated drive per site measured after the system has reached a steady state. An avalanche of size s is rescaled by the size of the largest observed avalanche in the time series, $s_{\max}$.

Figure 3.33(a) is the time series for the decoupled OFC model where $\alpha = 0$ on a square lattice of size $L = 8$. All avalanches are of size one, so that $s_{\max} = 1$. The time in between the avalanches is a fingerprint of the dynamical variable in the initial configuration being uniformly distributed over the interval $[0, 1]$ when the threshold $F^{\text{th}} = 1$. Since the dynamical variables on the sites are entirely decoupled, the avalanche time series repeats itself with period $t = 1$, irrespective of system size.

Figure 3.33(b) is the time series for the conservative OFC model where $\alpha = 1/4$ on a square lattice of size $L = 256$. The avalanche time series for $\alpha = 1/4$ is totally different from that for $\alpha = 0$ with no apparent periodicity. There is a broad range of avalanche sizes spanning six orders of magnitude with $s_{\max} = 2\,834\,103$. The frequency of avalanches decays with size and there is no apparent characteristic avalanche size other than $s_{\max}$, which is increasing with system size.

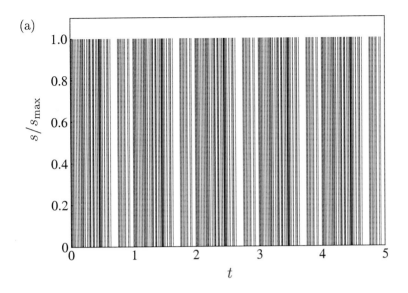

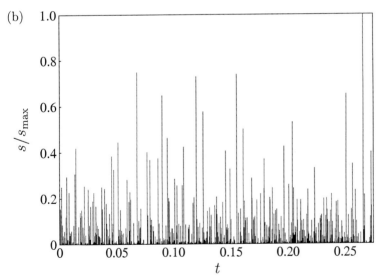

Fig. 3.33 Numerical results for the rescaled avalanche size, $s/s_{\max}$, versus the time, t, measured in terms of the drive per site (from an arbitrary initial time in a steady state) for the OFC model with threshold $F^{\text{th}} = 1$. (a) The time series for the decoupled model where $\alpha = 0$ on a square lattice of size $L = 8$. All avalanches are of size one and the series has period $t = 1$. (b) The time series for the conservative model where $\alpha = 1/4$ on a square lattice of size $L = 256$. The avalanche sizes span over six orders of magnitude, with the frequency of avalanches decaying with size. The largest avalanche recorded in a sequence of 10 000 avalanches is of size $s_{\max} = 2\,834\,103$.

Figure 3.34 displays the time series for the nonconservative OFC model with $\alpha = 0.20$ on a square lattice of size $L = 256$ in two different time intervals $0 \leq t \leq 2$ and $25 \leq t \leq 27$, respectively. The largest avalanche recorded within the two intervals $s_{\max} = 24\,584$.

Within the small time interval $0 \leq t \leq 2$, there appears to be a characteristic avalanche size with $s/s_{\max} \approx 0.4$ occurring periodically with period $t \approx 0.2$. Similarly, in the beginning of the time interval $25 \leq t \leq 27$, there appears to be a different characteristic avalanche size with $s/s_{\max} \approx 0.2$ although its periodicity is less clear cut. This pattern is disrupted by the largest avalanche recorded.

In fact, the temporal signal reflects the self-organisation of the dynamical variable into patches of approximately equal value [Middleton and Tang, 1995]. These patches of synchronised sites are responsible for the apparent periodic pulse behaviour. However, the patches are not robust: they have their own dynamics and interact with one another through avalanches, merging and breaking up. In some sense, the persistence of patches can be thought of as the system having a 'memory'.

The microscopic origin for the synchronisation is due to the relaxing sites decreasing to zero rather than by a fixed amount, see Equation (3.64). Consider the sites that have relaxed during an avalanche. After the avalanche has come to a halt, the approximate value of the dynamical variables on these sites will depend only on how many of their nearest-neighbours subsequently relaxed. Therefore, the values of the dynamical variables on these sites are distributed narrowly around multiples of α.

As mentioned previously, a synchronisation of the dynamical variable among sites may imply that the scaling exponent of the average drive with system size deviates from the trivial value, that is, $\delta(\alpha) < 2$. Therefore, for a non-conservative OFC model, it is the synchronisation that may allow the average avalanche size to diverge with system size. Hence, self-organised criticality may be present in non-conservative models.

This is surprising. The knowledge gained from the branching process seems to support the conjecture that the introduction of a non-conservative relaxation rule will leave the system sub-critical: a finite correlation length which does not scale with system size will appear [Hwa and Kardar, 1989; Grinstein et al., 1990]. However, we must bear in mind that a branching process only deals with systems without spatial or temporal correlations. Thus we cannot apply the results of branching processes to systems with spatio-temporal correlations.

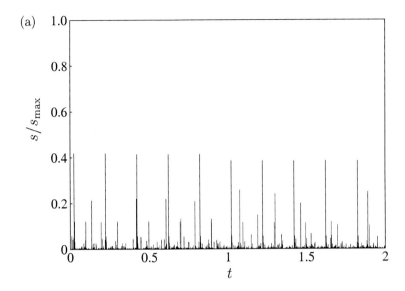

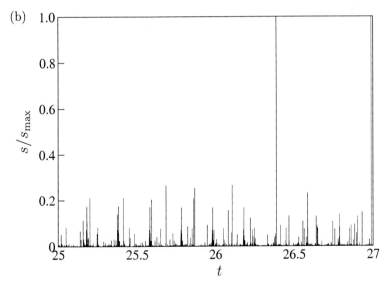

Fig. 3.34 Numerical result for the rescaled avalanche size, $s/s_{\max}$, versus the time, t, measured in terms of the drive per site (from an arbitrary initial time in a steady state) for the nonconservative OFC model with $\alpha = 0.20$ and threshold $F^{\text{th}} = 1$ on a square lattice of size $L = 256$. (a) The time interval $0 \leq t \leq 2$ consisting of 73 031 avalanches. (b) The time interval $25 \leq t \leq 27$ consisting of 63 079 avalanches. The largest avalanche recorded in the two sequences of avalanches is of size $s_{\max} = 24\,584$.

3.10.8 Avalanche-size probability

The avalanche-size probability is trivial for the decoupled OFC model where $\alpha = 0$. All avalanches are of size one if no sites are initiated with the same value of F.

In order to sample the avalanche-size probability for the OFC model when $0 < \alpha \leq 1/4$, we collect statistics of avalanche sizes when the system is in a steady state. We repeat simulations for a range of system sizes to measure the avalanche-size probability $P(s, \alpha; L)$.

Figure 3.35 displays $P(s, 1/4; L)$ versus the avalanche size s for the conservative OFC model where $\alpha = 1/4$ on square lattices of size $L = 32, 64, 128, 256$. The avalanche-size probability has a broad distribution.

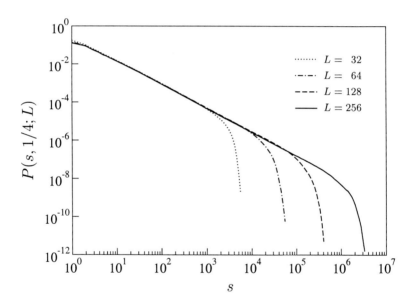

Fig. 3.35 Numerical results for the avalanche-size probability, $P(s, 1/4; L)$, versus the avalanche size, s, for the conservative OFC model where $\alpha = 1/4$ on lattices of size $L = 32, 64, 128, 256$ marked with lines of increasing dash length. There is no typical size of an avalanche except for a cutoff avalanche size which increases with system size. For large avalanche sizes, the avalanche-size probability is well approximated by a power-law decay, $P(s, 1/4; L) \propto s^{-\tau_s}$ for $1 \ll s \ll s_c$.

For large avalanches, the frequency decays with size approximately according to a power law before reaching a cutoff avalanche size which increases with system size.

To test whether the avalanche-size probability for the conservative OFC model satisfies the simple scaling ansatz in Equation (3.33), we estimate the critical exponents D and τ_s by applying the moment scaling analysis and then attempt a data collapse.

Following the standard recipe for moment analysis outlined in Section 3.7, we first determine $d\log\langle s^k\rangle/d\log L$ for $L \gg 1$ for each k which is an estimate of the exponent $D(1 + k - \tau_s)$, see Equation (3.40). Figure 3.36 displays the estimated exponents $D(1+k-\tau_s)$ versus the moment k. Linear regression of the resulting graph yields the critical exponents $D = 3.01$ and $\tau_s = 1.253$.

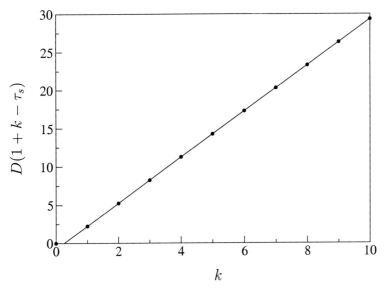

Fig. 3.36 Moment analysis of the avalanche-size probability for the OFC model with $\alpha = 1/4$ and F^{th} on lattices of size $L = 64, 128, 256$. The estimated exponents, $D(1+k-\tau_s)$, determined by $d\log\langle s^k\rangle/d\log L$ for $L \gg 1$ versus the moment, k. The slope of the straight line is $D = 3.01$, intersecting the k-axis at $k = 0.253$, implying $\tau_s = 1.253$.

Assuming that the avalanche-size probability for the conservative OFC model satisfies the simple finite-size scaling ansatz in Equation (3.33), we should be able to perform a data collapse using the critical exponents estimated from moment analysis. Figure 3.37(a) displays the transformed avalanche-size probability $s^{\tau_s}P(s, 1/4; L)$ versus the avalanche size s, using the critical exponent $\tau_s = 1.243$. The characteristic feature of each graph, namely the onset of the rapid decay of the avalanche-size probabilities, vertically align.

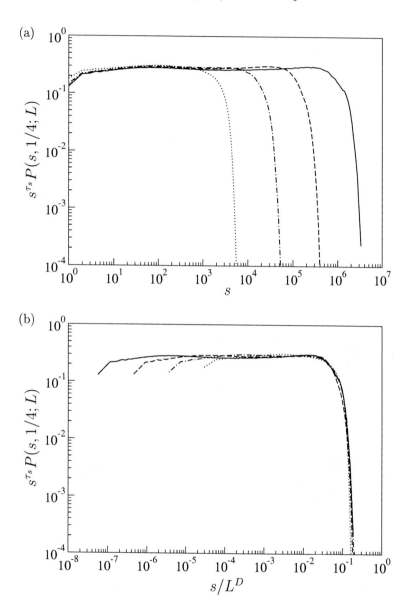

Fig. 3.37 Data collapse of the numerical results of the avalanche-size probabilities for the conservative OFC model where $\alpha = 1/4$ on lattices of size $L = 32, 64, 128, 256$ marked with lines of increasing dash length. (a) The transformed avalanche-size probability, $s^{\tau_s} P(s, 1/4; L)$, versus the avalanche size, s, using the critical exponent $\tau_s = 1.253$. (b) Transformed avalanche-size probability, $s^{\tau_s} P(s, 1/4; L)$, versus the rescaled avalanche size, s/s_c, where the cutoff avalanche size, $s_c(L) \propto L^D$, with $D = 3.01$. For $s \gg 1, L \gg 1$, the curves collapse onto the graph for the scaling function $\mathcal{G}^{\mathrm{OFC}}$.

Using $D = 3.01$, the avalanche size is then rescaled by the factor L^D, which horizontally aligns the characteristic feature of each graph, see Figure 3.37(b). Since the data collapse is of good quality, we conclude that the conservative OFC model displays self-organised criticality with the exponents $D \approx 3.01$ and $\tau_s \approx 1.253$.

There is a geometrical interpretation for the value of the critical exponent D which describes the scaling of the cutoff avalanche size with system size.

Figure 3.38 displays the spatial distribution of the number of topplings on a lattice of size $L = 64$ averaged over 500 avalanches with size $s \geq 40\,000$ for the conservative OFC model. Such avalanches belong to the tail of the avalanche-size probability for $L = 64$ in Figure 3.35. The volume under the surface represents the average total number of topplings. The scaling of the volume under the surface with system size L determines D. If the height of the surface scales linearly with system size, then $D = 3$.

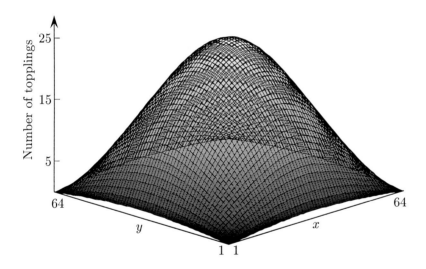

Fig. 3.38 Numerical results for the conservative OFC model where $\alpha = 1/4$ on a lattice of size $L = 64$. The spatial distribution of the number of topplings averaged over 500 avalanches with size $s \geq 40\,000$. The volume under the surface is a graphical representation of the cutoff avalanche size. The volume scales with system size like L^3, implying that the critical exponent $D = 3$ in the conservative OFC model.

Figure 3.39 displays the maximum height of the surface and the average number of topplings along the edge as a function of lattice size. The maximum height scales linearly with system size, consistent with $D = 3$. The average number of topplings at edge sites scales linearly with L. Since the total number of topplings scales like L^3, the fraction of edge sites slipping, $P_e \propto L/L^3 = 1/L^2$.

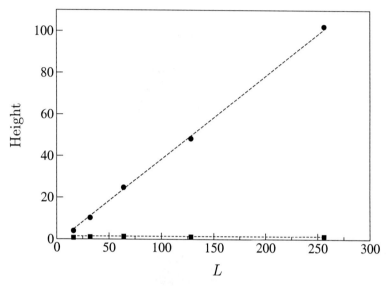

Fig. 3.39 Numerical results for the conservative OFC model where $\alpha = 1/4$ on square lattices of size $L = 16, 32, 64, 128, 256$. The maximum height (solid circles) and the average height along the edge (solid squares) of the surface displayed in Figure 3.38 as a function of system size. The maximum height scales linearly with system size for $L \gg 1$ while the average height along the edge is constant.

Figure 3.40(a) displays $P(s, 0.20; L)$ versus the avalanche size, s, for the non-conservative OFC model with $\alpha = 0.20$ on square lattices of size $L = 64, 128, 256, 512$. The avalanche-size probability is broad with an apparent power-law decay in the frequency with size, up to a cutoff that increases with system size.

To test whether the avalanche-size probabilities satisfy simple finite-size scaling, we once again attempt to extract estimates for the critical exponents D and τ_s by moment scaling analysis of the avalanche-size probability. Figure 3.40(b) displays the estimated exponents $D(1 + k - \tau_s)$ for the non-conservative OFC model on lattices of size $L = 128, 256, 512$, versus the moment k. Linear regression of the resulting graph yields the critical exponents $D = 2.2$ and $\tau_s = 1.8$. We notice that data fall outside this line for large k, indicating that simple finite-size scaling might not prevail.

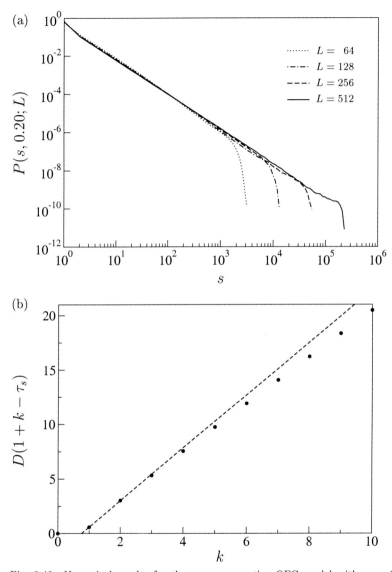

Fig. 3.40 Numerical results for the non-conservative OFC model with $\alpha = 0.20$ on square lattices. (a) The avalanche-size probability, $P(s, 0.20; L)$, versus the avalanches size, s on lattices of size $L = 64, 128, 256, 512$ marked with lines of increasing dash length. There is no apparent typical size of an avalanche except for a cutoff avalanche size which increases with system size. (b) Moment analysis of the avalanche-size probability on lattices of size $L = 128, 256, 512$. The estimated exponents, $D(1 + k - \tau_s)$, extracted as the slope $d\log\langle s^k\rangle/d\log L$ for $L \gg 1$, versus the moment, k. The slope of the straight line is $D = 2.2$, and the line intersects the k-axis at $k = 0.8$, implying $\tau_s = 1.8$.

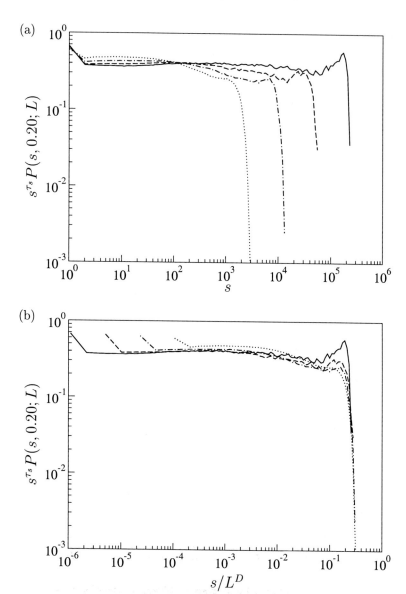

Fig. 3.41 Data collapse of the numerical results of the avalanche-size probabilities for the non-conservative OFC model with $\alpha = 0.20$ on lattices of size $L = 64, 128, 256, 512$ marked with lines of increasing dash length. (a) The transformed avalanche-size probability, $s^{\tau_s} P(s, 0.20; L)$, versus the avalanche size, s, using the critical exponent $\tau_s = 1.8$. (b) The transformed avalanche-size probability, $s^{\tau_s} P(s, 0.20; L)$, versus the rescaled avalanche size, s/s_c, where the cutoff avalanche size, $s_c(L) \propto L^D$, with $D = 2.2$. The data does not fall onto the graph for a scaling function $\mathcal{G}^{\mathrm{OFC}}$.

Assuming that the avalanche-size probability for the non-conservative OFC model satisfies the simple finite-size scaling ansatz in Equation (3.33), we should be able to perform a data collapse using the critical exponents estimated from moment analysis.

Figure 3.41(a) displays the transformed avalanche-size probability $s^{\tau_s} P(s, 0.20; L)$ versus the avalanche size s, using the critical exponent $\tau_s = 1.8$. Using $D = 2.2$, the avalanche size is then rescaled by the factor L^D, see Figure 3.41(b). Clearly, the data collapse is of rather poor quality even though the cutoff avalanche sizes are aligned horizontally.

Similar to the discussion related to the position α_c of the transition from non-diverging to diverging average avalanche behaviour, it is still an open question whether the OFC model displays self-organised criticality for $0 < \alpha < 1/4$.

3.11 Rainfall

Rain is a very familiar and important natural phenomenon, so much so that a large vocabulary exists for its description. One speaks of trickles, drizzle, bursts, showers, downpours, and torrents, or more prosaically, slight, moderate, heavy, and violent rain. This is somewhat reminiscent of the early Mercalli scale of earthquake magnitude based on eyewitness' reports of an earthquake's severity. This subjective scale was later replaced by the more objective and quantitative Richter scale, making the discovery of the Gutenberg-Richter law possible. Fortunately, recent advances in radar technology have also allowed for a more accurate quantitative study of rainfall.

The comparison between earthquakes and rain is deliberate rather than accidental. There are intriguing analogies between the mechanisms of the two phenomena, and we will shortly argue that there is, in fact, a rain equivalent of the Gutenberg-Richter law.

3.11.1 Rainfall mechanism

Energy arriving at the Earth from the Sun heats up the lower parts of the atmosphere and induces convective currents. These convective currents transport evaporated water to the atmosphere where it is stored intermediately in the form of clouds. When a saturation threshold is exceeded, a rain shower is triggered and the clouds release the water, see Figure 3.42.

Fig. 3.42 A rain 'event' over the Grand Canyon, Arizona, releasing energy from the atmosphere. Photo courtesy of Gary Jedlovec, NASA, U.S.A.

3.11.2 Rainfall time series

Rain data have been collected at the Baltic coast Zingst over the period 1.1.1999 – 1.7.1999 by the Max-Planck-Institute for Meteorology, Hamburg, using a vertically pointing Doppler radar developed by METEK. They measured the rain rate, $q(t)$, in mm/min in a particular volume of the sky 250 m above sea level with a one minute resolution. The sensitivity of the radar was 0.0001 mm/min and rain rates below this value were defined to be zero. In Figure 3.43(a), the observed rain rates at time t, measured in minutes from 00:00 on 1.1.1999 averaged over $\Delta t = 1$ min intervals, appear as peaks of height $q(t)$. Figure 3.43(b) shows a close-up of the time series of rain rates. There are two sequences of non-zero rain rates. We identify each of these sequences with a 'rain event'. The first event is of duration $T = 6$ min and the second event is of duration $T = 64$ min. Rain events are always separated by at least one minute of zero rain rate. In the same way that an earthquake is still in progress while the ground is trembling, or an avalanche is still in progress while grains are tumbling, so a rain event is still in progress while rain drops are falling.

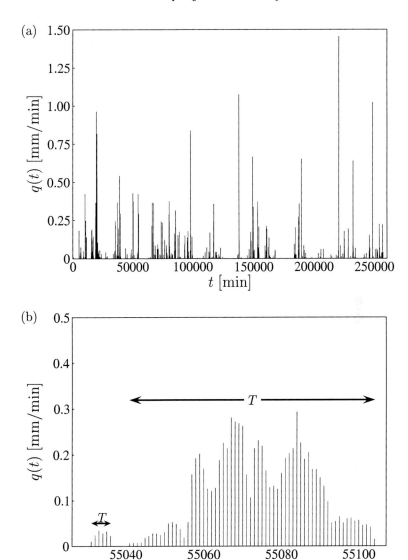

Fig. 3.43 (a) The rain rate, $q(t)$, in mm/min versus the time, t, measured in minutes from 00:00 on 1.1.1999 at the Baltic coast Zingst (54.43°N 12.67°E), Germany. The rain rates are averaged over $\Delta t = 1$ min intervals. (b) Two sequences of non-zero rain rates, each defining a 'rain event'. The event size, M, is the total released water measured in mm, that is, the time integral of the rain rates over an event. The first rain event is of duration $T = 6$ min and size $M = 0.15$ mm. The second rain event has duration $T = 64$ min and size $M = 7.74$ mm.

We now identify a proper observable for the size of a rain event, that is, the rain equivalent of the magnitude for an earthquake. Assume that the rain rate is zero at time t_0, non-zero from $(t_0 + 1)$ to $(t_0 + T)$, and zero at time $(t_0 + T + 1)$. The size, M, of the event of duration T, starting at $(t_0 + 1)$ and stopping at $(t_0 + T)$ is the total released water during that event, that is,

$$M = \sum_{t=t_0+1}^{t_0+T} q(t)\Delta t. \tag{3.76}$$

With the event size M as a measure for the energy released in the atmosphere, we can construct the time series of event sizes by plotting each event as a spike of height $M/M_{\max}$ at $(t_0 + T/2)$, where $M_{\max} = 33.17\,\text{mm}$ is the largest event observed in the complete series of $N = 1\,097$ rain events for the half year period, see Figure 3.44. The minimum rain event size for the series is $M_{\min} = 0.0001\,\text{mm}$. Therefore, the rain event sizes span over six orders of magnitude and, qualitatively, the frequency of events decays with size.

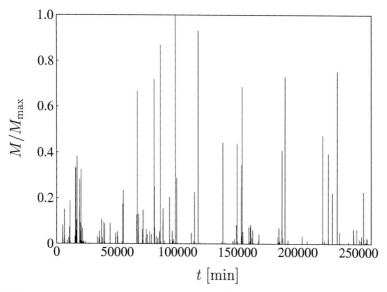

Fig. 3.44 Rain data for the rescaled event sizes, $M/M_{\max}$, versus the time, t, measured in minutes from 00:00 on 1.1.1999 at the Baltic coast Zingst, Germany. The largest event recorded in the sequence of $N = 1\,097$ events during the period 1.1.1999 − 1.7.1999 is of size $M_{\max} = 33.17\,\text{mm}$ and the smallest event detectable is of size $M_{\min} = 0.0001\,\text{mm}$. The event sizes span over six orders of magnitude, with the frequency of events decaying with size.

3.11.3 Rainfall-size number density

We proceed by counting the annual number of rain events in the interval $[M, M+dM]$ from the rain event time series, thereby obtaining the number density, $N(M)\,dM$, see Figure 3.45.

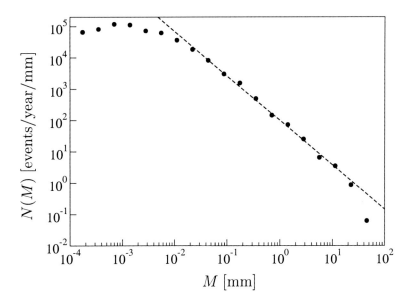

Fig. 3.45 The annual number density of rain events, $N(M)$, versus the event size, M, measured in mm (solid circles) based on data from METEK covering the period $1.1.1999 - 1.7.1999$ at the Baltic coast Zingst, Germany. There is no typical event size. For a range of event sizes, the annual number density of rain events with size M is well approximated by a power-law decay, $N(M) \propto M^{-\tau_M}$, for $0.01\,\text{mm} \lesssim M \lesssim 30\,\text{mm}$ with $\tau_M = 1.42$. The dashed straight line has slope -1.42.

The number density is consistent with a power-law decay,

$$N(M) \propto M^{-\tau_M} \quad \text{for } 0.01\,\text{mm} \lesssim M \lesssim 30\,\text{mm} \tag{3.77}$$

spanning more that three orders of magnitude in size M, and four orders of magnitude in number-frequency $N(M)$. The exponent describing the decay $\tau_M \approx 1.42$. This statistical law is the rain-equivalent of the Gutenberg-Richter law for earthquakes.

The deficit of events with size $M \lesssim 0.01\,\text{mm}$ indicates that a different physical process controls small events – it is not related to the sensitivity of the measuring device as in the case or earthquakes and avalanches in ricepiles.

There is also a deficit of events with size $M \gtrsim 30\,\text{mm}$, but it is not *a priori* clear whether this reflects insufficient statistics or a genuine cutoff in the data collected. One way to distinguish between these alternatives is to estimate the expected number of events, $N(M > 30)$, of size $M > 30\,\text{mm}$, assuming that the power-law decay in Equation (3.77) is valid for all sizes $M > 0.01\,\text{mm}$, and compare this to the observed number of events. First, we calculate the relative number of events of size $M > M_2$ and $M > M_1$,

$$\frac{N(M > M_2)}{N(M > M_1)} = \frac{\int_{M_2}^{\infty} M^{-\tau_M}\,dM}{\int_{M_1}^{\infty} M^{-\tau_M}\,dM} = \left(\frac{M_2}{M_1}\right)^{1-\tau_M}, \quad (3.78)$$

since $1 - \tau_M < 0$, such that the contribution from the upper limit vanishes for each integral. Rearranging for the expected number of events of size $M > M_2$,

$$N(M > M_2) = N(M > M_1)\left(\frac{M_2}{M_1}\right)^{1-\tau_M}. \quad (3.79)$$

Substituting $M_1 = 0.01\,\text{mm}$, $M_2 = 30\,\text{mm}$ and reading off $N(M > 0.01) \approx 756$ events per year from the data, we find that the expected number of events of size $M > 30\,\text{mm}$

$$N(M > 30) \approx 756\left(\frac{30}{0.01}\right)^{-0.42} \approx 26\,\text{events/year}. \quad (3.80)$$

However, only two such events were observed per year. Therefore, we conclude that the deficit of the events with size $M \gtrsim 30\,\text{mm}$ most likely reflects a genuine cutoff in the data collected.[4] We should have in mind, however, that this may not be a genuine cutoff of the *physical* process of energy release in the atmosphere, because the data are collected from a fixed volume, while the atmosphere may release energy before and after passing through the observation volume.

Whether the rain-equivalent of the Gutenberg-Richter law in Equation (3.77) holds worldwide is an open question that would be worthwhile pursuing.

---

[4]Such a procedure was not necessary in the discussion of the upper cutoff in earthquake data, since Figure 3.29 displays the annual number of earthquakes with size $S > s$ directly.

3.12 Summary

In this chapter, we have focused on non-equilibrium, slowly driven systems that display intermittent behaviour. Many natural phenomena displaying intermittent behaviour are characterised by being highly susceptible with large deviations away from the average response. Although extreme fluctuations away from the average response are very rare, they may, nonetheless, determine the average response.

In particular, we have examined avalanches in ricepiles, earthquakes in the crust of the Earth, and rain in the atmosphere. There are numerous other examples of non-equilibrium phenomena in Nature in which appropriately defined 'events' appear to obey a scale-free frequency-size distribution:

- Traffic jams: A traffic jam begins when cars get stuck at some instant in time. The accumulated time of all cars stuck in a jam is a measure of its size [Nagel and Paczuski, 1995].
- Relative price fluctuations of commodities: The price of commodities changes in time. The relative price fluctuation from month to month, say, is a measure of the size of the fluctuation [Mandelbrot, 1963].
- Extinction events in biology: An extinction event occurs whenever species become extinct. From fossil records one can estimate the size of extinction events by the percentage of species that became extinct during a given period, say, a few million years [Raup, 1986]. Furthermore, the extinction events apparently have no characteristic size and they occur intermittently. Therefore, fossil records indicate that evolution is not slow and gradual but rather rapid and intermittent, giving birth to the notion of 'punctuated equilibrium' [Eldrege and Gould, 1972].
- Solar flares: The Sun emits solar flares when intermittently stored magnetic energy in the solar atmosphere is suddenly released. The size of a solar flare is determined by the intensity of the associated X-rays [Dennis, 1985; Lu and Hamilton, 1991].

As opposed to critical phenomena in equilibrium systems, there is currently no existing all-encompassing theory of non-equilibrium slowly driven dissipative systems displaying criticality.

Self-organised criticality may offer an appropriate framework for dealing with intermittent relaxational processes, in which the statistics of events have no inherent scale. However, there is no clear understanding of the content and breadth of the classes of systems displaying self-organised criticality.

Let us reflect upon the general features shared by the ricepile, the crust of the Earth and the atmosphere, that we have considered in detail in this chapter. The general features of all three physical systems are summarised in Table 3.2.

Table 3.2 Avalanches, earthquakes and rain events are examples of relaxational processes of slowly driven systems. Analogies are drawn between three physical systems in terms of their mechanisms for adding, storing and releasing energy. The systems are capable of storing energy because of the threshold. Together with the slow driving rate, the threshold also gives rise to the intermittent release of energy.

System	Ricepile	Crust of the Earth	Atmosphere
Energy source	Addition of grains	Convective currents	Sun
Energy storage	Gravitational potential	Strain	Latent heat
Threshold origin	Friction	Friction	Saturation
Release of energy	Avalanche	Earthquake	Rain event

In all three physical systems, we were able to identify relaxational events, namely avalanches, earthquakes, and rain events, respectively.

All three systems are slowly driven, such that the average time in between events is much longer than the average duration of events. Put differently, since there is no energy dissipation in between events, the rate of energy input is slow compared to the rate of energy output during an event. In the ricepile, grains are added at a rate slow enough for the pile to relax in between additions. In the crust of the Earth, convective currents cause strain energy to build up across the tectonic plates at a rate which is compared to the rate of energy release during an earthquake. In the atmosphere, the Sun evaporates water at a rate which is slow compared to the rate of rain release during a rain event.

All three systems have the capacity to store a finite amount of energy because of the existence of a threshold for the relevant physical processes. In the ricepile, gravitational potential energy can be stored because of the friction among the grains. In the crust of the Earth, strain energy can be stored because of the friction among the tectonic plates. In the atmosphere, energy in the form of latent heat is stored because of the saturation threshold for condensation.

The combination of the slow drive and the thresholds enable all three systems to release the energy intermittently rather than continuously. In fact, it is exactly the intermittent release of energy that makes it possible to define events in the first place. Such intermittent behaviour is sometimes referred to as 'stick-slip' behaviour. The ricepile sticks as long as all local slopes are below the threshold slope, but slips when some local slopes exceed the threshold. The tectonic plates stick as long as the strain is below the static friction threshold, but slip when the strain exceeds the threshold. The atmosphere 'sticks' as long as the water vapour is below the saturation threshold, but 'slips' when the water vapour exceeds the saturation threshold.

Self-organised criticality offers a single unifying explanation for events of all sizes with no characteristic scale. Slowly driven non-equilibrium systems with threshold dynamics self-organise into a steady state, in which events of all sizes are caused by the same mechanism and decrease in frequency as a power law with size. There are no so-called 'freak' events that require special consideration.

Exercises

3.1 *Power-law probability density with exponent* -2.

You have been hired to build and manage a dam to prevent a river from overflowing. A natural spring guarantees that the level h of the river is always positive, that is, $h \geq h_{min} > 0$. Measurements of the level of the river once a day show that it is described by a power-law probability density,

$$P(h) = \begin{cases} Ah^{-2} & \text{for } h \geq h_{min} \\ 0 & \text{otherwise,} \end{cases} \quad (3.1.1)$$

where A is a non-zero constant. In the following, you may assume that there are no correlations between the daily measured levels of the river.

(i) Show that the probability density is normalised if $A = h_{min}$.

(ii) (a) What is the probability, on a given day, of measuring a level $h \geq h_{max}$ with $h_{max} \geq h_{min}$?

(b) On average, how many days would you have to wait to see one event with $h \geq h_{max}$?

(iii) Your job contract has a clause that if the dam overflows, you get fired.

(a) Can you build a dam that would guarantee your job forever? Explain your answer.

(b) You would like to keep your job for N years with probability p. How high should you build the dam?

(c) Find the height of the dam, assuming you want to keep your job for 10 years with 90% probability when $h_{min} = 0.01$ m.

(iv) (a) Find the average level $\langle h \rangle$ of the river.

(b) Discuss modification(s) to the probability density $P(h)$ that would make the model more physically realistic.

3.2 Olami-Feder-Christensen model.

A model system displaying self-organised criticality has an event size probability

$$P(s) = \begin{cases} s^{-\tau_s} & \text{for } s \geq 1 \\ 0 & \text{otherwise.} \end{cases} \quad (3.2.1)$$

(i) Derive an inequality for τ_s such that $P(s)$ can be normalised. Derive an inequality for τ_s such that the average avalanche size $\langle s \rangle = \sum_{s=1}^{\infty} s P(s)$ diverges.

Consider a model system where the level of conservation is defined by a parameter α and where the event size distribution is given by

$$P(s, \alpha) \propto \begin{cases} s^{-\tau_s} H(1 - s/s_\xi) & \text{for } s \geq 1 \\ 0 & \text{otherwise,} \end{cases} \quad (3.2.2a)$$

$$s_\xi(\alpha) \propto (\alpha_c - \alpha)^{-1/\sigma} \quad \text{for } \alpha < \alpha_c \quad (3.2.2b)$$

where s_ξ is a cutoff event size and Θ the Heaviside step function, that is, $\Theta(x) = 1$ for $x \geq 0$ and $\Theta(x) = 0$ for $x < 0$.

(ii) Make a sketch of $P(s, \alpha)$ for three different levels of conservation $\alpha_1 < \alpha_2 < \alpha_3 < \alpha_c$. What is the limiting function of $P(s)$ for $\alpha \to \alpha_c$?

(iii) Assume the scaling form of $P(s, \alpha)$ given in (ii). Derive a scaling relation between the scaling exponent γ_k describing the divergence of the kth moment $\langle s^k \rangle = \sum_{s=1}^{\infty} s^k P(s, \alpha)$ when $\alpha \to \alpha_c$ and the critical exponents τ_s and σ.

(iv) Define explicitly the dynamical rules for the two-dimensional Olami-Feder-Christensen (OFC) spring-block model on a square lattice in terms of the force F_i on block i and the parameter $\alpha = K/(4K+K_L)$, see Figure 3.30 on page 306. Discuss briefly the time scale of drive and the response, and the level of conservation.

(v) Define the concepts of the OFC model being (a) critical and (b) noncritical. How does the model behave for $\alpha = 0$? How would you expect the model to behave for $\alpha = 0.25$? Discuss briefly the possibility for a phase transition, that is, the existence of a critical value of α_c above which the model is critical and below which the model is noncritical.

3.3 Modified Bak-Tang-Wiesenfeld model on a tree-like lattice.

Consider a Bak-Tang-Wiesenfeld type sandpile model defined on a tree-like lattice with N sites, $i = 1, \ldots, N$, each having h_c downward neighbours, see Figure 3.3.1. Each site i is assigned an integer h_i. Grains are added at random sites $h_i \to h_i + 1$. When $h_i \geq h_c$, the site i topples and one grain is added to each of the h_c downward neighbours.

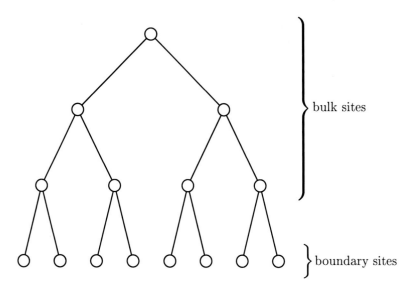

Fig. 3.3.1 A tree-like lattice structure with $h_c = 2$. The bottom most sites are boundary sites. The remaining sites are bulk sites.

The algorithm for the dynamics is:

1. Place the pile in an arbitrary stable configuration with $h_i < h_c$ for all sites i.
2. Add a grain at a random site i, that is, $h_i \to h_i + 1$.
3. If $h_i \geq h_c$, the site relaxes and

$$h_i \to h_i - h_c,$$
$$h_j \to h_j + 1 \quad \text{for the } h_c \text{ downward neighbours}$$

except when boundary sites topple where,

$$h_i \to h_i - h_c.$$

A stable configuration is reached when $h_i < h_c$ for all sites i.
4. Return to **2**.

(i) (a) How many stable configurations are there in total?
(b) Discuss briefly the concept of transient and recurrent configurations in sandpile models in general.
(c) Explain why both the minimally stable configuration with $h_i = h_c - 1$ for all sites i and the empty configuration with $h_i = 0$ for all sites i are recurrent states. Hence argue why all stable configurations are recurrent configurations on this tree-like structure.

(ii) (a) Let P_h denote the probability that a site contains h grains. The probability that the addition of a single grain to a stable configuration at a random site in the bulk will cause b of its h_c downward neighbours to topple is

$$p_b = \binom{h_c}{b} P_{h_c-1}^b (1 - P_{h_c-1})^{h_c-b} \quad b = 0, \ldots, h_c. \quad (3.3.1)$$

Justify this result.

(b) Argue why the average number of new topplings is

$$\langle b \rangle = h_c P_{h_c-1}. \quad (3.3.2)$$

(iii) You may assume that in a typical stable configuration

$$P_h = P_{h-1} \quad \text{for } h = 1, \ldots, h_c - 1. \quad (3.3.3)$$

Hence show that the average number of new topplings

$$\langle b \rangle = 1 \quad (3.3.4)$$

and comment on the result.

(iv) (a) Consider a tree with $h_c = 2$, see Figure 3.3.1. Discuss how this model is related to percolation on a Bethe lattice with coordination number $z = 3$.
(b) Using an argument similar to that for deriving the average cluster size in percolation on a Bethe lattice, show that the average avalanche size in an infinite tree with $h_c = 2$ is

$$\langle s \rangle = \frac{P_{h_c-1}}{1 - h_c P_{h_c-1}}. \quad (3.3.5)$$

3.4 *Oslo model and moments.*

Consider a one-dimensional granular pile on a finite horizontal base of length L confined between two glass plates. There is a vertical wall at the left boundary. Grains can leave the system at the right boundary.

(i) Using the metaphor of a one-dimensional granular pile driven by adding grains at the left boundary, explain briefly the concept of self-organised criticality.

(ii) (a) Define explicitly the dynamical algorithm for the one-dimensional Oslo rice pile model driven at the left boundary.

(b) For the Oslo rice pile model, explain the concept of recurrent configurations.

The one-dimensional Oslo model of linear size L displays self-organised criticality and the probability density of the avalanche size s satisfies

$$P(s; L) \propto s^{-\tau_s} \mathcal{G}(s/L^D) \quad \text{for } L \gg 1, s \gg 1, \quad (3.4.1)$$

where D is the avalanche dimension and τ_s the avalanche size exponent. The scaling function $\mathcal{G}(x)$ is a non-zero constant for $x \ll 1$ and decays rapidly for $x \gg 1$.

(iii) Assuming that Equation (3.4.1) is valid for all s, derive the scaling of the kth moment,

$$\langle s^k \rangle = \sum_{s=1}^{\infty} s^k P(s; L), \quad (3.4.2)$$

with system size $L \gg 1$ in terms of the exponents D and τ_s.

Numerical results of the kth moment in a variant of the Oslo model are show in Table 3.4.1.

Table 3.4.1 Numerical results of the moments $k = 1, 2, 4, 6$ of the avalanche-size probability in a variant of the Oslo model in lattices of size $L = 100, 400, 1\,600$.

Moment	$L = 100$	$L = 400$	$L = 1\,600$
$\langle s \rangle$	3.38×10^3	5.35×10^4	8.54×10^5
$\langle s^2 \rangle$	3.65×10^7	1.21×10^{10}	4.16×10^{12}
$\langle s^4 \rangle$	8.76×10^{15}	1.35×10^{21}	2.28×10^{26}
$\langle s^6 \rangle$	3.61×10^{24}	2.73×10^{32}	2.31×10^{40}

(iv) (a) Show that the numerical results of the first moment in this variant of the Oslo model are consistent with

$$\langle s \rangle \propto L^2 \quad \text{for } L \gg 1, \quad (3.4.3)$$

and hence derive a scaling relation between the critical exponents D and τ_s.

(b) Using the numerical measurements of your choice, determine the critical exponents D and τ_s.

3.5 *Moment ratios and universality.*

Consider a system of size L which displays self-organised criticality. Let s denote the avalanche size with a corresponding avalanche-size probability

$$P(s; L) = as^{-\tau_s} \mathcal{G}\left(s/bL^D\right) \quad \text{for } L \gg 1, s \gg 1, \qquad (3.5.1)$$

where τ_s and D are universal critical exponents and a and b are non-universal constants. The scaling function $\mathcal{G}(x)$ is a non-zero constant for $x \ll 1$ and decays rapidly for $x \gg 1$.

(i) Assume that the avalanche-size probabilities are measured in systems of sizes $L_4 \gg L_3 \gg L_2 \gg L_1 \gg 1$. Explain how to make a data collapse of the graphs for the avalanche-size probabilities. Illustrate your explanation with a sketch, clearly labelling your axes.

(ii) In the following you may assume that Equation (3.5.1) is valid for all avalanche sizes.

(a) Keeping track of the non-universal constants, show that the kth moment of the avalanche-size probability

$$\langle s^k \rangle = \sum_{s=1}^{\infty} s^k P(s; L) \to \Gamma_k L^{\gamma_k} \quad \text{for } L \to \infty, \qquad (3.5.2)$$

and identify the critical exponent γ_k and the critical amplitude Γ_k. Comment on whether they are universal.

(b) Express the moment ratio

$$g_k = \frac{\langle s^k \rangle \langle s \rangle^{k-2}}{\langle s^2 \rangle^{k-1}} \qquad (3.5.3)$$

in terms of the critical amplitudes and show that g_k is a universal quantity.

(iii) Consider three one-dimensional models A, B, and C displaying self-organised criticality in which g_3 has been measured numerically as a function of system size.

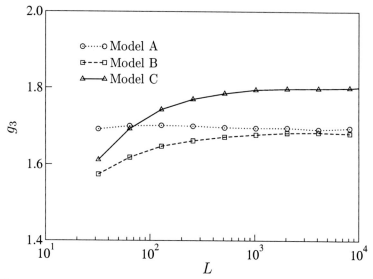

Fig. 3.5.1 The numerical results of moment ratio g_3 as a function of system size L for three models A, B, and C.

(a) Taking the numerical measurements of g_3 for Model C as an example, argue qualitatively why you would expect g_3 to be constant only for $L \to \infty$.

(b) Considering that g_3 is independent of non-universal constants, what can you conclude from the data about the three models with respect to universality classes?

Appendix A

Taylor Expansion

Let $f(x)$ be a function of one variable with a continuous nth derivative. The Taylor expansion of $f(x)$ about a point $x = x_0$ to order n is

$$f(x) \approx f(x_0) + \frac{f^{(1)}(x_0)}{1!}(x-x_0) + \frac{f^{(2)}(x_0)}{2!}(x-x_0)^2 + \cdots + \frac{f^{(n)}(x_0)}{n!}(x-x_0)^n$$

$$= \sum_{j=0}^{n} \frac{f^{(j)}(x_0)}{j!}(x-x_0)^j, \qquad (A.1)$$

where $f^{(j)}(x_0)$ is the jth derivative of the function $f(x)$ evaluated at $x = x_0$ with the convention that the 0th derivative $f^{(0)}(x_0) = f(x_0)$. Note that the factorial in the denominator ensures that the jth derivative of the Taylor expansion evaluated in $x = x_0$ equals $f^{(j)}(x_0)$. Therefore, the Taylor expansion Equation (A.1) is the polynomial of order n that best approximates the function $f(x)$ about the point $x = x_0$. The following Taylor expansions are about the point $x_0 = 0$ to order $n = 2$, see Figure A.1.

$$\ln(1+x) = x - \frac{x^2}{2} + \mathcal{O}(x^3), \qquad (A.2a)$$

$$\ln(1-x) = -x - \frac{x^2}{2} + \mathcal{O}(x^3), \qquad (A.2b)$$

$$\cosh^2 x = 1 + x^2 + \mathcal{O}(x^4), \qquad (A.2c)$$

$$\cosh x = 1 + \frac{x^2}{2} + \mathcal{O}(x^4), \qquad (A.2d)$$

$$\sinh x = x + \frac{x^3}{6} + \mathcal{O}(x^5), \qquad (A.2e)$$

$$\operatorname{sech} x = 1 - \frac{x^2}{2} + \mathcal{O}(x^4). \qquad (A.2f)$$

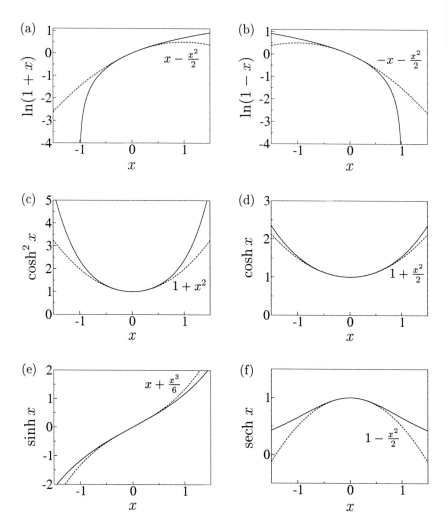

Fig. A.1 Taylor expansions (dashed lines) about the point $x_0 = 0$ to order $n = 2$ for the functions in Equation (A.2).

Appendix B

Hyperbolic Functions

The hyperbolic functions are defined as

$$\cosh x = \frac{\exp(x) + \exp(-x)}{2}, \tag{B.1a}$$

$$\sinh x = \frac{\exp(x) - \exp(-x)}{2}, \tag{B.1b}$$

$$\tanh x = \frac{\exp(x) - \exp(-x)}{\exp(x) + \exp(-x)}, \tag{B.1c}$$

$$\coth x = \frac{\exp(x) + \exp(-x)}{\exp(x) - \exp(-x)} \quad \text{for } x \neq 0, \tag{B.1d}$$

$$\operatorname{sech} x = \frac{2}{\exp(x) + \exp(-x)}, \tag{B.1e}$$

$$\operatorname{csch} x = \frac{2}{\exp(x) - \exp(-x)} \quad \text{for } x \neq 0. \tag{B.1f}$$

The graphs of the hyperbolic functions are shown in Figure B.1.

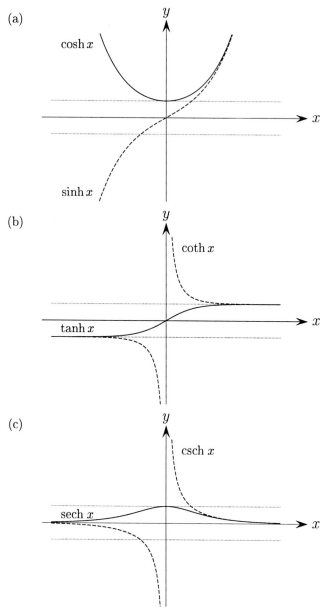

Fig. B.1 (a) Hyperbolic cosine (solid line) and sine (dashed line) of x. (b) Hyperbolic tangent (solid line) and cotangent (dashed lines) of x. (c) Hyperbolic secant (solid line) and cosecant (dashed lines) of x. The horizontal dotted lines are $y = 1$ and $y = -1$.

Appendix C

Homogeneous and Scaling Functions

A function $f(x)$ of one variable $x > 0$ is a homogeneous function if

$$f(\lambda^a x) = \lambda f(x) \quad \text{for all } \lambda > 0, \tag{C.1a}$$

and it is a power law with exponent $1/a$ if

$$f(x) = x^{1/a} f(1), \tag{C.1b}$$

where $f(1)$ is the value of the function at $x = 1$.

We will show that $f(x)$ is a homogeneous function if and only if $f(x)$ is a power law.

Assume that $f(x)$ is a homogeneous function. Since Equation (C.1a) is satisfied for all values of λ, we may choose $\lambda^a = 1/x$ and find, after simple rearrangement, that

$$\begin{aligned} f(x) &= \frac{1}{\lambda} f(1) \\ &= x^{1/a} f(1), \end{aligned} \tag{C.2a}$$

so that $f(x)$ is a power law.

Assume that $f(x)$ is a power law. Then

$$\begin{aligned} f(\lambda^a x) &= (\lambda^a x)^{1/a} f(1) \\ &= \lambda x^{1/a} f(1) \\ &= \lambda f(x), \end{aligned} \tag{C.2b}$$

so that $f(x)$ is a homogeneous function.

Note that a homogeneous function or, equivalently, a power law has the unique property that the relative change

$$\frac{f(\lambda^a x)}{f(x)} = \lambda \qquad (C.3)$$

is a constant and therefore independent of x. Since $(\log \lambda^a x - \log x) = \log \lambda^a$, the distance between the arguments of the numerator and denominator in Equation (C.3) is constant on a logarithmic scale. Therefore, for a function $f(x)$ to satisfy Equation(C.3), it must be a straight line on a double logarithmic plot, see Figure C.1(a).

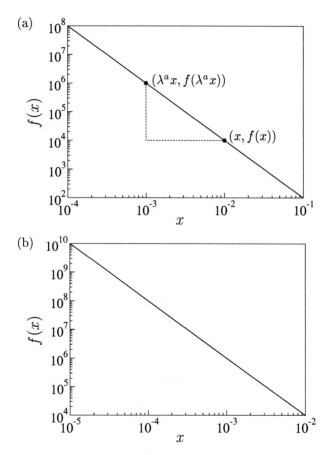

Fig. C.1 A power law $f(x) = x^{-2}$ corresponding to $a = -1/2$. (a) The graph of the function $f(x)$ for $x \in [10^{-4}, 10^{-1}]$. That the ratio $f(\lambda^a x)/f(x) = \lambda$ is independent of x is illustrated for $\lambda = 100$. (b) The graph of the function $f(x)$ for $x \in [10^{-5}, 10^{-2}]$ when the y-axis is rescaled by the factor $\lambda = 100$.

In addition, the graph of the function $f(x)$ for $x \in [x_0, x_1]$ is identical to the graph of the function $f(x)$ for $x \in [\lambda^a x_0, \lambda^a x_1]$ *provided* that the y-axis is rescaled by the factor λ. In this sense, a homogeneous function has no characteristic scale.

Figure C.1(a) displays a power law $f(x) = x^{-2}$, that is, $a = -1/2$, for $x \in [10^{-4}, 10^{-1}]$. Choosing $\lambda = 100$ so that $\lambda^a = 10^{-1}$, and plotting the power law for $x \in [10^{-5}, 10^{-2}]$, an identical graph appears when rescaling the y-axis by the factor $\lambda = 100$, see Figure C.1(b).

A function $f(x, y)$ of two variables is a generalised homogeneous function if

$$f(\lambda^a x, \lambda^b y) = \lambda f(x, y) \quad \text{for all } \lambda > 0, \tag{C.4a}$$

and it satisfies a scaling form with exponent $1/a$ if

$$f(x, y) = |x|^{1/a} \mathcal{G}_\pm\left(y/|x|^{b/a}\right), \tag{C.4b}$$

where $\mathcal{G}_\pm\left(y/|x|^{b/a}\right) = f\left(\pm 1, y/|x|^{b/a}\right)$ is the value of the function f at $x = \pm 1$ and the rescaled variable $y/|x|^{b/a}$. The function $\mathcal{G}_\pm$ is the so-called scaling function.

We will show that $f(x, y)$ is a generalised homogeneous function if and only if it can be recast in the scaling form Equation (C.4b).

Assume that $f(x, y)$ is a generalised homogeneous function. Since Equation (C.4a) is satisfied for all values of λ, we may choose $\lambda^a = 1/|x|$, and find after simple rearrangement that

$$\begin{aligned} f(x, y) &= \frac{1}{\lambda} f(\pm 1, \lambda^b y) \\ &= |x|^{1/a} f\left(\pm 1, y/|x|^{b/a}\right) \\ &= |x|^{1/a} \mathcal{G}_\pm\left(y/|x|^{b/a}\right), \end{aligned} \tag{C.5a}$$

where the scaling function, $\mathcal{G}_\pm$, is a function of the rescaled variable $y/|x|^{b/a}$ only, so that $f(x, y)$ satisfies the scaling form in Equation (C.4b).

Assume that $f(x, y)$ satisfies the scaling form in Equation (C.4b). Then

$$\begin{aligned} f(\lambda^a x, \lambda^b y) &= (\lambda^a |x|)^{1/a} \mathcal{G}\left(\lambda^b y/(\lambda^a |x|)^{b/a}\right) \\ &= \lambda |x|^{1/a} \mathcal{G}\left(y/|x|^{b/a}\right) \\ &= \lambda f(x, y), \end{aligned} \tag{C.5b}$$

so that $f(x, y)$ is a generalised homogeneous function.

Note that a generalised homogeneous function has the unique property that the relative change

$$\frac{f(\lambda^a x, \lambda^b y)}{f(x,y)} = \lambda \qquad (C.6)$$

is independent of x. Therefore, the graph of the function $f(x,y)$ for $(x,y) \in [x_0, x_1] \times [y_0, y_1]$ is identical to the graph of the function $f(x,y)$ for $(x,y) \in [\lambda^a x_0, \lambda^a x_1] \times [\lambda^b y_0, \lambda^b y_1]$ *provided* that the z-axis is rescaled by the factor λ. In this sense, a generalised homogeneous function has no characteristic scale.

Figure C.2(a) displays a generalised homogeneous function $f(x,y) = x^3 + y^2$, that is, $a = 1/3, b = 1/2$, for $(x,y) \in [1, 10] \times [1, 10]$. Choosing $\lambda = 27$ so that $\lambda^a = 3, \lambda^b \approx 5.2$, and plotting the graph for $(x,y) \in [3, 30] \times [5.2, 52]$, an identical graph appears when rescaling the z-axis by the factor $\lambda = 27$, see Figure C.2(b).

In practice, how can we show whether a function of two variables is a generalised homogeneous function? If we have the exact functional form of $f(x,y)$, we can simply test whether there exist two exponents a and b such that Equation (C.4a) is satisfied. Generally, we only have measurements of $f(x,y)$ as function of x for various choices of y without access to the exact functional form of $f(x,y)$. However, it is still possible to argue that the underlying function of two variables is a generalised homogeneous function by performing a data collapse. According to Equation (C.4b), plotting the transformed function $x^{-1/a} f(x,y)$ versus the rescaled argument $y/x^{b/a}$, all the data collapses onto the curve representing the graph of the scaling function $\mathcal{G}$.[1] Since Equation (C.4b) is equivalent to Equation (C.4a), the data collapse is a unique property of a generalised homogeneous function.

For example, Figure C.3(a) displays $f(x,y) = x^3 + y^2$ for $x \in [1, 10]$ and $y = 1, 3, 5, 7, 9$. The reader will recognise these curves from the surface displayed in Figure C.2(a). Figure C.3(b) displays the transformed function $x^{-3} f(x,y) = 1 + y^2/x^3$ for $x \in [1, 10]$. The data collapse in Figure C.3(c) reveals that $f(x,y) = 1 + \left(y/x^{3/2}\right)^2$ is only a function of the rescaled variable $y/x^{3/2}$.

---

[1] Note that for a homogeneous function of one variable, the transformed function, $x^{-1/a} f(x)$ is a constant, namely $f(1)$.

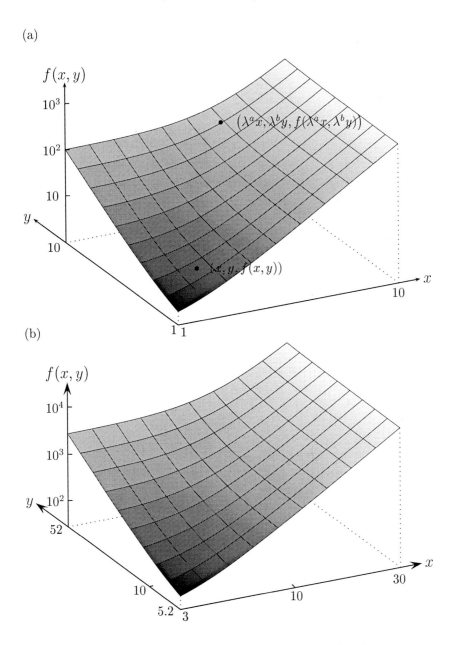

Fig. C.2 The homogeneous function of two variables $f(x,y) = x^3 + y^2$. (a) The graph of the function $f(x,y)$ for $(x,y) \in [1,10] \times [1,10]$. (b) The graph of the function $f(x,y)$ for $(x,y) \in [3,30] \times [5.2,52]$ when the z-axis is rescaled by the factor $\lambda = 27$.

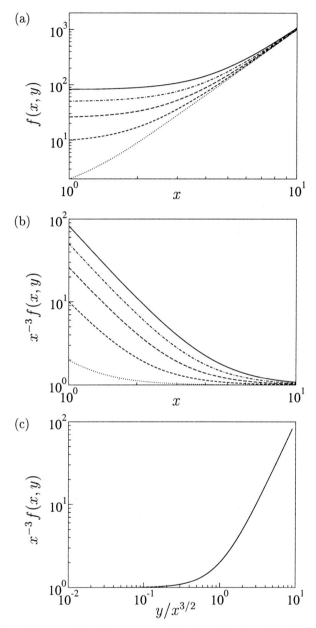

Fig. C.3 The generalised homogeneous function $f(x,y) = x^3 + y^2$ for $y = 1, 3, 5, 7, 9$ shown with lines of increasing dash length. (a) The function $f(x,y)$ for $x \in [1, 10]$. (b) The transformed function $x^{-3}f(x,y)$ for $x \in [1, 10]$. (c) The transformed function $x^{-3}f(x,y)$ as a function of the rescaled variable $y/x^{3/2}$.

Appendix D

Fractals

A fractal is a self-similar geometrical structure that looks alike on all length scales. There is no *a priori* way of identifying a characteristic scale nor is it possible to determine at which scale a fractal is being viewed.

The property of self-similarity implies that the mass $M(\ell)$ of a fractal geometrical structure grows with its linear size ℓ, according to

$$M(\ell) \propto \ell^D \quad \text{for } \ell \gg 1, \tag{D.1}$$

where the exponent D is the fractal dimension of the structure.[2] The density of a fractal geometrical structure

$$\rho(\ell) = \frac{M(\ell)}{\ell^d} \propto \ell^{D-d} \quad \text{for } \ell \gg 1 \tag{D.2}$$

will decay with the length scale on which it is observed when $D < d$, approaching zero for $\ell \to \infty$. For a homogeneous geometrical structure $D = d$ and the density approaches a constant for $\ell \to \infty$.

In practice, the fractal dimension of a geometrical structure is determined by measuring its mass as a function of the linear scale

$$D = \lim_{\ell \to \infty} \frac{\log M(\ell)}{\log \ell}. \tag{D.3}$$

Graphically, the fractal dimension is the slope of the graph of the logarithm of the mass $\log M(\ell)$ as a function of the logarithm of the linear scale $\log \ell$ for $\ell \to \infty$.

Fractals can be either deterministic or random. A deterministic fractal is created by applying a rule of some sort iteratively and indefinitely. Figure D.1 shows the first four iterations in creating a deterministic fractal known as the Sierpinski carpet.

---

[2] For a more rigorous discussion of fractal dimensions in general, see [Falconer, 2003].

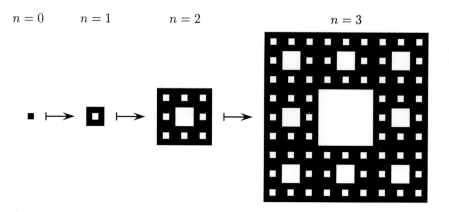

Fig. D.1 Iterative construction of the Sierpinski carpet. At each iteration n, the linear size ℓ is enlarged by a factor 3 and a black unit square is replaced with 3×3 unit squares in which the centre square is white, while a white unit square is replaced with 3×3 white unit squares.

After n iterations, the linear size of the geometrical structure $\ell = 3^n$. If the mass of a unit square (black or white) is 1, then at iteration n the mass of the geometrical structure of black squares $M_{\text{black}} = 8^n$ while the mass of the geometrical structure of white squares $M_{\text{white}} = 9^n - 8^n$.

Table D.1 At iteration n, the linear size $\ell = 3^n$ with a total number of 9^n unit squares of which 8^n are black. The remaining $9^n - 8^n$ unit squares are white.

Iteration	n	0	1	2	3
Linear size ℓ	3^n	1	3	9	27
Total number of unit squares	9^n	1	9	81	729
Number of black unit squares	8^n	1	8	64	512
Number of white unit squares	$9^n - 8^n$	0	1	17	217

The fractal dimension of the geometrical structure of black squares and white squares, respectively

$$D_{\text{black}} = \lim_{n \to \infty} \frac{n \log 8}{n \log 3} = \frac{\log 8}{\log 3} \approx 1.89, \tag{D.4a}$$

$$D_{\text{white}} = \lim_{n \to \infty} \frac{\log(9^n - 8^n)}{n \log 3} = \frac{\log 9}{\log 3} = 2. \tag{D.4b}$$

The geometrical structure of black squares is a fractal with fractal dimension $D_{\text{black}} \approx 1.89 < d$ which is less than the embedding dimension while the geometrical structure of white squares is compact with dimension $D_{\text{white}} = d$.

The densities of the geometrical structure of black squares and white squares are

$$\rho_{\text{black}} \propto \ell^{-0.11} \quad \text{for } \ell \gg 1, \tag{D.5a}$$

$$\rho_{\text{white}} \propto \ell^0 = \text{constant} \quad \text{for } \ell \gg 1. \tag{D.5b}$$

Hence, the density of the geometrical structure of black squares will decay with length scale and approach zero for $\ell \to \infty$ while the density of the geometrical structure will approach a constant.

Since the density of a fractal decreases with length scale, this has to be taken into account if one wants to estimate the density of a geometrical structure on a length scale ℓ_2, based on the density measured at another length scale ℓ_1. For example, consider an oil reservoir where the oil resides in connected pores inside a porous fractal material in $d = 3$. Samples of volumes 0.001 m^3, 0.008 m^3 and 0.064 m^3 from an oil field have densities of oil 250 kg m^{-3}, 177 kg m^{-3} and 125 kg m^{-3}, respectively. How much oil can be recovered from an oil field of volume 10^3km^3?

The graph of $\log \rho_{\text{oil}}(\ell)$ as a function of $\log \ell$ is a straight line with slope

$$D_{\text{oil}} - d = -0.5. \tag{D.6}$$

The length scale of the $d = 3$ oil field is $\ell_2 = 10000$ m. In order to calculate $M(\ell_2)$, we first note that the ratio of densities

$$\frac{\rho_{\text{oil}}(\ell_2)}{\rho_{\text{oil}}(\ell_1)} = \left(\frac{\ell_2}{\ell_1}\right)^{D_{\text{oil}} - d} \tag{D.7}$$

such that the density at length scale ℓ_2 is given by

$$\rho_{\text{oil}}(\ell_2) = \left(\frac{\ell_2}{\ell_1}\right)^{D_{\text{oil}} - d} \rho_{\text{oil}}(\ell_1) = 100000^{-0.5}\, 250 \text{ kg m}^{-3} \approx 0.79 \text{ kg m}^{-3}. \tag{D.8}$$

The amount of oil that can be recovered from the reservoir is

$$M_{\text{oil}}(\ell_2) = \ell_2^3 \rho_{\text{oil}}(\ell_2) = 10000^3 \text{ m}^3 0.79 \text{ kg m}^{-3} = 7.9\, 10^{11} \text{ kg}. \tag{D.9}$$

which is orders of magnitudes less than what one would have estimated by erroneously using one of the densities from the samples.

For a general review of fractals we recommend [Feder, 1988; Falconer, 2003].

Appendix E

Data Binning

When data are sampled in Nature or in numerical simulations, we have to analyse data with finite statistics. Assume, for example, that we try to estimate an underlying avalanche size probability, $P(s)$, by sampling a finite series of N avalanches $s_1, s_2, \ldots, s_N$. We define

$$P_N(s) = \frac{\text{No. of avalanches with } s_i = s}{N}. \tag{E.1}$$

As the sample size N increases, the sampled avalanche size probability $P_N(s)$ tends to the underlying avalanche size probability $P(s)$, that is,

$$P(s) = \lim_{N \to \infty} P_N(s)$$
$$\approx P_N(s) \quad \text{for } N \gg 1. \tag{E.2}$$

Let us sample a series of avalanches from a branching process with an underlying avalanche size probability $P(s)$, given by Equation (3.26). Choosing $p = 0.4950$, the avalanche size probability $P(s)$ decays like a power law with exponent $-3/2$ for $1 \ll s \ll s_c$, where the cutoff avalanche size $s_c \approx 10\,000$.

The left column in Figure E.1 from top to bottom displays $P_N(s)$ for increasing sample size $N = 10^4, 10^6, 10^8$. We observe that with increasing statistics, $P_N(s)$ probes the underlying avalanche size probability $P(s)$ for larger and larger avalanche sizes. Also, the approximation to the underlying avalanche size probability improves for small avalanches, with a decrease in scatter. However, for a sample of $N = 10^4$ avalanches, the statistics of $P_N(s)$ is insufficient to reveal that $P(s)$ actually has a finite cutoff. Only with increasing statistics does the finite cutoff reveal itself.

Data binning can, in this case, be applied to extract information on $P(s)$ that is present in the noisy tail of $P_N(s)$ where the statistics are poor. We divide the s-axis into bins labelled $j = 0, 1, \ldots$, where the jth bin covers the interval $[a^j, a^{j+1}[$. With $a > 1$, the bins are exponentially increasing in length. Let $s^j_{\min}$ and $s^j_{\max}$ denote the minimum and maximum integer avalanche sizes in bin j. Then $\Delta s^j = (s^j_{\max} - s^j_{\min} + 1)$ is the number of integers in the interval. We now count the number of avalanches that fall within bin j, that is, in the interval $\left[s^j_{\min}, s^j_{\max} \right]$ and define

$$\tilde{P}_N(s^j) = \frac{\text{No. of avalanches in bin } j}{N \Delta s^j} \qquad \text{(E.3a)}$$

$$s^j = \sqrt{s^j_{\min} s^j_{\max}}, \qquad \text{(E.3b)}$$

where s^j is the geometric mean of the avalanche sizes in bin j.

The right column in Figure E.1 from top to bottom displays the binned data $(s^j, \tilde{P}_N(s^j))$ of $P_N(s)$ for $N = 10^4, 10^6, 10^8$ using the factor $a = 1.1$. Clearly, data binning reduces the noise for large avalanche sizes and approximates the underlying avalanche size probability well. Note, however, that there is no one-to-one correspondence between $P_N(s)$ and $\tilde{P}_N(s^j)$. After assigning an avalanche size s_i to a bin j, we only know that the avalanche is somewhere in the interval $\left[s^j_{\min}, s^j_{\max} \right]$ but we lose information on exactly where.

Whenever appropriate, we have applied the procedure of data binning throughout the book.

Data Binning 357

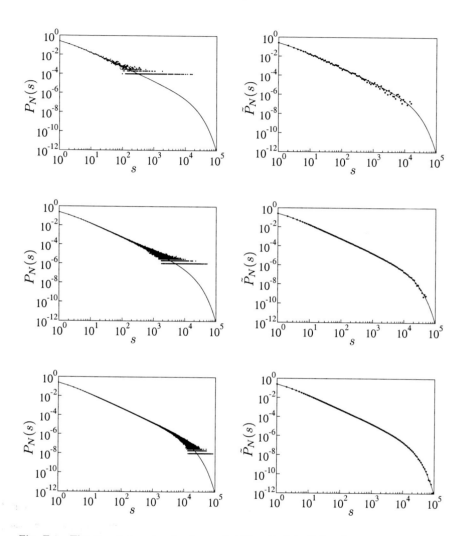

Fig. E.1 The sampled avalanche-size probability, $P_N(s)$, (left column, solid circles) and the binned avalanche-size probability, $\tilde{P}_N(s^j)$, (right column, solid circles) for $N = 10^4, 10^6, 10^8$ (top to bottom) compared with the exact avalanche-size probability, $P(s)$, (solid line) in a branching process with branching ratio $\langle b \rangle = 0.99$. The factor $a = 1.1$ has been applied.

Appendix F

Boltzmann Distribution

Consider a large isolated system at equilibrium with constant energy E^{tot} at temperature T. We now divide this system into a subsystem and a reservoir, such that the number of states available to the reservoir is vastly greater than that available to the subsystem. The subsystem can exchange energy with the reservoir, but the total energy must remain constant,

$$E_r + E'_r = E^{\text{tot}} = \text{constant}, \tag{F.1}$$

where E_r is the energy of the subsystem in state r, and E'_r is the corresponding energy of the reservoir. In addition, in equilibrium the subsystem and the reservoir have a common temperature T. Our task is to determine the probability that the subsystem has energy E_r.

Suppose that the subsystem is in a specified state with energy E_r. The reservoir can be in any microstate consistent with constant energy $E^{\text{tot}} - E_r$, and all such microstates are assumed to be equally likely. Therefore, the probability p_r that the subsystem is at energy E_r is proportional to the number $\Omega'(E^{\text{tot}} - E_r)$ of microstates that the reservoir can assume, consistent with that energy. That is,

$$p_r \propto \Omega'\left(E^{\text{tot}} - E_r\right). \tag{F.2}$$

Now since the reservoir is much larger than the subsystem, $E^{\text{tot}} \gg E_r$. This suggests expanding the number of states accessible to the reservoir in powers of E_r. For reasons of convergence, the expansion is performed on the logarithm of the number of states,

$$\ln \Omega'\left(E^{\text{tot}} - E_r\right) = \ln \Omega'\left(E^{\text{tot}}\right) + \left(\frac{\partial \ln \Omega'}{\partial E_r}\right)_{E_r=0} E_r + \cdots. \tag{F.3}$$

The energy of the whole system is fixed; therefore the first term on the right-hand side is a constant. The partial derivative in the second term is nothing other than the statistical mechanical definition of inverse temperature, $\beta' = 1/k_B T'$, for the reservoir. The subsystem and reservoir are at the same temperature, $\beta = \beta'$, so that the sought after probability that the subsystem has energy E_r is given by

$$p_r \propto \exp\left(-\beta E_r\right), \qquad (\text{F.4})$$

or, after normalisation,

$$p_r = \frac{\exp\left(-\beta E_r\right)}{\sum_r \exp\left(-\beta E_r\right)}, \qquad (\text{F.5})$$

where the sum runs over all states accessible to the subsystem.

Appendix G

Free Energy

Recall that in the canonical ensemble the free energy,

$$F = \langle E \rangle - TS, \tag{G.1}$$

is minimised in equilibrium when the temperature and the external field are fixed. To see this, we examine the sign of the change of the free energy when the system exchanges a small amount of heat, $đQ$, with its surrounding heat bath. In differential form, the free energy

$$dF = d\langle E \rangle - TdS - SdT. \tag{G.2}$$

In the canonical ensemble the term SdT is zero, because the temperature is fixed. Also, no work is performed by the system since the external field is fixed. Therefore, all the exchanged heat goes into increasing the system's internal energy, so that

$$dF = đQ - TdS. \tag{G.3}$$

The second law of thermodynamics states that

$$dS \geq đQ/T. \tag{G.4}$$

Therefore the change in the free energy

$$dF \leq 0, \tag{G.5}$$

which implies that the free energy is indeed minimised once the system has attained equilibrium.

Appendix H

Metropolis Algorithm

The definition of the Ising model contains no information on its dynamics. However, we know that in an ensemble of identical systems at temperature T and external field H, the fraction of systems in a particular microstate $\{s_i\}$ are given by the Boltzmann distribution Equation (2.8). A Metropolis algorithm is designed to sample the Boltzmann distribution by artificially imposing dynamics on the Ising model. Let $\{s_i\}$ denote a microstate of the Ising model with energy $E_{\{s_i\}}$ and $\{s_i'\}$ a microstate with energy $E_{\{s_i'\}}$. Let the probability of a transition from $\{s_i\}$ to $\{s_i'\}$ be given by

$$P(\{s_i\} \to \{s_i'\}) = \begin{cases} \exp[-\beta(E_{\{s_i'\}} - E_{\{s_i\}})] & \text{for } E_{\{s_i'\}} > E_{\{s_i\}} \\ 1 & \text{for } E_{\{s_i'\}} \leq E_{\{s_i\}}. \end{cases} \quad \text{(H.1)}$$

The Metropolis algorithm [Metropolis et al., 1953] is then given as follows:

1. *Initialisation.* Prepare the system in an arbitrary microstate $\{s_i\}$.
2. Calculate the energy $E_{\{s_i\}}$. Choose a random site j and calculate the energy $E_{\{s_i'\}}$ of the microstate $\{s_i'\}$ resulting from flipping the jth spin (i.e., $s_j \to -s_j$) in microstate $\{s_i\}$. With probability $P(\{s_i\} \to \{s_i'\})$ accept the new microstate.
3. Return to **2**.

Consider an ensemble with $N_{\{s_i\}}$ systems in microstate $\{s_i\}$ and $N_{\{s_i'\}}$ systems in microstate $\{s_i'\}$. Assume, without loss of generality, that $E_{\{s_i'\}} > E_{\{s_i\}}$. Then the number of transitions

$$N(\{s_i\} \to \{s_i'\}) \propto N_{\{s_i\}} \exp[-\beta(E_{\{s_i'\}} - E_{\{s_i\}})], \quad \text{(H.2a)}$$

$$N(\{s_i'\} \to \{s_i\}) \propto N_{\{s_i'\}}. \quad \text{(H.2b)}$$

First we show that the system reaches a steady state where the number of transitions $N(\{s_i\} \to \{s_i'\})$ equals $N(\{s_i'\} \to \{s_i\})$. The net flow out of microstate $\{s_i\}$ is

$$\Delta N(\{s_i\} \to \{s_i'\}) = N(\{s_i\} \to \{s_i'\}) - N(\{s_i'\} \to \{s_i\})$$
$$\propto N_{\{s_i\}} \exp[-\beta(E_{\{s_i'\}} - E_{\{s_i\}})] - N_{\{s_i'\}}$$
$$\propto N_{\{s_i\}} \left(\frac{\exp(-\beta E_{\{s_i'\}})}{\exp(-\beta E_{\{s_i\}})} - \frac{N_{\{s_i'\}}}{N_{\{s_i\}}} \right). \quad \text{(H.3)}$$

Assume that $\Delta N(\{s_i\} \to \{s_i'\}) > 0$: there is a net flow from microstate $\{s_i\}$ into microstate $\{s_i'\}$. This will increase the ratio $N_{\{s_i'\}}/N_{\{s_i\}}$ until $\Delta N(\{s_i\} \to \{s_i'\}) = 0$.

Assume that $\Delta N(\{s_i\} \to \{s_i'\}) < 0$: there is a net flow into microstate $\{s_i\}$ from microstate $\{s_i'\}$. This will decrease the ratio $N_{\{s_i'\}}/N_{\{s_i\}}$ until $\Delta N(\{s_i\} \to \{s_i'\}) = 0$.

Therefore, the system reaches a steady state with $\Delta N(\{s_i\} \to \{s_i'\}) = 0$ where

$$\frac{N_{\{s_i\}}}{N_{\{s_i'\}}} = \frac{p_{\{s_i\}}}{p_{\{s_i'\}}} = \frac{\exp(-\beta E_{\{s_i\}})}{\exp(-\beta E_{\{s_i'\}})} \quad \text{(H.4)}$$

which is the desired ratio derived from the Boltzmann probability distribution.

The choice of the transition probability in Equation (H.1) is not unique. Other choices exist that also correctly sample the microstates of the Ising model according the Boltzmann probability distribution. Finally, we note that the Metropolis algorithm is particularly inefficient in the vicinity of the critical point $(T_c, 0)$, and that other 'cluster-flipping' (rather than 'spin-flipping') algorithms are more efficient and should be used. Whenever appropriate, we have applied the so-called Wolff algorithm [Wolff, 1989].

Bibliography

Adler, J., Meir, Y., Aharony, A., and Harris, A. B. (1990). Series Study of Percolation Moments in General Dimension. *Phys. Rev. B*, 41:9183.
Amit, D. J. (1989). *Modeling Brain Function: the World of Attractor Neural Networks*. Cambridge University Press.
Anderson, P. W. (1991). Is Complexity Physics? Is it Science? What is it? *Phys. Today*, 44:9.
Arisue, H., Fujiwara, T., and Tabata, K. (2004). Higher Orders of the High-Temperature Expansion for the Ising Model in Three Dimensions. *Nucl. Phys. B (Proc. Suppl.)*, 129&130:774.
Bak, P. (1996). *How Nature Works: the Science of Self-Organized Criticality*. Springer-Verlag, New York.
Bak, P., Tang, C., and Wiesenfeld, K. (1987). Self-Organized Criticality: An Explanation of $1/f$ Noise. *Phys. Rev. Lett.*, 59:381.
Bak, P., Tang, C., and Wiesenfeld, K. (1988). Self-Organized Criticality. *Phys. Rev. A*, 38:364.
Ball, Z., Phillips, H. M., Callahan, D. L., and Sauerbrey, R. (1994). Percolative Metal-Insulator Transition in Excimer-Laser Irradiated Polyimide. *Phys. Rev. Lett.*, 73:2099–2102.
Ballesteros, H. G., Fernández, L. A., Martín-Mayor, V., Muñoz Sudupe, A., Parisi, G., and Ruiz-Lorenzo, J. J. (1997). Measures of Critical Exponents in the Four-Dimensional Site Percolation. *Phys. Lett. B*, 400:346.
Ballesteros, H. G., Fernández, L. A., Martín-Mayor, V., Muñoz Sudupe, A., Parisi, G., and Ruiz-Lorenzo, J. J. (1999). Scaling Corrections: Site Percolation and Ising Model in Three Dimensions. *J. Phys. A*, 32:1.
Barber, M. (1983). Finite Size Scaling. In Domb, C. and Lebowitz, J., editors, *Phase Transitions and Critical Phenomena*, volume 8. Academic Press, London.
Barouch, E., McCoy, B. M., and Wu, T. T. (1973). Zero-Field Susceptibility of the Two-Dimensional Ising Model near T_c. *Phys. Rev. Lett.*, 31:1409.
Baxter, R. J. (1982). *Exactly Solved Models in Statistical Mechanics*. Academic Press.
Binder, K. and Luijten, E. (2001). Monte Carlo Tests of Renormalization-Group

Predictions for Critical Phenomena in Ising Models. *Phys. Rep.*, 344:179.
Brush, S. G. (1967). History of the Lenz-Ising Model. *Rev. Mod. Phys.*, 39:883.
Burridge, R. and Knopoff, L. (1967). Model and Theoretical Seismicity. *Bull. Seismol. Soc. Am.*, 57:341.
Callen, H. B. and Welton, T. A. (1951). Irreversibility and Generalized Noise. *Phys. Rev.*, 83:34.
Cardy, J. L., editor (1988). *Finite-Size Scaling*. North Holland, Amsterdam.
Cardy, J. L. (1996). *Scaling and Renormalization in Statistical Physics*. Cambridge University Press.
Cardy, J. L. and Grassberger, P. (1985). Epidemic Models and Percolation. *J. Phys. A*, 18:L267.
Christensen, K., Corral, A., Frette, V., Feder, J., and Jøssang, T. (1996). Tracer Dispersion in a Self-Organized Critical System. *Phys. Rev. Lett.*, 77:107.
Christensen, K. and Olami, Z. (1993). Sandpile Models with and without an Underlying Spatial Structure. *Phys. Rev. E*, 48:3361.
Chua, A. and Christensen, K. (2002). Exact Enumeration of the Critical States in the Oslo Model. *cond-mat/0203260*.
Cohen, R., ben Avraham, D., and Havlin, S. (2002). Percolation Critical Exponents in Scale-Free Networks. *Phys. Rev. E*, 66:036113–036117.
Corral, A. (2004). Calculation of the Transition Matrix and of the Occupation Probabilities for the States in the Oslo Sandpile Model. *Phys. Rev. E*, 69:026107.
De Menech, M., Stella, A. L., and Tebaldi, C. (1998). Rare Events and Breakdown of Simple Scaling in the Abelian Sandpile Model. *Phys. Rev. E*, 58:R2677.
den Nijs, M. P. M. (1979). A Relation between the Temperature Exponents of the Eight-Vertex and q-state Potts Models. *J. Phys.*, 12:1857.
Dennis, B. R. (1985). Solar Hard X-Ray Bursts. *Solar Phys.*, 100:465.
Dhar, D. (2004). Steady State and Relaxation Spectrum of the Oslo Rice-Pile Model. *Physica A*, 340:535.
Drossel, B. (2000). Scaling Behavior of the Abelian Sandpile Model. *Phys. Rev. E*, 61:R2168.
Eldrege, N. and Gould, S. J. (1972). Punctuated Equilibria: an Alternative to Phyletic Gradualism. In Schopf, T. J. M., editor, *Models in Paleobiology*, page 82. San Francisco: Freeman, Cooper & Co.
Falconer, K. (2003). *Fractal Geometry*. John Wiley & Sons, 2nd edition.
Feder, J. (1988). *Fractals*. New York: Plenum Press, 2nd edition.
Fisher, M. E. (1998). Renormalization Group Theory: Its basis and Formulation in Statistical Physics. *Rev. Mod. Phys.*, 70:653.
Fisher, M. E. and Essam, J. W. (1961). Some Cluster Size and Percolation Problems. *J. Math. Phys.*, 2:609.
Francesco, P. D., Mathieu, P., and Sénéchal, D. (2001). *Conformal Field Theory*. Springer-Verlag, New York, 2nd edition.
Frette, V., Christensen, K., Malthe-Sørenssen, A., Feder, J., Jøssang, T., and Meakin, P. (1996). Avalanche Dynamics in a Pile of Rice. *Nature*, 379:49.
Gaunt, D. S. and Sykes, M. F. (1973). Estimation of Critical Indices for the Three-Dimensional Ising Model. *J. Phys. A: Math., Nucl. Gen.*, 6:1517.

Goldenfeld, N. (1992). *Lectures on Phase Transitions and the Renormalisation Group*. Perseus Publishing.
Griffiths, R. B. (1972). Rigorous Results and Theorems. In Domb, C. and Green, M., editors, *Phase Transitions and Critical Phenomena*, volume 1, page 7. Academic Press, London.
Grimmett, G. (1999). *Percolation*. Springer-Verlag, Berlin Heidelberg, 2nd edition.
Grinstein, G., Lee, D., and Sachdev, S. (1990). Conservation Laws, Anisotropy, and 'Self-Organized Criticality' in Noisy Nonequilibrium Systems. *Phys. Rev. Lett.*, 64:1927.
Gutenberg, B. and Richter, C. F. (1944). Frequency of Earthquakes in California. *Bull. Seismol. Soc. Am.*, 34:185.
Herrmann, H. J. and Roux, J., editors (1990). *Statistical Models for the Fracture of Disordered Media*. North Holland, Amsterdam.
Hopfield, J. J. (1982). Neural Networks and Physical Systems with Emergent Collective Computational Abilities. *Proc. Natl. Acad. Sci.*, 79:2554.
Hsu, H.-P., Nadler, W., and Grassberger, P. (2005). Simulations of Lattice Animals and Trees. *J. Phys. A*, 38:775.
Hwa, T. and Kardar, M. (1989). Dissipative Transport in Open Systems: An Investigation of Self-Organized Criticality. *Phys. Rev. Lett.*, 62:1813.
Ising, E. (1925). Beitrag zur Theorie des Ferromagnetismus. *Z. Phys.*, 31:253.
Jensen, H. J. (1998). *Self-Organised Criticality*. Cambridge University Press.
Jensen, I. (2001). Enumerations of Lattice Animals and Trees. *J. Stat. Phys.*, 102:865.
Kadanoff, L. P. (1966). Scaling Laws for Ising Models Near T_c. *Physics*, 2:263.
Kesten, H. (1980). The Critical Probability of Bond Percolation on the Square Lattice Equals 1/2. *Commun. Math. Phys.*, 74:41.
King, P. R., Buldyrev, S. V., Dokholyan, N. V., Havlin, S., Lee, Y., Paul, G., and Stanley, H. E. (1999). Applications of Statistical Physics to the Oil Industry: Predicting Oil Recovery using Percolation Theory. *Physica A*, 274:60.
Kramers, H. A. and Wannier, G. H. (1941). Statistics of the Two-dimensional Ferromagnet. Part I. *Phys. Rev.*, 60:247.
Landau, L. D. (1937). *Phys. Z. Sowjetunion*, 11:26.
Lenz, W. (1920). Beitrag zum Verständnis der Magnetischen Erscheinungen in festen Körpern. *Phys. Z.*, 21:613.
Lorenz, C. D. and Ziff, R. M. (1998a). Precise Determination of the Bond Percolation Thresholds and Finite-Size Scaling Corrections for the SC, FCC, and BCC Lattices. *Phys. Rev. E*, 57:230.
Lorenz, C. D. and Ziff, R. M. (1998b). Universality of the Excess Number of Clusters and the Crossing Probability Function in Three-Dimensional Percolation. *J. Phys. A*, 31:8147.
Lu, E. T. and Hamilton, R. J. (1991). Avalanches and the Distribution of Solar Flares. *Astrophys. J.*, 380:L89.
Majewski, J., Li, H., and Ott, J. (2001). The Ising Model in Physics and Statistical Genetics. *Am. J. Hum. Genet.*, 69:853.

Mandelbrot, B. (1963). The Variation of Certain Speculative Prices. *J. Business*, 36:394.
Maris, H. J. and Kadanoff, L. P. (1978). Teaching the Renormalization Group. *Am. J. Phys.*, 46:652.
McKenzie, D. P. and Parker, R. L. (1967). The North Pacific: An Example of Tectonics on a Sphere. *Nature*, 216:1276.
Metropolis, N., Rosenbluth, A. W., Rosenbluth, M. N., Teller, A. H., and Teller, E. (1953). Equation of State Calculations by Fast Computing Machines. *J. Chem. Phys.*, 21:1087.
Middleton, A. A. and Tang, C. (1995). Self-Organized Criticality in Nonconserved Systems. *Phys. Rev. Lett.*, 74:742.
Nagel, K. and Paczuski, M. (1995). Emergent Traffic Jams. *Phys. Rev. E*, 51:2909.
Nelson, D. R. and Fisher, M. E. (1975). Soluble Renormalization Groups and Scaling Fields for Low-Dimensional Ising Systems. *Ann. Phys.*, 91:226.
Newman, M. E. J. and Ziff, R. M. (2000). Efficient Monte Carlo Algorithm and High-Precision Results for Percolation. *Phys. Rev. Lett.*, 85:4104.
Nienhuis, B. (1982). Exact Critical Point and Critical Exponents of $\mathcal{O}(n)$ Models in Two Dimensions. *Phys. Rev. Lett.*, 49:1062.
Olami, Z., Feder, H. J. S., and Christensen, K. (1992). Self-Organized Criticality in a Continuous, Nonconservative Cellular Automaton Modeling Earthquakes. *Phys. Rev. Lett.*, 68:1244.
Onsager, L. (1944). Crystal Statistics. I. A Two-Dimensional Model with an Order-Disorder Transition. *Phys. Rev.*, 65:117.
Pathria, R. K. (1996). *Statistical Mechanics*. Butterworth-Heinemann, 2nd edition.
Paul, G., Ziff, R. M., and Stanley, H. E. (2001). Percolation Threshold, Fisher Exponent, and Shortest Path Exponent for Four and Five Dimensions. *Phys. Rev. E*, 64:026115.
Peierls, R. E. (1936). On Ising's Model of Ferromagnetism. *Proc. Cambridge Philos. Soc.*, 32:477.
Peters, O., Hertlein, C., and Christensen, K. (2002). A Complexity View of Rainfall. *Phys. Rev. Lett.*, 88:018701.
Privman, V., editor (1990). *Finite Size Scaling and Numerical Simulations of Statistical Systems*. World Scientific, Singapore.
Raup, D. M. (1986). Biological Extinction in Earth History. *Science*, 213:1528.
Reynolds, P. J., Stanley, H. E., and Klein, W. (1977). A Real-Space Renormalization Group for Site and Bond Percolation. *J. Phys. C*, 10:L167.
Roman, H. E., Bunde, A., and Dieterich, W. (1986). Conductivity of Dispersed Ionic Conductors: a Percolation Model with two Critical Points. *Phys. Rev. B*, 34:3439–3445.
Sahimi, M. (1994). *Applications of Percolation Theory*. Taylor & Francis, London.
Sivardière, J. (1997). Simple Mechanical Systems Exhibiting Instabilities. *Eur. J. Phys.*, 18:384.
Sornette, D. (2003). *Why Stock Markets Crash? Critical Events in Complex Financial Systems*. Princeton University Press.
Stapleton, M. (2004). *Private communication*.

Stauffer, D. and Aharony, A. (1994). *Introduction to Percolation Theory.* Taylor & Francis, London, 2nd edition.
Stauffer, D. and Ziff, R. M. (2000). Reexamination of Seven-Dimensional Site Percolation Thresholds. *Int. J. Mod. Phys. C*, 11:205.
Sykes, M. F. and Essam, J. W. (1964). Exact Critical Percolation Probabilities for Site and Bond Problems in Two Dimensions. *J. Math. Phys.*, 5:1117.
Sykes, M. F., Gaunt, D. S., Roberts, P. D., and Wyles, J. A. (1972). High Temperature Series for the Susceptibility of the Ising Model II. Three Dimensional Lattices. *J. Phys. A: Math., Gen. Phys.*, 5:640.
van der Marck, S. C. (1997). Percolation Thresholds and Universal Formulas. *Phys. Rev. E*, 55:1514.
van der Marck, S. C. (1998). Calculation of Percolation Thresholds in High Dimensions for FCC, BCC and Diamond Lattices. *Int. J. Mod. Phys. C*, 9:529.
Wegener, A. (1924). *The Origin of Continents and Oceans.* Dutton, New York.
Weidlich, W. (2001). *Sociodynamics: A Systematic Approach to Mathematical Modelling in the Social Sciences.* Harwood Academic Publishers.
Widom, B. (1965). Equation of State in the Neighborhood of the Critical Point. *J. Chem. Phys.*, 43:3898.
Wierman, J. C. (1981). Bond Percolation on the Honeycomb and Triangular Lattices. *Adv. Appl. Prob.*, 13:298.
Wilson, K. G. (1971a). Renormalization Group and Critical Phenomena. I. Renormalization Group and the Kadanoff Scaling Picture. *Phys. Rev. B*, 4:3174.
Wilson, K. G. (1971b). Renormalization Group and Critical Phenomena. II. Phase-Space Cell Analysis of Critical Behavior. *Phys. Rev. B*, 4:3184.
Wolff, U. (1989). Collective Monte Carlo Updating for Spin Systems. *Phys. Rev. Lett.*, 62:361.
Yang, C. N. (1952). The Spontaneous Magnetization of a Two-Dimensional Ising Model. *Phys. Rev.*, 85:808.
Young, A. P. and Stinchcombe, R. B. (1975). A Renormalization Group Theory for Percolation Problems. *J. Phys. C: Solid St. Phys.*, 8:L535.

List of Symbols

The list of abbreviations, functions, and symbols, their definition and the page of first appearance.

BTW	Bak-Tang-Wiesenfeld	254
fct(s).	Function(s)	371
nnn	Next-nearest-neighbour	217
OFC	Olami-Feder-Christensen	254
SCSN	Southern California Seismographic Network	303
csch	Hyperbolic cosecant	129
cos	Cosine	186
cosh	Hyperbolic cosine	125
coth	Hyperbolic cotangent	184
e	Base of natural logarithm	8
exp	Exponential fct. to the base e	8
$\Gamma(x)$	Gamma fct.	29
ln	Natural logarithm	8
log	Logarithm to the base 10	33
π	Pi	79
sech	Hyperbolic secant	129
sign(x)	Sign of x	133
sin	Sine	79
sinh	Hyperbolic sine	127
tan	Tangent	186
tanh	Hyperbolic tangent	127
$|x|$	Absolute value of x	24
$\sqrt{x}$	Square root of x	19

Chapter 1: Percolation

a	Lattice spacing	3
B	Contribution to average cluster size from a branch	17
b	Rescaling factor	66
D	Fractal dimension	57
d	Euclidean dimension	5
d_u	Upper critical dimension	81
$\mathcal{G}, \mathcal{G}(s/s_\xi)$	Scaling fct. for the cluster number density	41
$\mathcal{G}'(0)$	Derivative of the scaling fct. of the cluster number density	41
$\mathcal{G}''(0)$	Derivative of the scaling fct. of the cluster number density	41
$\tilde{\mathcal{G}}$	Finite-size scaling fct. of the cluster number density	76
$\mathcal{G}_\pm$	Scaling fcts. for the cluster number density for $p \to p_c^\pm$	41
$\mathcal{G}_{1\mathrm{d}}$	Scaling fct. for the cluster number density in $d=1$	40
$\mathcal{G}_{2\mathrm{d}}$	Scaling fct. for the cluster number density in $d=2$	42
$\mathcal{G}_{\mathrm{Bethe}}$	Scaling fct. for the cluster number density in Bethe lattice	42
$g(i,j)$	Site-site correlation fct. in Bethe lattice	29
$g(\mathbf{r}_i, \mathbf{r}_j)$	Site-site correlation fct.	13
$g(s,t)$	Number of different clusters of size s and perimeter t	22
i	Site index	13
k	Order of moment	51
L	Lattice size	3
L'	Lattice length	3
l	Generation number	15
$l_\xi, l_\xi(p)$	Characteristic generation	30
ℓ	Window size	55
$M_k(p)$	kth moment of cluster number density	51
$M_k(p; L)$	kth moment of cluster number density in lattice of size L	73
$M(s, p, \ell)$	Mass of a large finite cluster in window of size ℓ	58
$M_\infty(p; \ell)$	Mass of percolating cluster in window of size ℓ	55
$M_\infty(\xi; \ell)$	Mass of percolating cluster in window of size ℓ	63
$M_\infty(p; L)$	Mass of percolating cluster in lattice of size L	71
$m(\ell/R_s)$	Crossover fct. of mass of large finite cluster	61
$m_\infty(\ell/\xi)$	Crossover fct. of mass of percolating cluster	66
N_{clu}	Number of clusters	8
N_{occ}	Number of occupied sites	8
$N(i;l)$	Average number of sites $l \geq 1$ generations away	29
$N(s, p; L)$	Cluster size frequency	3
$n(l)$	Number of sites in the lth generation	29

List of Symbols

$n(s,p)$	Cluster number density	7
$n(s,p;L)$	Cluster number density in lattice of size L	76
$P_\infty(p)$	Probability of a site in percolating cluster	13
$P_\infty(p;\ell)$	Probability of a site in percolating cluster in window of size ℓ	55
$P_\infty(p;L)$	Probability of a site in spanning cluster in lattice of size L	71
$\mathcal{P}$	Scaling fct. for the probability of a site in percolating cluster	106
p	Occupation probability	3
p_c	Critical occupation probability	5
$p^\star$	Fixed point in p-space	86
$Q_\infty(p)$	Probability a branch does not connect to percolating cluster	19
$R_b, R_b(p)$	Renormalisation group transformation	86
$R(s)$	Radius of gyration of an s-cluster	60
R_s	Radius of gyration of s-clusters, ensemble average	59
$\mathbf{r}_{\mathrm{cm}}$	Position vector to centre of mass of a cluster	59
$\mathbf{r}_i$	Position vector of site i	13
r	Distance between $\mathbf{r}_i$ and $\mathbf{r}_j$	13
s	Cluster size	3
s_k	Size of cluster k	8
$s_\xi, s_\xi(p)$	Characteristic cluster size	7
$T_b, T_b(p)$	Rescaling transformation	85
t	Number of perimeter sites of a cluster	22
u	Integration variable	28
z	Coordination number	15
β	Critical exponent of order parameter	31
γ	Critical exponent of average cluster size	33
ν	Critical exponent of correlation length	63
$\xi, \xi(p)$	Correlation length	14
$\Pi_\infty(p;L)$	Probability of having a percolation cluster in lattice of size L	111
$\Pi_\infty(\xi;L)$	Probability of having a percolation cluster in lattice of size L	111
Π	Scaling fct. for the probability of having a percolation cluster	111
σ	Critical exponent of characteristic cluster size	37
τ	Critical exponent of cluster number density	26
$\chi(p)$	Average cluster size	10
$\chi(p;L)$	Average cluster size in lattice of size L	73
$\mathcal{X}, \mathcal{X}(L/\xi)$	Scaling fct. for the average cluster size	109

Chapter 2: Ising Model

A	$n \times n$ matrix	142
A_{ij}	Entries in matrix **A**	142
$A_{\{s_i\}}$	Observable of a microstate $\{s_i\}$	119
$\langle A \rangle$	Average of observable	119
a_2, a_4	Coefficients to expansion of mean-field free energy	171
b	Rescaling factor	202
B	$n \times n$ matrix	143
$\mathcal{C}_\pm$	Scaling fcts. for the specific heat for $t \to 0^\pm$	191
$c, c(T, H)$	Specific heat	122
d	Euclidean dimension	123
$\det(\mathbf{A})$	Determinant of $n \times n$ matrix **A**	142
E_{ext}	Total external energy	118
E_{int}	Total internal interaction energy	117
$E_{\{s_i\}}$	Total energy of a microstate $\{s_i\}$	118
$E'_{\{s_I\}}$	Total energy of a microstate $\{s_I\}$ in renormalised lattice	207
$\Delta E_{\{s_i\}}$	Difference in total energy and mean-field energy	200
$\langle E \rangle$	Average total energy	120
$E(\kappa)$	Complete elliptic integral of the second kind	184
$\mathbf{e}_i(b)$	Eigenvector of $\mathbf{M}(b)$	232
$\mathcal{F}_\pm$	Scaling fcts. for the free energy per spin for $t \to 0^\pm$	190
$F, F(T, H)$	Total free energy	120
F_{boundary}	Free energy boundary contribution	123
F_{bulk}	Free energy bulk contribution	123
$F_{1\text{-dom}}$	Total free energy of spin system with one domain	155
$F_{2\text{-dom}}$	Total free energy of spin system with two domains	155
$F(\kappa)$	Complete elliptic integral of the first kind	184
$f, f(T, H)$	Free energy per spin	121
$f(t, h)$	Free energy per spin as a fct. of reduced variables	208
$f(T, H; \phi)$	Free energy per spin in Landau theory	177
f_0	Entropic part of the free energy per spin	170
$f_r(t, h)$	Regular part of the free energy per spin	188
$f_s, f_s(t, h)$	Singular part of the free energy per spin	188
$\mathcal{G}_\pm$	Scaling fcts. for the spin-spin correlation fct. for $t \to 0^\pm$	192
$g(\mathbf{r}_i, \mathbf{r}_j)$	Spin-spin correlation fct.	134
H	External field	117
H_c	Critical external field	154
h	Reduced external field	188

List of Symbols

h'	Reduced external field in renormalised lattice	204
I	Identity matrix	142
I	Block of spins	202
i	Site index	116
$\langle ij \rangle$	Distinct nearest-neighbour pairs ij	118
J	Constant interaction energy between nearest-neighbour spins	118
J_i	Interaction energy between spins s_i and s_{i+1} in $d=1$	118
J_{ij}	Interaction energy between spins s_i and s_j	117
K	Coupling constant	179
$\delta \mathbf{K}$	Deviation of $\mathbf{K}$ from fixed point $\mathbf{K}^\star$	231
$\delta \mathbf{K}'$	Deviation of $\mathbf{K}'$ from fixed point $\mathbf{K}^\star$	231
$\mathbf{K}$	Coupling constant	223
$\mathbf{K}'$	Renormalised coupling constant	226
$\mathbf{K}^\star$	Fixed point in coupling space	227
$\mathbf{K}_c$	Critical coupling constant	225
$K^\star$	Fixed point of coupling constant	213
K_c	Critical value of coupling constant	186
K_1	Nearest-neighbour coupling constant	211
K'_0	Renormalised coupling constant	212
K'_1	Renormalised nearest-neighbour coupling constant	212
$K_1^{(n)}$	Renormalised nearest-neighbour coupling constant, n iterations	222
K'_2	Renormalised nnn coupling constant	217
K'_3	Renormalised quadruple coupling constant	217
$\tilde{K}'_1$	Effective renormalised nn coupling constant	220
k	Positive constant related to reduced 'temperature' in $d=1$	196
k_B	Boltzmann's constant	118
L	Lattice size	116
$\mathbf{M}(b)$	Linearised renormalisation group transformation	231
$\mathcal{M}_\pm$	Scaling fcts. for the magnetisation for $t \to 0^\pm$	189
$M_{\{s_i\}}$	Total magnetisation of a microstate $\{s_i\}$	120
$\langle M \rangle$	Average total magnetisation	120
$m, m(T,H)$	Average magnetisation per spin	121
$m(t,h)$	Average magnetisation per spin as a fct. of reduced variables	189
$m_0(T)$	Average magnetisation per spin in zero external field	132
m_I	Magnetisation per spin within block I	209
N	Number of spins	116
N'	Number of spins in renormalised lattice	207
$n_\pm$	Number of microstates within block consistent with $S_I = \pm 1$	206

$p_{\{s_i\}}$	Probability of system to be in microstate $\{s_i\}$	119
$R_b, R_b(\mathbf{K})$	Renormalisation group transformation	226
$\mathbf{r}_i$	Position vector of site i	134
$\mathbf{r}_I$	Position vector of block I	209
r	Distance between $\mathbf{r}_i$ and $\mathbf{r}_j$	192
r'	Distance between $\mathbf{r}_I$ and $\mathbf{r}_J$ in renormalised lattice	209
S	Total entropy	120
s_i	Spin at position $\mathbf{r}_i$	116
s_I	Block spin at position $\mathbf{r}_I$	202
$\langle s_i \rangle$	Average value of spin s_i	134
$\{s_i\}$	Spin microstate configuration $\{s_1, s_2, \ldots, s_N\}$	118
$\mathbf{T}$	Transfer matrix	141
$T_{s_i s_{i+1}}$	Entries in transfer matrix	141
T	Temperature	118
T_c	Critical temperature	132
$\mathrm{Tr}\,(\mathbf{A})$	Trace of matrix $\mathbf{A}$	142
t	Reduced temperature	188
t'	Reduced temperature in renormalised lattice	204
$\mathbf{U}$	Unitary 2×2 matrix	142
u_i	Scaling field	233
u_i'	Renormalised scaling field	233
y_i	Eigenvalue associated with scaling field u_i	234
y_h	Eigenvalue associated with external field scaling field	206
y_t	Eigenvalue associated with temperature scaling field	206
Z	Partition fct.	119
$Z(T, H, N)$	Partition fct.	120
$Z(K_1, N)$	Partition fct. of $d = 1$ Ising model	211
z	Coordination number	157

$\alpha, \alpha^\pm$	Critical exponent of specific heat for $t \to 0^\pm$	134
$\alpha_k(T, H)$	Coefficients in expansion of free energy per spin	177
$\tilde{\alpha}_k$	Coefficients in expansions of $\alpha_k(T, H)$	178
β	Critical exponent of order parameter at $H = 0$	133
β	Inverse temperature	120
β_c	Inverse temperature at $T = T_c$	166
$\Gamma^\pm$	Critical amplitudes for $t \to 0^\pm$	164
γ	Critical exponent of susceptibility	133
$\gamma^\pm$	Critical exponent of susceptibility for $t \to 0^\pm$	164

Δ	Gap exponent	189
δ	Critical exponent or order parameter at $t = 0$	133
$\varepsilon, \varepsilon(T, H)$	Energy per spin	121
η	Critical exponent of correlation fct.	135
ϕ	General order parameter	177
κ, κ'	Parameters	179
$\lambda_\pm$	Eigenvalues of transfer matrix	142
$\lambda_t(b)$	Proportionality constant	204
$\lambda_i(b)$	Eigenvalue of $\mathbf{M}(b)$	232
ν	Critical exponent of correlation length	134
$\xi, \xi(T, H)$	Correlation length	134
ξ'	Correlation length in renormalised lattice	206
$\xi(\mathbf{K})$	Correlation length associated with $\mathbf{K}$ in coupling space	224
$\chi, \chi(T, H)$	Susceptibility per spin	121
$\mathcal{X}_\pm$	Scaling fcts. for the susceptibility for $t \to 0^\pm$	192
φ	Integration variable	179
$\Omega(n)$	Number of different interfaces separating n spins	186

Chapter 3: Self-Organised Criticality

B	Exponent of the Gutenberg-Richter law	304
b	Number of induced relaxations	266
$\langle b \rangle$	Average branching ratio	266
D	Avalanche dimension	273
E	Energy dissipated by avalanche	287
$E_{\text{pot}}^{\text{add}}$	Added potential energy	287
$E_{\text{pot}}^{\text{i}}$	Initial potential energy	287
$E_{\text{pot}}^{\text{f}}$	Final potential energy	287
$E_{\max}$	Maximum energy release	289
$\mathbf{e}_i$	Unit position vector	279
δF	Infinitesimal increase in force on blocks during driving	309
$F_{i,j}$	Total force exerted by springs on block (i, j)	307
F_{nn}	Total force on nearest neighbour blocks	308
F^{th}	Threshold frictional force	307
$\delta F_{i,j}$	Change in force on block (i, j) when neighbouring block slips	308
F_L	Force exerted on block by load spring	307
F_{b}^{r}	Force on bulk sites before slipping	310
F_{e}^{r}	Force on edge sites before slipping	310

F_c^r	Force on corner sites before slipping	310
$f_{i,j}^{i\pm1,j\pm1}$	Force exerted by springs in plane on block (i,j)	307
$f_{i,j}^L$	Force exerted by leaf spring on block (i,j)	307
$\mathcal{G}$	Scaling fct. for avalanche-size probability	273
$\mathcal{G}_{1d}^{BTW}$	Scaling fct. for $d=1$ BTW model	261
$\mathcal{G}^{BP}$	Scaling fct. for the branching process	261
$\mathcal{G}^{OFC}$	Scaling fct. for the OFC model	320
$\mathcal{G}^{Oslo}$	Scaling fct. for the Oslo model	295
$\mathcal{G}_{bulk}^{Oslo}$	Scaling fct. for the bulk-driven Oslo model	299
g	Acceleration due to gravity	289
$g(s,t)$	Number of different trees of size s and perimeter t	268
H_c	Critical external field	263
h_i	Height at position i	255
$\langle\text{influx}\rangle$	Average influx	258
K	Elastic constant	307
K_L	Elastic constant of load spring	307
L	Lattice size in units of lattice spacing	255
$\langle\ell\rangle$	Average length of rice grain	289
M	Rain event size	328
M_{max}	Maximum rain event size	328
M_{min}	Minimum rain event size	328
m	Earthquake magnitude	252
$\langle m\rangle$	Average mass of rice grain	289
$N(S>s)$	Annual number of earthquakes with size $S>s$	304
$N(M)$	Annual number density of rain events with size M	329
N_F	Number of different initial F values in the OFC model	311
$N_\mathcal{R}$	Number of recurrent configurations	292
$N_\mathcal{S}$	Number of stable configurations	265
n	Integer variable	258
$\langle\text{outflux}\rangle$	Average outflux	258
P_b	Fraction of bulk sites slipping	310
P_e	Fraction of edge sites slipping	310
P_c	Fraction of corner sites slipping	310
$P(E;L)$	Avalanche-size density in lattice of size L	289
$P(s;L)$	Avalanche-size probability in lattice of size L	260
$P(s,p)$	Avalanche-size probability	269
$P(s,\alpha;L)$	Avalanche-size probability with α in lattice of size L	318
P_z	Probability that site contains z units	265
p_c	Critical occupation probability	263

List of Symbols

$q_{i,j}$	Coordination number of block (i,j) in the OFC model	308
$q(t)$	Rain rate at time t	326
$\mathbf{R}$	Position vector	279
R	Length of position vector $\mathbf{R}$	279
$\mathcal{R}$	Set of recurrent configurations	257
$\mathcal{R}_j$	Recurrent configuration	258
$\mathcal{S}$	Set of stable configurations	257
$\mathcal{S}_j$	Stable configuration	257
s	Earthquake energy	252
s	Avalanche size	258
$s_c, s_c(p)$	Characteristic avalanche size	269
s_k	Size of finite tree k	263
$s_{\max}$	Maximum avalanche size	259
$\langle s \rangle$	Average avalanche size	258
$\langle s^k \rangle$	kth moment of avalanche-size probability	263
$\mathcal{T}$	Set of transient configurations	257
$\mathcal{T}_j$	Transient configuration	258
T	Rain event duration	326
T_c	Critical temperature	263
t	Perimeter	268
t	Time	259
t_0	Initial time	328
V	Relative velocity	307
$x_{i,j}$	Displacement of block (i,j)	307
$\tilde{x}_{i,j}$	Zero force displacement of block (i,j) after slip	308
z_i	Slope at position i	255
z^{th}	Threshold slope	255
$\langle z \rangle$	Average height	265
α	Dissipation parameter in the OFC model	308
α_n	Discrete parameter in random neighbour BTW model	264
$\delta(\alpha)$	Scaling exponent in the OFC model	312
Δt	Time interval	326
ϕ	Golden mean	292
$\Theta(x)$	Heaviside step fct.	261
τ_s	Avalanche-size exponent	270
τ_E	Avalanche-size exponent in pile of rice	290
τ_M	Event-size exponent in rainfall	329

Appendix

a	Number	345
b	Number	347
$d\langle E\rangle$	Infinitesimal change in total energy	361
dF	Infinitesimal change in total free energy	361
$đQ$	Infinitesimal heat transfer	361
dS	Infinitesimal change in total entropy	361
dT	Infinitesimal change in temperature	361
$E_{\{s_i\}}$	Total energy of a microstate $\{s_i\}$	363
$\langle E\rangle$	Average total energy	361
F	Total free energy	361
$f(x)$	Function of one variable x	341
$f^{(j)}(x_0)$	jth derivative of the fct. $f(x)$ evaluated in $x = x_0$	341
$f(x,y)$	Function of two variables x and y	347
$\mathcal{G}_\pm$	Scaling fcts. for generalised homogeneous fct.	347
j	Bin label	356
N	Sample size	355
$N_{\{s_i\}}$	Number of systems in ensemble in microstate $\{s_i\}$	363
$\Delta N(\{s_i\} \to \{s_i'\})$	Net flow out of microstate $\{s_i\}$	364
$P(s)$	Avalanche size probability	355
$P_N(s)$	Avalanche size probability from sample of size N	355
$\tilde{P}_N(s^j)$	Estimate of avalanche size probability density $P(s)\,ds$	356
$P(\{s_i\} \to \{s_i'\})$	Transition probability from $\{s_i\}$ to $\{s_i'\}$	363
$p_{\{s_i\}}$	Probability of system to be in microstate $\{s_i\}$	364
S	Total entropy	361
s	Avalanche size	355
s_c	Characteristic avalanche size	355
s_i	Avalanche size of avalanche i	355
s_j	Spin on position $\mathbf{r}_j$	363
$s_{\min}^j$	Minimum integer avalanche size in bin j	356
s^j	Geometric mean of avalanche sizes in bin j	356
$s_{\max}^j$	Maximum integer avalanche size in bin j	356
$\{s_i\}$	Spin microstate configuration $\{s_1, s_2, \ldots, s_N\}$	363
T	Temperature	361
T_c	Critical temperature	364
λ	Rescaling factor	345

Index

Algorithm
  Ising model
    Metropolis, 179, 363
    Wolff, 179, 364
  self-organised criticality
    BTW model $d = 1$, 256
    BTW model $d = 2$, 278
    BTW random neighbour, 264
    OFC model, 309
    Oslo model, 291
Appendix, *341-364*
Atmosphere
  rain event, 326
  rain event duration, 326
  rain event size, 328
  rain rate, 326
  rainfall time series, 326
  rainfall-size number density, 329
  saturation threshold, 325
Avalanche
  time series
    BTW model $d = 1$, 259
    BTW model $d = 2$, 280
    OFC model, 314
    Oslo model, 293
Avalanche dimension, 273
Avalanche size, 258
Avalanche-size exponent, 273
Avalanche-size probability
  branching process
    exact, 268
    scaling form, 270

BTW model $d = 1$, 260
BTW model $d = 2$, 281
  critical exponent D, 273
  critical exponent τ_s, 273
  kth moment, 275
  OFC model, 318
  Oslo model, 294
  scaling ansatz, 273
  scaling function, 273
Average avalanche size, 258
  branching process, 266, 272
  BTW model $d = 1$, 263
  BTW model $d = 2$, 280
  BTW random neighbour, 266
  OFC model, 310
Average branching ratio, 266, 268
Average cluster size, 10
  Bethe lattice, 16, 19
  critical exponent γ, 33
  finite-size scaling, 73
  percolation $d = 1$, 11
  percolation $d = 2$, 33
Average energy, 120
Average energy per spin, 121
  Ising model $d = 1$, 149
  Ising model $d = 2$, 184
  mean-field theory, 166
  non-interacting spins, 129
Average magnetisation, 120
Average magnetisation per spin, 121
  block, 209
  critical exponent β, 133

critical exponent δ, 133
Ising model $d = 1$, 145
Ising model $d = 2$, 180
  mean-field theory, 158
  non-interacting spins, 127
  scaling ansatz, 189, 191
  scaling functions, 189, 192
Average of observable, 119
Average value of spin, 134

Basin of attraction
  critical surface, 229
  fixed point, 229
Bethe lattice, 15
  average cluster size, 16
  characteristic cluster size, 23
  characteristic generation, 30
  chemical distance, 16
  cluster number density, 22, 25
    data collapse, 44
    scaling form, 28
    scaling function, 44
  coordination number, 15
  correlation function, 29
  critical exponent
    β, order parameter, 20
    γ, average cluster size, 19
    σ, characteristic cluster size, 25
    τ, cluster number density, 28, 29
  critical occupation probability, 16
  generation number l, 15
  order parameter, 19
  sum rule, 30
Binning data, 355
Block spin, 202, 209
  transformation, 202
Boltzmann distribution, 119, 359
Boltzmann's constant, 118
Bond percolation, 78
Branching process, 267
  avalanche-size probability
    exact, 268
    scaling form, 270
  average avalanche size, 266, 272
  average branching ratio, 268
  branching ratio, 267

rooted tree, 267
BTW mean-field theory, 267
  avalanche-size probability
    exact, 268
    scaling form, 270
  average avalanche size, 266, 272
  average branching ratio, 268
  branching ratio, 267
  rooted tree, 267
BTW model $d = 1$, 255
  algorithm, 256
  avalanche time series, 259
  avalanche-size probability, 260
  average avalanche size, 263
  configuration
    recurrent, 257
    stable, 255
    transient, 257
  local slope, 255
  separation of time scales, 257
  threshold slope, 255, 257
BTW model $d = 2$
  algorithm, 278
  avalanche time series, 280
  avalanche-size probability, 281
  average avalanche size, 280
  threshold, 278
BTW random neighbour model
  algorithm, 264
  average avalanche size, 266
  average branching ratio, 266
  stable configuration, 265
  threshold, 264

Canonical ensemble, 119
Cayley tree, 15
Centre of mass, 59
Characteristic cluster size, 8, 37
  Bethe lattice, 23, 25
  critical exponent σ, 37
  percolation $d = 1$, 8
  percolation $d = 2$, 37
Characteristic equation, 142
Characteristic generation, 30
Cluster, 3, 79
  average size, 10

centre of mass, 59
characteristic size, 37
geometric properties, 55–69
large finite at p_c, 58
number density, 7
perimeter sites, 22
radius of gyration, 59
size, 3, 79
size frequency, 6
Cluster number density, 7
Bethe lattice, 22, 25
  critical exponent τ, 28, 29
  data collapse, 44
  scaling form, 28
  scaling function, 44
characteristic cluster size, 41
critical exponent τ, 26, 41
data collapse, 41
exact, 35
finite-size scaling, 76
  ansatz, 76
  moment, 51
percolation $d = 1$, 6
  critical exponent τ, 40
  data collapse, 42
  scaling function, 40
percolation $d = 2$, 35
  critical exponent τ, 37, 49
  data collapse, 49
  scaling function, 49
scaling ansatz, 39–49
scaling function, 41
Coarse graining, 85, 202
Configuration
  stable, 255
Constant coupling term, 223, 236
Continuous phase transition, 20, 132
  Landau theory, 175
    coefficient expansion, 178
    coefficients, 177
    critical exponents, 178
    free energy per spin, 177
    order parameter, 177
    symmetry arguments, 177
Coordination number, 15
Correlation function

Ising model, 134
  $d = 1$ lattice, 152
  at critical point, 135
  critical exponent η, 135
  generalised homogeneous, 209
  renormalised, 209
  scaling ansatz, 192, 210
  scaling functions, 192
percolation, 13
  $d = 1$ lattice, 13
  Bethe lattice, 29
Correlation length
Ising model, 134
  $d = 1$ lattice, 152
  critical exponent ν, 134
percolation, 62
  $d = 1$ lattice, 14
  critical exponent ν, 63
  fixed point equation, 83
  self-similarity, 82
Coupling constant, 118
  renormalised, 226
Coupling space, 222
  critical manifold, 229
  critical surface, 229
  renormalisation group flow, 230
Critical amplitude
  Ising model $d = 2$, 182
  mean-field theory, 164
  universal ratio, 169
Critical exponent
Ising model
  scaling relations, 195
  table, 199
  universality, 199
α, specific heat, 134
β, order parameter, 133
γ, susceptibility per spin, 133
δ, order parameter, 133
η, correlation function, 135
ν, correlation length, 134
Δ, gap exponent, 189
percolation
  scaling relations, 51–54
  table, 81
  universality, 81

β, order parameter, 31
γ, average cluster size, 33
ν, correlation length, 63
σ, characteristic cluster size, 37
τ, cluster number density, 26
self-organised criticality
  D, avalanche-size probability, 273
  τ_s, avalanche-size probability, 273
Critical external field, 135
  reduced variable, 188
Critical manifold, 229
Critical occupation probability, 5
  Bethe lattice, 16
  non-universality, 78
  percolation $d = 1$, 14
  percolation $d = 2$, 31
  table, 80
Critical point, 135
  critical external field, 135
  critical temperature, 135
    Ising model $d = 1$, 154
    Ising model $d = 2$, 180, 186
    mean-field theory, 159
    table, 199
  reduced variables, 188
Critical surface, 229
  fixed point, 229
Critical temperature, 135
  Ising model $d = 1$, 154
    Peierls' argument, 154
  Ising model $d = 2$, 180
    Peierls' argument, 186
  mean-field theory, 159
  reduced variable, 188
Crossover
  cluster number density, 39, 44
  function for mass
    large finite cluster at p_c, 61
    percolating cluster, 55, 66

Data binning, 355
Data collapse
  cluster number density, 41
    Bethe lattice, 44
    percolation $d = 1$, 42
    percolation $d = 2$, 49

scaling ansatz, 39
scaling function, 41
Decimation rule, 204
Degeneracy factor, 22
Density, percolating cluster, 55
Distance
  chemical, 16
  lattice sites, 13
  pair of block spins, 209
  pair of spins, 135, 151
Distinct nearest-neighbour pairs, 118

Earthquake time series, 303
Earthquake-size frequency, 304
Effective external field, 157
Effective nn coupling constant, 220
Eigenvalue
  external scaling field, 206
  temperature scaling field, 206
  transfer matrix, 142
Eigenvalues, 232
Eigenvectors, 232
Energy
  external, 118
  internal interaction, 117
  Ising model $d = 1$, 140
  mean-field theory, 157
  microstate Ising model, 118
  renormalised, 207
  variance, 122
    fluctuation-dissipation thm, 122
    Ising model $d = 1$, 149
    Ising model $d = 2$, 184
    non-interacting spins, 131
Ensemble average, 119
Entropy, 120
Ergodic theorem, 119
Exercises
  Ising model, *241–247*
  percolation, *104–114*
  SOC, *334–340*
External energy, 118
External field, 117
  effective, 157
External magnetic field, 117

Index 385

Finite-size scaling
  percolation, 69–78
    average cluster size, 73
    cluster number density, 76
    moment of order k, 73
    order parameter, 71
Fixed point
  p-space, 86
  basin of attraction, 229
  correlation length, 83
  coupling constant, 213
  equation
    correlation length, 83
    RG transformation, 86
  high temperature, 214
  non-trivial, 229
  renormalisation group, 227
  self-similarity, 83
  strong-coupling, 229
  weak-coupling, 214, 229
Flow
  correlation length, 84
  coupling space, 222
  occupation probability, 84
  reduced external field, 206
  reduced temperature, 206
  renormalisation group, 222
Fluctuation-dissipation thm, 122
Fractal dimension, 55, 57
Fractals, 351
Free energy, 121, 361
  boundary contribution, 123
  bulk contribution, 123
  Ising model $d = 1$, 143
    single domain, 155
    two domains, 155
  Ising model $d = 2$
    one domain, 186
    two domains, 186
  minimisation, 120
  non-interacting spins, 126
Free energy per spin, 121
  entropic part, 170
  Landau theory
    continuous phase transition, 177
    Ising, 171

  mean-field theory, 158
  regular part, 188
  singular part, 188, 190
  scaling ansatz, 190

Gamma function, 29
Gap exponent, 189
Generalised homogeneous fct., 345
  correlation function, 209
Generalised Ising model, 225
  renormalised, 226
Generalised reduced energy, 225
  renormalised, 226
  symmetry, 225
Generation number, 15
Geometric properties, 55–69
Ginzburg criterion, 200
Golden mean, 292
Gutenberg-Richter law, 252

Heat capacity per spin, 122
Heaviside step function, 261
Height, 255
Helmholtz free energy, 120
Homogeneous function, 345
Hyperbolic functions, 343

Incipient infinite cluster, 5
  fractal dimension, 57
  mass, 57
Interaction strength, 117, 118
Internal energy, 117
Inverse temperature, 120
Irrelevant scaling field, 234
Ising model, *115–240*
Ising model
  generalised, 225
  Landau theory, 169
    coefficients, 171
    critical exponent, 172
    critical exponent, 175
    α^-, specific heat, 175
    β, order parameter, 174
    $\gamma^{\pm}$, susceptibility per spin, 174
    δ, order parameter, 175
    free energy per spin, 171

magnetisation per spin, 172
order parameter, 172
specific heat, 175
susceptibility per spin, 174
Ising model $d = 1$, 140
　average energy per spin, 149
　average mag. per spin, 145
　　scaling form, 197
　correlation function, 152, 193
　correlation length, 152
　　scaling form, 198
　critical exponent, 195
　　α, specific heat, 196
　　β, order parameter, 197
　　δ, order parameter, 197
　　η, correlation function, 198
　　γ, susceptibility per spin, 197
　　ν, correlation length, 198
　critical point, 154
　　critical external field, 154
　　critical temperature, 154
　critical temperature, 154
　energy, 140
　free energy, 143
　　scaling form, 196
　partition function, 141, 143
　reduced 'temperature, 196
　specific heat, 149
　sum rule, 154
　susceptibility per spin, 147
　　scaling form, 197
　transfer matrix, 141
　variance of energy, 149
　variance of magnetisation, 147
　Widom scaling form, 195
Ising model $d = 2$, 179
　average energy per spin, 184
　average mag. per spin, 180
　critical exponent
　　α, specific heat, 186
　　β, order parameter, 180
　　γ, susceptibility per spin, 182
　critical temperature, 180
　　Peierls' argument, 186
　order parameter, 180
　partition function, 179

specific heat, 184
susceptibility per spin, 182
variance of energy, 184
variance of magnetisation, 182
Ising model mean field, 156
　average energy per spin, 166
　average mag. per spin, 158
　correlation function, 193
　critical exponent
　　$\alpha^\pm$, specific heat, 169
　　β, order parameter, 161
　　$\gamma^\pm$, susceptibility per spin, 164
　　δ, order parameter, 166
　critical temperature, 159
　energy, 157
　free energy per spin, 158
　order parameter, 161
　partition function, 157
　specific heat, 169
　susceptibility per spin, 164
　variance of magnetisation, 164

Kadanoff, 202
　block spin, 202
　block spin transformation, 202
Kramers, 186
kth moment
　avalanche-size probability, 275
　cluster number density, 51
　finite-size scaling, 73

Landau theory
　continuous phase transition, 175
　coefficient expansion, 178
　coefficients, 177
　critical exponents, 178
　free energy per spin, 177
　order parameter, 177
　symmetry arguments, 177
　Ising model, 169
　　coefficients, 171
　　critical exponent, 172
　　α^-, specific heat, 175
　　β, order parameter, 174
　　$\gamma^\pm$, susceptibility per spin, 174
　　δ, order parameter, 175

free energy per spin, 171
magnetisation per spin, 172
order parameter, 172
specific heat, 175
susceptibility per spin, 174
Lattice length, 3
Lattice size, 3
Lattice spacing, 3

Magnetisation
  microstate, 120
  variance, 122
  fluctuation-dissipation thm, 122
  Ising model $d = 1$, 147
  Ising model $d = 2$, 182
  mean-field model, 164
  non-interacting spins, 129
Majority rule, 204
Marginal scaling field, 234
Mass
  crossover function
    large finite cluster at p_c, 61
    percolating cluster, 66
  incipient infinite cluster, 57
  large finite cluster at p_c, 58
  percolating cluster, 55, 63, 66
Mean-field theory, 156
  average energy per spin, 166
  average mag. per spin, 158
  correlation function, 193
  critical exponent
    $\alpha^{\pm}$, specific heat, 169
    β, order parameter, 161
    $\gamma^{\pm}$, susceptibility per spin, 164
    δ, order parameter, 166
  critical temperature, 159
  energy, 157
  free energy per spin, 158
  order parameter, 161
  partition function, 157
  specific heat, 169
  susceptibility per spin, 164
  variance of magnetisation, 164
Metropolis algorithm, 179, 363
Microstate, 118
Minimisation of free energy, 120

Moment of order k
  avalanche-size probability, 275
  cluster number density, 51
  finite-size scaling, 73

Nearest-neighbour
  interactions, 118
Nearest-neighbour (nn)
  coupling constant, 211, 215
Next-nearest-neighbour (nnn)
  coupling, 216
Non-equilibrium steady state, 249
Non-interacting spins, 124
  average energy per spin, 129
  average mag. per spin, 127
  free energy, 126
  partition function, 125
  specific heat, 129
  susceptibility per spin, 129
  total energy, 124
  variance of energy, 131
  variance of magnetisation, 129

Observable, 119
  average, 119
Occupation probability, 3
OFC model
  algorithm, 309
  avalanche time series, 314
  avalanche-size probability, 318
  average avalanche size, 310
  threshold, 307
One-dimensional BTW model, 255
  algorithm, 256
  avalanche time series, 259
  avalanche-size probability, 260
  average avalanche size, 263
  configuration
    recurrent, 257
    stable, 255
    transient, 257
  local slope, 255
  separation of time scales, 257
  threshold slope, 255, 257
One-dimensional Ising model, 140
  average energy per spin, 149

average mag. per spin, 145
  scaling form, 197
 correlation function, 152, 193
 correlation length, 152
  scaling form, 198
 critical exponent, 195
  α, specific heat, 196
  β, order parameter, 197
  δ, order parameter, 197
  η, correlation function, 198
  γ, susceptibility per spin, 197
  ν, correlation length, 198
 critical point, 154
  critical external field, 154
  critical temperature, 154
 critical temperature, 154
 energy, 140
 free energy, 143
  scaling form, 196
 partition function, 141, 143
 reduced 'temperature', 196
 specific heat, 149
 sum rule, 154
 susceptibility per spin, 147
  scaling form, 197
 transfer matrix, 141
 variance of energy, 149
 variance of magnetisation, 147
 Widom scaling form, 195
One-dimensional model
 configuration
  recurrent, 292
One-dimensional Oslo model
 algorithm, 291
 avalanche time series, 293
 avalanche-size probability, 294
 configuration
  stable, 292
  transient, 292
 threshold slope, 291
One-dimensional percolation, 5
 average cluster size, 11
 characteristic cluster size, 8
 cluster number density, 6, 7
 cluster size frequency, 6
 correlation function, 13

correlation length, 14
critical exponent
 β, order parameter, 13
 γ, average cluster size, 11
 ν, correlation length, 14
 σ, characteristic cluster size, 8
 τ, cluster number density, 40
critical occupation probability, 14
data collapse, 42
order parameter, 13
renormalisation group, 91
scaling function, 40
sum rule, 10, 13, 14
Onsager, 179
Order parameter
 Ising model, 132
  critical exponent β, 133
  critical exponent δ, 133
  Landau theory, 177
 percolation, 31
  $d = 1$ lattice, 13
  $d = 2$ lattice, 31
  Bethe lattice, 19
  critical exponent β, 31
  finite-size scaling, 71
Oslo model
 algorithm, 291
 avalanche time series, 293
 avalanche-size probability, 294
 configuration
  recurrent, 292
  stable, 292
  transient, 292
 threshold slope, 291

Partition function, 120
 Ising $d = 2$, 179
 Ising model $d = 1$, 141, 143
 mean-field theory, 157
 non-interacting spins, 125
Peierls, 186
Peierls' argument
 Ising $d = 1$, 154
 Ising $d = 2$, 186
Percolating cluster, 5
 crossover function of mass, 66

density, 55
mass, 55, 63
Percolation, *1–103*
Percolation
 correlation function, 13
Percolation $d = 1$, 5
 average cluster size, 11
 characteristic cluster size, 8
 cluster number density, 6, 7
 cluster size frequency, 6
 correlation function, 13
 correlation length, 14
 critical exponent
  β, order parameter, 13
  γ, average cluster size, 11
  ν, correlation length, 14
  σ, characteristic cluster size, 8
  τ, cluster number density, 40
 critical occupation probability, 14
 data collapse, 42
 order parameter, 13
 renormalisation group, 91
 scaling function, 40
 sum rule, 10, 13, 14
Percolation $d = 2$, 30
 average cluster size, 33
 characteristic cluster size, 37
 cluster number density, 35–42
 critical exponent
  β, order parameter, 31
  γ, average cluster size, 33
  σ, characteristic cluster size, 37
  τ, cluster number density, 37
 critical occupation probability, 31
 data collapse, 49
 order parameter, 31
 renormalisation group
  square lattice, 98
  triangular lattice, 95
 scaling ansatz, 39
 scaling function, 49
Percolation Bethe lattice, 15
 average cluster size, 16
 characteristic cluster size, 23
 characteristic generation, 30
 chemical distance, 16

cluster number density, 22, 25
data collapse, 44
scaling form, 28
scaling function, 44
coordination number, 15
correlation function, 29
critical exponent
 β, order parameter, 20
 γ, average cluster size, 19
 σ, characteristic cluster size, 25
 τ, cluster number density, 28, 29
critical occupation probability, 16
generation number l, 15
sum rule, 30
Perimeter sites, 22

Radius of gyration
 s-cluster, 60
 ensemble of s-clusters, 59
Rain
 rain rate, 326
 rainfall time series, 326
 saturation threshold, 325
Rain event, 326
 duration, 326
 size, 328
 size number density, 329
Random neighbour BTW model
 algorithm, 264
 average avalanche size, 266
 average branching ratio, 266
 threshold, 264
Random-neighbour BTW model
 stable configuration, 265
Reduced
 control parameter, 207
 energy, 220
 external field, 188, 205
 nn coupling constant, 211, 215
 number of spins, 213
 temperature, 188
Reduced energy
 generalised, 225
Relevant scaling field, 234
Renormalisation group
 flow, 222

coupling space, 230
near fixed point, 231
Ising model, 202–221
$d = 1$ lattice, 211
$d = 2$ square lattice, 204, 215
block spin, 209
decimation rule, 204
flow, (t, h), 204
invariant systems, 204
majority rule, 204
transformation, 202
percolation, 82–100
$d = 1$ lattice, 91
$d = 2$ square lattice, 98
$d = 2$ triangular lattice, 95
procedure, 87
scaling field, 233
irrelevant, 234
marginal, 234
relevant, 234
renormalised, 233
theory, 222
transformation, 86, 222, 226
analytic, 226
correlation length, 227
fixed point, 227
linearised, 231
operator identity, 226
recursion relation, 226
Renormalised
correlation function, 209
coupling constant, 212
energy, 207
generalised reduced energy, 226
nn coupling constant, 216
nnn coupling constant, 216
partition function, 207
quadruple coupling constant, 216
reduced control parameter, 207
reduced external field, 205
reduced temperature, 205
Rescaling factor, 66, 202
Rescaling transformation, 83, 85
Response function, 121, 122, 133
Review of statistical mechanics, 119
Ricepile

avalanche time series, 287
experiment, 285
modelling, 290
Ricepile model
algorithm, 291
avalanche time series, 293
avalanche-size probability, 294
configuration
recurrent, 292
stable, 292
transient, 292
threshold slope, 291

Sandpile metaphor, 250
Scaling ansatz, 188–193
avalanche-size probability, 273
average mag. per spin, 189
cluster number density, 39–49
free energy per spin, 190
Scaling field, 233
irrelevant, 234
marginal, 234
relevant, 234
renormalised, 233
Scaling function, 345
cluster number density, 41
percolation Bethe lattice, 44
percolation $d = 1$, 42
percolation $d = 2$, 49
Ising model
correlation function, 192
free energy per spin, 190
magnetisation per spin, 189
specific heat, 191
susceptibility per spin, 192
self-organised criticality
avalanche-size probability, 273
Scaling relations
Ising model, 193–195
α, β, and γ, 193
α, ν, and d, 194
β, γ, and δ, 193
γ, η, and ν, 195
percolation, 51–54, 62–69
β, ν, D, and d, 65
β, σ, and τ, 54

γ, σ, and τ, 52
ν, σ, and D, 63
Seismic system, 252
  earthquake time series, 303
  earthquake-size frequency, 304
  static friction threshold, 252, 307
Self-organised criticality, *249-333*
Self-similarity, 55
  correlation length, 82
  fixed point, 83
Site percolation, 3, 78
Size
  cluster, 3, 79
Specific heat, 122
  critical exponent α, 134
  Ising model $d = 1$, 149
  Ising model $d = 2$, 184
  Landau theory, 175
  mean-field theory, 169
  non-interacting spins, 129
Spin, 116
  average value, 134
  down, 116
  microstate, 118
  up, 116
Stable configuration, 255
Steady state
  non-equilibrium, 249
Sum rule
  Ising model, 135
  Ising model $d = 1$, 154
  percolation $d = 1$, 10, 13, 14
  percolation Bethe lattice, 30
Susceptibility per spin, 121
  critical exponent γ, 133
  Ising model $d = 1$, 147
  Ising model $d = 2$, 182
  Landau theory, 174
  mean-field theory, 164
  non-interacting spins, 129
Symbols
  Ising model, *374-377*
  percolation, *372-373*
  SOC, *377-380*
Symmetry breaking, 138

Table
  Ising model
    critical exponent, 199
    critical temperature, 199
    pair spin configurations, 117
    thermodynamic relations, 123
  percolation
    critical exponent, 81
    critical occupation probability, 80
Taylor expansion, 341
Temperature, 118
Temporal average, 119
Thermodynamic limit, 123
Thermodynamic relations, 123
  table, 123
Time series
  avalanche
    BTW model $d = 1$, 259
    BTW model $d = 2$, 280
    OFC model, 314
    Oslo model, 293
    ricepile, 287
  earthquake, 303
  rainfall, 326
Toppling, 255
Transfer matrix, 141
  entries, 141
  method, 141
Two-dimensional BTW model
  algorithm, 278
  avalanche time series, 280
  avalanche-size probability, 281
  average avalanche size, 280
  threshold, 278
Two-dimensional Ising model, 179
  average energy per spin, 184
  average mag. per spin, 180
  critical exponent
    α, specific heat, 186
    β, order parameter, 180
    γ, susceptibility per spin, 182
  critical temperature, 180
  Peierls' argument, 186
  order parameter, 180
  partition function, 179
  specific heat, 184

susceptibility per spin, 182
variance of energy, 184
variance of magnetisation, 182
Two-dimensional percolation, 30
  average cluster size, 33
  characteristic cluster size, 37
  cluster number density, 35–42
  critical exponent
    β, order parameter, 31
    γ, average cluster size, 33
    σ, characteristic cluster size, 37
    τ, cluster number density, 37
  critical occupation probability, 31
  data collapse, 49
  order parameter, 31
  renormalisation group
    square lattice, 98
    triangular lattice, 95
  scaling ansatz, 39
  scaling function, 49

Universal critical exponents
  Ising model, 199
  percolation, 81
Universality, 81, 231
Upper critical dimension, 81

Variance
  energy, 122
    fluctuation-dissipation thm, 122
    Ising model $d = 1$, 149
    Ising model $d = 2$, 184
    non-interacting spins, 131
  magnetisation, 122
    fluctuation-dissipation thm, 122
    Ising model $d = 1$, 147
    Ising model $d = 2$, 182
    mean-field model, 164
    non-interacting spins, 129

Wannier, 186
Widom scaling ansatz, 188
  free energy per spin, 190, 208
  magnetisation per spin, 189, 191
  renormalisation group, 235
  specific heat, 191

susceptibility per spin, 192
Window size, 55
Wolff algorithm, 179, 364

SCI QC 173.4 .C74 C48 2005

Christensen, Kim, 1962-

Complexity and criticality